T0332005

Rivers
and
Sustainable
Development

Rivers and Sustainable Development

Alternative Approaches and Their Implications

S. NAZRUL ISLAM

OXFORD
UNIVERSITY PRESS

OXFORD
UNIVERSITY PRESS

Oxford University Press is a department of the University of Oxford. It furthers
the University's objective of excellence in research, scholarship, and education
by publishing worldwide. Oxford is a registered trade mark of Oxford University
Press in the UK and certain other countries.

Published in the United States of America by Oxford University Press
198 Madison Avenue, New York, NY 10016, United States of America.

Library of Congress Cataloging-in-Publication Data
Names: Islam, S. Nazrul, author.
Title: Rivers and sustainable development : alternative approaches
and their implications/ S. Nazrul Islam.
Description: New York : Oxford University Press, [2020] |
Includes bibliographical references and index.
Identifiers: LCCN 2019053615 (print) | LCCN 2019053616 (ebook) |
ISBN 9780190079024 (hardback) | ISBN 9780190079048 (epub) |
ISBN 9780190079031 (updf) | ISBN 9780190079055 (oso)
Subjects: LCSH: Rivers. | Environmental policy. | Sustainable development. |
Water resources development.
Classification: LCC GB1203.2 .I85 2020 (print) | LCC GB1203.2 (ebook) |
DDC 333.91/62—dc23
LC record available at https://lccn.loc.gov/2019053615
LC ebook record available at https://lccn.loc.gov/2019053616

1 3 5 7 9 8 6 4 2

Printed by Integrated Books International, United States of America

To
River scholars and activists of all countries

Contents

Figures

Tables

Preface

Rivers are one of the most important components of our natural environment. For some countries, they are the most important component. The policies adopted toward rivers are therefore of crucial importance for the environment and development of a country.

In 2015, the world community adopted Sustainable Development Goals (SDGs), several of which concern rivers *directly*. Some other SDGs depend on rivers *indirectly*. To achieve SDGs, it is therefore necessary to adopt correct policies toward rivers. This book discusses alternative approaches to rivers and their implications for sustainable development. This discussion may help countries adopt river-related policies that are more conducive to sustainable development.

The ideas presented in this book gelled over time. Coming from the Bengal Delta, formed by the three great rivers of the world (Ganges, Brahmaputra, and Meghna), I became aware of the importance of rivers and river-related policies early in my life. The flood of 1974—during which I did risky relief work among people marooned in remote villages—had a deep impact on my understanding of the raw power of rivers and the limited effectiveness of frontal stands against this power. My understanding of river and irrigation related issues was broadened and deepened through my work on the Irrigation Study Project in the early eighties at the Agro-Economic Research (AER) section of Bangladesh's Ministry of Agriculture. My views—emerging from the experience of the 1974 flood—were reinforced by Bangladesh's Great Flood of 1988, which led to my article, "Let the Delta Be a Delta! An Essay in Dissent on the Flood Problem of Bangladesh" that drew considerable attention.

After getting my PhD in economics from Harvard University in 1993, I worked for a while at the then Harvard Institute for International Development (HIID), where I conducted research into the *general* relationship between environmental quality and a country's policies, showing that the latter plays an important role in determining the former. This finding agreed well with my earlier observations in the *particular* case of rivers. Bangladesh's Great Flood of 1998 prompted me to refocus attention on river-related issues, viewing them from a policy perspective. In the process, I developed several important concepts helpful for conceptualizing river policy issues. The most important among these are the *Commercial approach* to rivers and its opposite, the *Ecological approach*. Each of these has a *frontal* and a *lateral* version, applicable, generally speaking, to upper and lower reaches of rivers, respectively. Dams, barrages, and other cross-sectional interventions embody the frontal version of the Commercial approach. On the other hand, embankments, levees, dykes, and other interventions from the side embody its lateral version, which, to effectively convey its spirit and for ease and parsimony of discussion, I call the *Cordon* approach. The Ecological approach, by contrast, calls for minimization of both frontal and lateral interventions. For similar reasons, I call the lateral version of the Ecological approach the *Open* approach.

Earlier, I published a good number of journal articles expounding these ideas and concepts. This book brings together in an integrated framework these concepts and ideas and the supporting empirical information. Its main contribution lies in advancing the conceptualization of river-related policies. It puts forward the basic proposition that the Ecological approach and its derivative the Open approach to rivers are more conducive to sustainable development than the Commercial approach and its derivative the Cordon approach. The book supports this proposition by drawing upon a wide range of experiences regarding river policies from all across the world. To illustrate the contrast between the Cordon and Open approaches, it uses the experience of the Bengal Delta, the largest and most active delta in the world. The conceptualization and analysis offered in this book can help scholars studying the concrete experiences of rivers in other parts of the world.

The discussion of this book has been kept as non-technical as possible. This is to facilitate a public discussion regarding river policies. For a long time, these policies have remained in the grip of "experts," comprising mainly civil engineers. Together with politicians and bureaucrats, they form a triad that makes major decisions regarding rivers. Unfortunately, this technicalization and hiding from the public view of river policy formulation have not served us well. Experience has shown that river issues are multidisciplinary and protection of rivers requires *public* participation. The discussion of river issues therefore needs to be moved to the public arena and be decided through open, public discussion. It is my hope that by being accessible to the general public, this book will help to foster that discussion. In places where the use of certain technical terms proved unavoidable, they have been explained. Also, a simple glossary of these terms has been provided at the end of the book.

In writing and publishing this book, I incurred debts to many. To begin with, I would like to thank the reviewers who read the manuscript carefully and offered helpful suggestions that helped to improve the book significantly. In particular, I would like to thank Prof. Khalequzzaman of Lock Haven University for his thorough review and many helpful comments. Thanks are also due to the editors and publishers of the journals and books in which my earlier articles were published.

A new book always stands on the shoulders of previous works on the subject. I would like to express my gratitude to river scholars from whose works I learnt a lot. Particular mention in this regard should be made of the very informative report, *Dams and Development*, prepared by the World Commission on Dams (2000) and the hugely informative book, *Silenced Rivers*, by Patrick McCully (2001). Various other reports by the United Nations and the European Union also proved helpful. Many observations made and ideas expressed in this book have already been put forward by previous scholars, often using different terminology.

I would also like to thank river activists across the world who are fighting for protection of rivers, and without whose efforts rivers in many places would have been in worse shape than they are now. I would like to thank my colleagues in the environment movement, whose work has been a source of both information and inspiration. Special thanks are due to Dr. Mustafizur Rahman of Centre for Policy Dialogue (CPD) for his help in getting hold of the Master Plan, prepared by the International Engineering Company (IECO) in 1964 for water and power development in Bangladesh—a document that played a vital role in the water development history of the country.

Taking this opportunity, I would like to thank Ms. Bronwyn Geyer and others of Oxford University Press and all of Newgen Knowledge Works for their sincere work in publishing this book.

Finally, I would like to express my gratitude to my family—my wife Dr. Tanvira, daughter Nusaybah, and son Rahul for allowing me all the hours that could be their. I would also like to thank members of my extended family (my brothers and sisters and their families) and my friends for their support and encouragement.

The views presented in this book are my personal and need not be ascribed to any of the organizations to which I belong.

S. Nazrul Islam
New York
February, 2020

Acknowledgments

The following individuals and organizations are gratefully acknowledged for their cooperation in using some of their figures and images in this book. The numbers of figures in this book that benefited from their cooperation are mentioned after their names. Except where mentioned, the figures are reproductions from respective sources.

National Aeronautics and Space Administration (NASA) Socioeconomic Data and Applications Center (SEDAC) (Figures 2.2, 2.3, 2.4, 2.5, 2.7, 2.8)

India Water Portal and National Institute of Hydrology, available at https://projectupsc.wordpress.com/2018/06/18/seven-rivers-and-three-interlinking-projects/ (Figure 2.6)

C. Franko and M. Johnson and maps.com (Figure 4.1)

Discover Murray River (Figure 4.2)

USAID; used by Kristina Schneider (2001), Civil Engineering Department, University of Texas; and currently available in US Wiki Commons at https://commons.wikimedia.org/wiki/File:The_Caucasus_and_Central_Asia_-_Political_Map.jpg (Figure 4.3)

Landsat program of United States Geological Services (USGS) and NASA and Global Resource Information Database (GRID)-Sioux Falls of the United Nations Environmental Programme (UNEP) (Figure 4.4)

Agnes Stienne and *Le Monde Diplomatique* (August 2013) (Figure 4.5)

Gill Braulik, Uzma Noureen, Masood Arshad, and Randall R. Reeves (2015) (Figure 4.6)

Miah Adel, University of Arkansas at Pine Bluff, Department of Chemistry and Physics (Figure 5.1)

Sandra Postel, *Pillar of Sand*, W.W. Norton, 1999 (Figure 5.2).

Animesh K. Gain and Carlo Giupponi (2014) (Figure 5.3)

Shamsuddoza Sajen and *The Daily Star* (May 26, 2016), Bangladesh (Figure 5.4)

Amanur Aman and *The Daily Star* (August 29, 2016), Bangladesh (Figure 5.5)

Gauri Noolkar-Oak and Both Ends (2017) (Figure 5.6)

Rashed Shuman and Pinaki Roy, *The Daily Star* (May 25, 2013) (Figure 5.8)

Subir Bhaumik (2009) and *Bengal Newz*, July 17 (Figure 5.9)

Uwe Dedering at German Wikipedia / CC BY-SA, modified by Nusaybah Islam (Figure 5.10)

Kalendra Sejuwal and *my Republica* (August 17, 2017) (Figure 5.11)

International Rivers and Central Electricity Authority India (Figure 5.12)

Nirmala Ganapathy and *The Strait Times Graphics* (September 27, 2016) (Figure 5.13)

Sandeep Dikshit and *The Hindu* (June 13, 2016) (Figure 5.14)

Diganta and The New Horizon (June 28, 2009) and Ritesh Ghosh (2017) "Is China to divert Brahmaputra water after its failure at Doklam?" https://bengali.oneindia.com/news/features/after-doklam-standoff-defeat-china-again-tries-test-india-s-patience-brahmaputra-river/articlecontent-pf31825-025736.html (Figure 5.15)

American Rivers (Figure 8.1)

Ajit Bajaj and Downtoearth.org.in (May 18, 2018) (Figure 8.2)

International Rivers Network (Figure 8.3)

Avantika Tewari and *Feminism in India* (August 10, 2017) (Figure 8.4)

Mississippi River Commission and Mississippi Valley Division Corps of Engineers (Figure 9.3)

USGS Central Midwest Water Science Center (Figure 9.4)

China Highlights Travel (Figure 9.5)

China Connection Tours (CCT) (Figure (9.6)

M. Getzner, M. Jungmeier, T. Kostl, and S. Weiglhofer (2011) and American Society of Civil Engineers (ASCE) (Figure 10.1)

Suvendu Roy, Department of Geography, Kalipada Ghosh Tarai Mahavidyalaya, West Bengal, India (Figure 11.3)

Himanshu Thakkar, South Asian Network on Dams, Rivers, and People (SANDRP) (Figure 11.4)

International Engineering Company (IECO), San Francisco, USA (Figures 11.5, 11.6, 11.8, 11.9, 11.10, 11.11, 11.14, 11.15, 11.17, 12.4)

Center for Geographical and Environmental Information Services (CGEIS), Dhaka, Bangladesh (Figure 11.7)

Anisur Rahman and *The Daily Star* (January 1, 2016), Bangladesh (Figure 11.13)

Food and Agriculture Organization (FAO) and Rafiqul Islam (Figure 11.16)

Mustafa Alam, Asia Pacific Memo (APM) of Institute of Asian Research (IAR), University of British Columbia. Available at https://apm.sites.olt.ubc.ca/files/2014/01/Mustafa-Alam_Bangladesh-Water-Resource_image01.jpg (Figure 11.18)

UN-HABITAT (Figure 11.19)

Crelis F. Remmelt, Zahed Md. Masud, and Arvid Masud (2018) (Figure 12.1)

Shampa (Institute of Water and Flood Management (IWFM), Bangladesh University of Engineering and Technology (BUET), and Ibne Mayaz Md. Pramanik (2012) (Figure 12.2)

Loek Weijts and Martijn van Staveren (Figure 12.3)

Brian D. Smith, Benazir Ahmed, Muhammad Edrise Ali, and Gill Braulik (2001) (Figure 12.6)

Abbreviations

ABD	Adaptation By Design
ADAB	Association of Development Agencies in Bangladesh
ADB	Asian Development Bank
AER	Agro-Economic Research
AfDB	African Development Bank
ASCI	Areas of Special Conservation Interest
BADC	Bangladesh Agriculture Development Corporation
BAPA	Bangladesh Poribesh Andolon
BAU	Business As Usual
BDP	Bangladesh Delta Plan
BEN	Bangladesh Environment Network
BJP	Bharatiya Janata Party
BRHE	Brahmaputra Right Hand Embankment
BRI	Belt and Road Initiative
BuRec	Bureau of Reclamation
BWDB	Bangladesh Water Development Board
CADP	Command Area Development Project
CBD	Convention on Biodiversity
CCD	Convention on Combatting Desertification
CEGIS	Centre for Environmental and Geographical Information Services
CEN	Coalition of Environmental NGOs
CEP	Coastal Embankment Project
CERP	Coastal Embankment Rehabilitation Project
CIP	Chandpur Irrigation Project
CSD	Commission on Sustainable Development
CO2	Carbon dioxide
Cumec	Cubic meter per second
Cusec	Cubic foot per second
DIFPP	Dhaka Integrated Flood Protection Project
DND	Dhaka-Narayanganj-Demra
DVC	Damodor Valley Corporation
DWP	Directive on Water Policy
EBSATA	East Bengal State Acquisition and Tenancy Act
EC	European Community
ECOSOC	Economic and Social Council
ECRR	European Center for River Restoration
EEA	European Environmental Agency
EEC	European Economic Community

EGAT	Electricity Generating Authority of Thailand
EIA	Environmental Impact Assessment
EIRR	Economic Internal Rate of Return
EKC	Environmental Kuznets Curve
EMINWA	Environmentally Sound Management of Water
EPWAPDA	East Pakistan Water and Power Development Authority
EU	European Union
EWS	Eastern Waters Study
FAP	Flood Action Plan
FAO	Food and Agriculture Organization
FC	Flood Control
FCD	Flood Control and Drainage
FCDI	Flood Control, Drainage, and Irrigation
FLOBAR	Floodplain Biodiversity and Restoration
FPCO	Flood Plan Coordination Committee
FRMP	Flood Risk Management Plan
FYP	Five Year Plan
GDP	Gross Domestic Product
GEMS	Global Environment Monitoring System
GHG	Greenhouse Gas
GHI	Gross Happiness Index
G-K	Ganges-Kobadak
GNH	Gross National Happiness
GNI	Gross National Income
GNP	Gross National Product
GoB	Government of Bangladesh
GoN	Government of the Netherlands
GRA	Grievance Redressal Authority
GW	Gigawatts
GWP	Global Warming Potential
HIID	Harvard Institute for International Development
HIMSAR	South Asian Network for Himalayan Rivers
HLPF	High Level Political Forum
HMRA	Hazard Mitigation and Relocation Act
HYV	High Yielding Variety
IBRD	International Bank for Reconstruction and Development
ICBEN	International Conference on Bangladesh Environment
ICOLD	International Commission on Large Dams
ICRCTR	International Conference on Regional Cooperation on Transboundary Rivers
IDA	International Development Association
IDWSSD	International Drinking Water Supply and Sanitation Decade
IECO	International Engineering Company
IER	Income-Environment Relationship

IFCDR	Institute of Flood Control and Drainage Research
IFMRC	Interagency Floodplain Management Review Committee
IFPRI	Internation Food Policy Research Institute
ILO	International Labor Organization
IPCC	Intergovernmental Panel on Climate Change
IRLP	Indian River Linking Project
IRN	International Rivers Network
IRT	Indus River Treaty
IUCN	International Union for Conservation of Nature
IWMI	International Water Management Institute
JICA	Japan International Cooperation Agency
KIA	Kachin Independence Army
KC	Kuznets Curve
KJDRP	Khulna-Jessore Drainage Rehabilitation Project
KWh	Kilowatt hour
LACPR	Louisiana Coastal Protection and Restoration
LDCs	Least Development Countries
LGED	Local Government Engineering Division
LGRD	Local Government and Rural Development
LIFE	L'Instrument Financier pour l'Environnement
LLP	Low Lift Pump
LWDAB	Land and Water Advisory Board
LWRSS	Land and Water Resources Sector Study
M-D	Meghna-Dhonagoda
MDBA	Murray-Darling Basin Auhtority
MDBC	Murray-Darling Basin Commission
MDBP	Murray-Darling Basin Plan
MDGs	Millennium Development Goals
MOEE	Ministry of Electricity and Energy
MPO	Master Plan Organization
MRC	Mekong River Commission
MR-GO	Mississippi River—Gulf Outlet
MRAG	Marine Resources Assessment Group
MW	Megawatts
NAPM	National Alliance for Peoples Movement
NASA	National Aeronautics and Space Administration
NBA	Narmada Bachao Andolon
NBI	Nile River Initiative
NCA	Narmada Control Authority
NCAR	National Center for Atmospheric Research
NDA	National Democratic Alliance
NDC	Netherlands Development Cooperation
NEDECO	Netherlands Development Cooperation
NEEPC	North Eastern Electric Power Corporation

NFC	National Flood Council
NGO	Non-Government Organization
NHPC	National Hydro Power Company
NLD	National League for Democracy
NOAA	National Oceanic and Atmospheric Administration
NRB	Non-Resident Bangladeshi
NRCC	National River Conservation Commission
NVDA	Narmada Valley Development Authority
NWC	National Water Council
NWRM	National Water Retention Measures
NWMP	National Water Management Plan
OECD	Organization for Economic Cooperation and Development
OED	Operations and Evaluation Department
OM	Organic Matter
OWC	Optimum Water Control
O&M	Operation and Maintenance
PEM	Post Evaluation Mission
PFRA	Preliminary Flood Risk Assessment
PIRDP	Pabna Irrigation and Rural Development Project
PoE	Panel of Experts
POROSH	Poribesh Rokkhya Shopoth
RBMP	River Basin Management Plan
RC	Review Committee
RESTORE	Rivers Engaging, Supporting, and Transferring Knowledge for Restoration in Europe
RFREV	Report on Flood Risk and Environmental Vulnerability
RLP	River Linking Project
RoR	Resolution on Rivers
RRI	River Research Institute
SAC	Special Areas of Conservation
SAAPM	South Asian Alliance for Peoples' Movements
SANDRP	South Asian Network on Dams, Rivers, and People
SASRP	South Asian Solidarity for Rivers and People
SCI	Sites of Community Interest
SDGs	Sustainable Development Goals
SEA	Strategic Environmental Assessment
SEDAC	Socioeconomic Data and Applications Center
SFRM	Strategic Flood Risk Management
SMEC	Snowy Mountain Engineering Corporation
SPA	Special Protection Area
SPIC	State Power Investment Corporation
SRDI	Soil Resources Development Institute
SSD	Sardar Sarovar Dam
STIFPP	Secondary Towns Integrated Flood Protection Project

SWT	Shallow Tube Well
TC	Technical Committee
TGPC	Three Gorges Project of China
TRM	Tidal River Management
TVA	Tennessee Valley Authority
UK	United Kingdom
UN	United Nations
UNCED	United Nations Conference on Environment and Development
UNCHE	United Nations Conference on Human Environment
UNCSD	United Nations Conference on Sustainable Development
UN-DESA	United Nations Department of Economic and Social Affairs
UNDP	United Nations Development Programme
UNEP	United Nations Environment Program
UNESCO	United Nations Education Scientific and Cultural Organization
UNTAA	United Nations Technical Assistance Administration
UNTAP	United Nations Technical Assistance Programme
UNWCM	United Nations Water Control Mission
UPA	United Progressive Alliance
USA	Unites States of America
USACE	United States Army Corps of Engineers
USAID	United States Agency for International Development
USSC	United States Supreme Court
USSR	Union of Soviet Socialist Republics
WARPO	Water Resources Planning Organization
WASA	Water and Sewerage Authority
WCD	World Commission on Dams
WCED	World Commission on Environment and Development
WMA	Water Management Association
WRSA-CC	Water Resources of South Asian Conflict to Cooperation
WSSD	World Summit on Sustainable Development
WWF	World Wildlife Foundation
WWT	Wildlife and Wetlands Trust

1

Rivers, sustainable development, and policies

1.1 Introduction

Rivers are an important component of our environment and an important natural resource. From time immemorial, rivers provided human beings with water for drinking, bathing, and irrigation. They provided fish to eat and served as an important means for transportation. Rivers played an important role in developing the economies, determining the settlement patterns, and shaping the social makeups. However, in many places rivers are in a bad state and have become a source of contention and conflict, and there is an ongoing debate about the best way to use them as a natural resource.

This book aims at advancing the conceptualization of river-related issues. For this purpose, it puts forward several important concepts. Among these are the *Commercial* and *Ecological* approaches to rivers, both of which have their *frontal* and their *lateral* versions. Dams and barrages embody the frontal version of the Commercial approach, while embankments, levees, and dykes represent its lateral version. By contrast, the Ecological approach urges for minimization of both frontal and lateral interventions in rivers. To convey their spirit effectively and for ease and parsimony of discussion, I call the lateral version of the Commercial approach the *Cordon* approach and the lateral version of the Ecological approach the *Open* approach. The main proposition of this book is that the Ecological and Open approaches are more conducive to sustainable development than the Commercial and Cordon approaches. The book supports this proposition by drawing upon world experience regarding river policies.

The examination of river policies has assumed renewed importance in view of the fact that the world community has now formally accepted the goal of sustainable development. In September 2015, the leaders of 193 countries gathered at the United Nations General Assembly in New York and adopted the 2030 agenda for sustainable development, formulated in the form of 17 Sustainable Development Goals (SDGs) and 169 targets. The SDGs strive to integrate more closely the three dimensions of sustainable development, namely economic growth, social development, and environmental protection.[1] As a result, SDGs are more interconnected, and these interconnections both pose a challenge and

Rivers and Sustainable Development. S. Nazrul Islam, Oxford University Press (2020). © Oxford University Press.
DOI: 10.1093/oso/9780190079024.001.0001

create an opportunity. They pose a challenge, because efforts toward achieving a particular goal have to take into account their consequences for other goals. However, they present an opportunity too, because, with proper understanding of these interconnections, it should be possible to make progress toward several goals simultaneously. In fact, these interconnections have created the possibility of identifying certain nexuses where different goals intersect more closely, so that focusing on these nexuses can provide an effective way to achieve several goals simultaneously.[2] It so happens that, in many countries, rivers provide one such nexus. By adopting the correct approach to rivers, it is therefore possible to advance the cause of many SDGs in these countries.

This opening chapter provides an introduction to the basic concepts proposed in the book, shows why river-related policies are important for sustainable development, and gives an overview of the remaining chapters. It begins with a brief discussion of the concept of sustainable development and notes its various interpretations (Section 1.2). It also notes the implications of recent developments—in particular, of climate change—for these interpretations. The chapter next shows how rivers can influence the three dimensions of sustainable development (Section 1.3). While most observers focus on links between rivers and environmental sustainability, it is important to realize that links between rivers and economic growth and social development are also important. The chapter then discusses how policies in general can play an important role in determining the environmental quality of a country (Section 1.4). This general proposition paves the way for the particular proposition that correct policies regarding rivers can play an important role in achieving sustainable development.

The chapter next introduces the main concepts put forward in this book, highlighting their relationships (Section 1.5). It shows that there are broadly two opposing approaches to rivers. One is the Commercial approach that originated with the Industrial Revolution and proceeds from the basic premise that "Any river water that passes to the sea is a waste!" Opposed to this approach is the Ecological approach that views rivers' main function in carrying precipitation water to the seas, so as to maintain the earth's hydrological cycle and sustain river basin ecology. Both these approaches have their corresponding frontal and lateral versions. The lateral version of the Commercial approach is the Cordon approach, according to which floodplains should be cordoned off from river channels through construction of embankments. By contrast, the lateral form of the Ecological approach is the Open approach, which advocates keeping floodplains open to river channels for the benefit of both. The goal of the book is to show that adoption of the Ecological approach—and its derivative, the Open approach—is more conducive to achieving sustainable development, particularly in countries where rivers are an important part of the ecology.

1.2 Sustainable development

The concept of sustainable development received a big boost from the report by the World Commission on Environment and Development (WCED 1987).[3] It defined sustainable development as development that "meets the needs of the current generation without sacrificing the needs of the future generations." This definition captures succinctly the essence of the concept, pointing to the basic intergenerational equity argument behind sustainable development. The report also makes clear that the proposition of sustainable development applies to both developed and developing countries.[4]

1.2.1 Natural capital versus produced capital

Since the WCED report, there have been efforts to concretize and operationalize the definition of sustainable development, leading to two interpretations of the concept: strict and lax. Both interpretations agree that, under sustainable development, future generations should have at least as much capital per capita as members of the current generation have. However, they disagree on the elements of capital to which this intergenerational equity principle should apply.[5]

Capital can be classified in many ways. From one perspective, capital may be classified into two types, namely natural and produced (Figure 1.1). Natural capital refers to all resources that exist in nature, prior to interventions by conscious human labor. Primary forests, unaffected by direct human interventions, provide an example of natural capital. Produced capital, on the other hand, comprises objects that are the outcome of human labor (directed generally on items that are available in nature). Produced capital in turn can be of two types, namely physical and human capital. The fact that producing physical capital requires human labor is self-evident. However, accumulation of human capital also requires effort by the person who acquires the human capital and by others (parents, teachers, instructors, colleagues, health workers, etc.) who help the person to acquire human capital through education, training, healthcare, and so on. Thus, human capital is also produced capital.

The difference between the strict and lax definitions of sustainable development hinges on the question of whether to allow substitutability between natural and produced capital in considering the capital bequeathed to future generations. The proponents of the strict definition do not allow such substitutability. They think that the definition of sustainable development applies specifically to natural capital. Accordingly, they demand that the members of the current generation should bequeath the future generation at least the same amount of natural capital per capita as it had inherited from the past generation. The proponents of

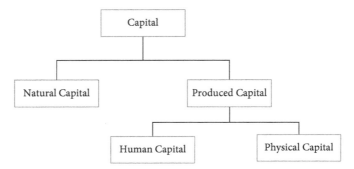

Figure 1.1 Classification of capital
Source: Author

the lax definition, on the other hand, allow substitution of natural capital by produced capital. In their view, development will be sustainable even if there is some loss of natural capital per capita, as long as that loss is more than offset by a gain in produced capital per capita.

There are however several problems with the lax interpretation of sustainable development, including its operationalization, because it requires measurement and valuation of different types of capital. While produced capital generally can be evaluated using market prices, many types of natural capital are not marketed and hence do not have market prices. Even if some components of natural capital are marketed (e.g., timber extracted from primary forests), their market prices generally do not capture the positive externalities associated with them (i.e., various positive functions they perform for nature and human beings). Also, these externalities are not only contemporaneous (for the present generation only) but extend to future times and generations. Since these future externalities generally do not find adequate reflection in current market prices, they are hard to be captured through formal cost-benefit analysis. Monetary valuation of natural capital is therefore difficult and controversial. Some of these measurement and evaluation issues also apply to produced capital, in particular to human capital, which also has many positive externalities. However, the magnitude of this problem for natural capital is much greater than it is for produced capital. As a result of these measurement and evaluation problems, it is difficult to know whether the loss of natural capital is adequately compensated by the gain in produced capital.

The growing environmental crisis illustrates the importance of positive externalities of natural capital. It shows that natural capital has intrinsic values that proposed monetary measures cannot capture adequately. (Alternatively, one may say that their value is infinite.) For example, the aggravation of climate change shows that, no matter how much physical capital the current generation may bequeath future generations, it will not be of much help to them if catastrophic consequences of climate change make human civilization untenable on the earth. As a result, more people now think that the strict interpretation of sustainable development is the more appropriate interpretation.

1.2.2 Three dimensions of sustainable development

Sustainable development, as noted, is generally considered to have three dimensions, namely economic growth, social development, and environmental protection (figure 1.2). This multidimensionality is a reflection of several factors. First, large parts of the world population still remain stuck at a very low material standard of living, with a low level of education and low life expectancy. For them, raising the material standard of living is a priority and a precondition for sustainable development, because the question of sustainability of development arises only when there *is* development. However, at the stage at which these parts of the world population are in, there can be hardly any development without economic growth. It is because of this on-the-ground reality that economic growth is recognized as an important dimension of sustainable development.[6]

Aggregate economic growth alone, however, often proves insufficient for improving the material standard of living of people who need it most. Fair distribution of the gains of economic growth, adequate access to various basic public services and utilities, opportunities for meaningful participation (voice) in making decisions affecting their lives, and so on, are also needed. In recent decades, it has become conventional to use the expression "social development" to denote progress in these particular areas (i.e., distribution, access, and voice).[7] As a result, social development is recognized as another dimension of sustainable development.

The rationale for environmental protection to be a dimension of sustainable development is obvious. Without protection of the environment, the current generation cannot bequeath to the future generation as much per capita natural

Figure 1.2 Three dimensions of sustainable development
Source: Author

capital as it inherited from the previous generation. Even under the lax interpretation, there are limits to which produced capital can substitute for natural capital so that development can be called sustainable. Thus, protection of the environment has to be an important dimension of sustainable development, irrespective of which interpretation is adopted.

The recognition of the three dimensions of sustainable development is therefore, in part, a reflection of the uneven stages of development across different parts of the world. The economic growth and social development dimensions of sustainable development are more important for developing countries, large parts of whose populations are still deprived of the minimum acceptable material standard of living. By contrast, people in developed countries, by and large, have reached the threshold material standard of living necessary for fulfilling the potentialities of human life, though many issues of social development remain pertinent for them too.

The issue of environmental protection, on the other hand, is important for all countries. In some respects, this issue is more important for developed countries than for developing countries, because in their (successful) pursuit of a high material standard of living, developed countries did considerable damage to the environment, both local and global. Through efforts in recent decades, they have been successful at repairing some of the local damage, but they are yet to fulfill adequately their responsibility for repairing the damage done to the global environment. On the other hand, developing countries, as they try to achieve fast economic growth, run the risk of damaging their local environment in addition to damaging the global environment. Yet, as the climate change phenomenon shows, the global environment cannot take any further damage without catastrophic consequences.

Thus, no matter which interpretation of sustainable development one adopts, and which part of the world one lives in, protection of the environment and conservation of natural capital remain a vital task. To the extent that rivers are an important part of our natural capital, it is simply not possible to bequeath natural capital to the next generation adequately and achieve sustainable development without protecting the rivers.[8] Protection of rivers is therefore an important task for all countries of the world.

1.3 Sustainable Development Goals (SDGs) and rivers

The importance of rivers and water issues in achieving sustainable development found reflection in the SDGs both directly and indirectly, as already noted. A good number of SDGs relate to river and water issues directly. For example, SDG-6 calls for "sustainable management of water," and many of its targets are related directly to rivers.[9] For example, target 6.4 calls for sustainable withdrawal of water, which is a key question of river management. Similarly, target 6.5 calls for transboundary cooperation regarding rivers. This cooperation, as we shall see, depends crucially on the approach adopted toward rivers by co-riparian countries and communities. Target 6.6 calls for protection and restoration of water-related ecosystems, including mountains, forests, wetlands, rivers, aquifers, and lakes. Needless to say, achieving these objectives requires correct approaches and policies toward rivers.

Similarly, SDG-14 calls for "sustainable use of oceans, seas, and marine resources." In particular, it calls for sustainable management and protection of marine and coastal ecosystems to avoid significant adverse impacts, and to achieve healthy and productive oceans (target 14.2). Clearly, coastal and marine ecosystems, particularly near river mouths, depend crucially on the flow of rivers that reaches the sea. The quantity and quality of this flow in turn depend on the approach adopted toward rivers. Thus, achieving target 14.2 requires the correct approach to rivers.

In the same vein, SDG-15 urges the world to "protect, restore, and promote sustainable use of terrestrial ecosystems, sustainably manage forests, combat desertification, halt and reverse land degradation, and halt biodiversity loss."[10] In particular, it calls for protection of "inland freshwater ecosystems and their services" (target 15.1). Obviously, rivers are the most important component of inland freshwater ecosystems and wetlands. Hence, conservation, restoration, and sustainable use of the latter depends crucially on the approach toward rivers. Target 15.3 calls for fighting against land degradation due to desertification, drought, and floods. Floods obviously depend on river management, and withdrawal of river water is an important cause of downstream desertification. At the same time, excessive irrigation, made possible through diversion of river water, often causes salinity through evapotranspiration and capillary action in arid areas. Similarly, excessive use of chemical fertilizers and pesticides—that usually accompanies irrigation—often leads to toxification of the soil. Thus, approach to rivers has a multipronged role in achieving the objectives included in target 15.3. Meanwhile, the role of proper river management for preservation of aquatic biodiversity—called for by target 15.5—hardly needs any explanation. Since the vegetation of river basins depends crucially on rivers themselves, much of the non-aquatic (i.e., land-based) biodiversity also depends on river management. Thus, SDGs 6, 14, and 15 all depend directly on policies and approaches adopted toward rivers.

There are however other SDGs that depend indirectly on river-related policies and approaches. For example, SDG-12 calls for switching to "sustainable consumption and production patterns." In particular, target 12.2 calls for sustainable management and efficient use of natural resources by 2030. Rivers being one of the most important components of the world's natural resources, this target cannot be achieved without adopting the right approach toward rivers. Similarly, SDG-2 calls for "sustainable agriculture." Its target 2.3 asks for adoption of agricultural practices that help maintain ecosystems, strengthen capacity for adaptation to climate change, extreme weather, drought, flooding, and other disasters, and help to improve land and soil quality.[11] Clearly, the right approach to rivers is needed to achieve these objectives.

The above review of the SDGs shows that, overall, a large number of them depend—either directly or indirectly—on the correct approach to rivers. In fact, given the interrelationships among various goals, the success of the entire SDG effort in many countries depends to a significant degree on the approach and policies adopted toward rivers.

1.4 Environment, rivers, and policies

The importance of policies for the protection of rivers is just one example of the importance of policies for environmental protection in general. The Income-Environment Relationship (IER) is a useful way to visualize this importance: it describes the generic dynamic relationship between the level of economic development and the quality of environment in a country. A particular proposition about the nature of this relationship is the Environmental Kuznets Curve (EKC) hypothesis, according to which the pollution level in a country will first increase as per capita income increases and then decrease after the country reaches a high level of per capita income.[12] With per capita income level on the horizontal axis and pollution level on the vertical axis, the EKC will therefore have an inverted-U shape (Figure 1.3).

The EKC is sometimes used to suggest policy irrelevance and therefore policy inaction, because it may be viewed as promoting a sense of inevitability and automaticity. For example, the rising part of the EKC curve may suggest that an increase in pollution is an inevitable outcome (by-product) of economic growth, prompting countries not to do much to curb pollution. Similarly, the falling part of the curve may suggest that after reaching a relatively high level of income (the threshold level, Y^*, in Figure 1.3), the level of pollution will decrease automatically, without requiring policy action.

A closer look however shows that this interpretation is not correct. Even proponents of the EKC have pointed out that the declining part of the EKC embodies policy action and is neither automatic nor guaranteed. For example, once a country reaches a high level of income, there is a greater demand for environmental quality and hence for pollution-curbing policies. Also, at a higher income level, a country acquires the financial and technological resources and

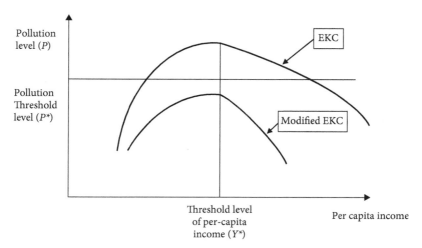

Figure 1.3 Environmental Kuznets Curve
Source: Author

institutional capability to curb pollution. Thus, both demand and supply side factors can play potential roles in bringing down the pollution level. However, to what extent this potential will be realized depends on conscious action, including appropriate government policies. Similarly, the EKC allows policy influence on its rising part too. For example, countries witnessing economic growth may use policies to flatten the rising portion of the curve. In particular, policies are necessary to ensure that environmental deterioration does not cross the threshold level, beyond which the damages become irreversible (P^* in Figure 1.3). As a result, it is possible to have a further policy-modified EKC, with a lower gradient in the rising part, a lower peak, and a quicker descent in the falling part, as compared to a benchmark EKC (Figure 1.3). Thus, instead of suggesting policy inaction, the EKC actually allows environmental policies to play a significant role at all stages of economic growth.

The role of policies in determining environmental outcomes is also clear from the empirical evidence showing that the IER in practice does not always conform to the EKC: the EKC held true for some pollutants, but not all. Also, the IER has been found to differ across regions and time periods. These variations suggest that there is no iron law that the pollution level has to first increase before decreasing. Instead, research suggests that the main role in determining the shape of the actual IER belongs to policies.[13]

These findings regarding protection of the environment in general apply equally to the role of policies in protecting rivers. In fact, the issue of irreversibility of damage is particularly relevant for rivers. Once a river is destroyed through inappropriate approaches and policies, it is often impossible to recover it, even if such a desire arises later. That is why knowing what policies and approaches are appropriate for protection of rivers is so important.

Also important is the fact that, unlike with many other environmental issues, the correct policies to be adopted regarding rivers are not always apparent. For example, in the case of climate change, there is little doubt that greenhouse gas (GHG) emissions need to be reduced, and it is also generally quite clear what needs to be done to reduce GHGs. This is however not the case with rivers. What is the best way to ensure sustainable use of rivers as a resource? Should rivers be dammed or let free to run? Should rivers be contained within embankments or allowed to overflow onto their floodplains? Answers to these questions are not always obvious. A more detailed understanding of the pros and cons of various approaches and policy options is needed to decide on such matters. The goal of this book is to help develop that understanding. It hopes to help countries achieve sustainable development by making more informed choices regarding approaches and policies toward rivers.

1.5 Alternative approaches to rivers

In advancing the conceptualization of river-related policies and issues, this book puts forward two pairs of fundamental concepts, namely the Commercial approach and its opposite, the Ecological approach; and the Cordon approach and its opposite, the Open approach.

The Commercial approach has two basic characteristics. First, it is focused on commercial gains, ignoring long-term ecological and ultimately human consequences. Second, it advocates the use of industrial era technologies to intervene in rivers to achieve those commercial gains. Meeting these commercial purposes generally requires construction of structures on rivers. The Commercial approach therefore is closely associated with what is known as the Structural approach. While the term commercial refers to the intention of the approach, the term structural refers to the method of realizing that intention. However, it is not accurate to equate the Commercial approach with the Structural approach, because some structures may be necessary even under the alternative, Ecological approach, as we shall see.

The interventions inspired by the Commercial approach can be classified into two broad types, namely frontal and lateral. Frontal structures are those that run across river channels, and hence are sometimes also called cross-sectional. These structures include dams, barrages, and weirs. Lateral structures intervene river flows from their sides, generally trying to constrict river flows to their channels so that they do not spread beyond their banks. Among lateral structures are embankments and floodwalls.

Based on the type of intervention involved, the Commercial approach can therefore assume two versions, namely frontal and lateral. The frontal version of the Commercial approach proceeds from the premise that "Any river water that passes to the sea is a waste!" This premise therefore promotes impounding and abstraction of river water for various commercial purposes, such as irrigation, electricity generation, and industrial and municipal use. Fulfilling these commercial purposes generally requires large-scale frontal interventions, such as construction of dams and barrages. The frontal version of the Commercial approach generally plays a more important role in the upper reaches of rivers where they pass through mountains and hills.

The lateral version of the Commercial approach proceeds from the premise that river flow should remain confined to its channels all the time and not overflow on to floodplains. It is thought that protection of floodplains from river overflows will increase their commercial value by allowing more crops to be grown and more development (meaning construction of dwellings and structures) to be undertaken. Achieving this commercial goal also requires large-scale lateral interventions in the form of embankments, flood-walls, and so on. Since the general goal of lateral interventions is to cordon off floodplains from river channels, I call the lateral version of the Commercial approach the Cordon approach. Clearly, this approach generally plays a larger role in the middle and lower reaches of rivers, where they pass through floodplains and deltas. Figure 1.4 presents these propositions in a schematic form.

In view of the discussion above, the book uses the term Commercial approach in two senses. The first is broader and includes both frontal and lateral versions; the second is narrower and refers to the frontal version only. The context of the discussion will make it clear in which sense the term has been used at a particular place in the book.

While the Commercial and Cordon approaches yielded many intended benefits, it was realized over time that these benefits were often less than expected; entailed substantial financial, environmental, and human costs; and could be achieved using alternative methods. This growing realization gave rise to the Ecological approach, which emphasizes the role of rivers as the main link in the earth's hydrological cycle, carrying precipitation water back to the seas, and creating, in the process, unique ecologies and cultures in their respective basins. Instead of viewing the passing of water by rivers to the seas as a waste, the Ecological approach views it as the basic function of rivers, necessary for maintaining the hydrological cycle of the earth and the ecology of river basins, including estuaries. The Ecological approach therefore discourages large-scale abstraction of river water and argues for fewer structural interventions in the volume and direction of river flows.

The Ecological approach also has two versions, frontal and lateral. The frontal version proceeds from the premise that carrying precipitation water to the seas is the basic hydrological function of rivers and, by performing this function, rivers

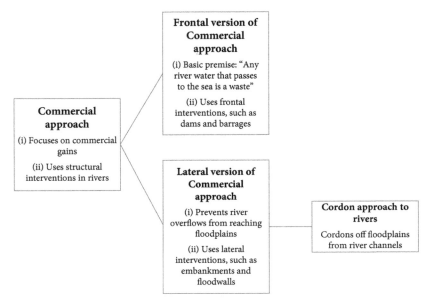

Figure 1.4 Commercial approach to rivers
Source: Author

also sustain the unique ecologies of their basins. The Ecological approach there-fore argues for preservation of the natural volume and direction of river flows, as much as possible, and hence discourages frontal interventions, such as dams and barrages. Clearly, the frontal version of the Ecological approach is more impor-tant for the upper reaches of rivers.

The lateral version of the Ecological approach proceeds from the premise that floodplains are as much a part of rivers as their channels, and river overflows on to floodplains are a natural and desirable phenomenon that performs vital ecological functions, necessary for both floodplains and river channels. The Ecological approach therefore argues for keeping floodplains open to river channels and discourages construction of continuous embankments that pre-vent rivers from overflowing onto them. To express the essence of the lateral ver-sion of the Ecological approach better, I have given it a separate name—the Open approach. It is clear that the Open approach is more relevant for middle and lower reaches of rivers, where they pass through floodplains and deltas. Figure 1.5 presents these propositions in a schematic form.

As is the case with the Commercial approach, the book also uses the term Ecological approach in two senses. The first, broader sense includes both frontal and lateral versions; the second, narrower sense refers to the frontal version only. The context will make it clear in which sense the term has been used in a partic-ular place of the book.

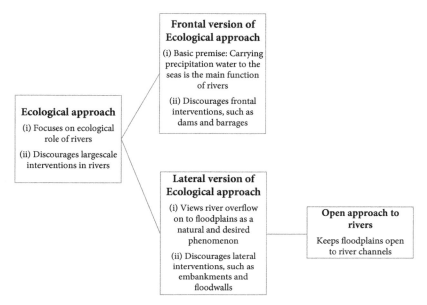

Figure 1.5 Ecological approach to rivers
Source: Author

Having proposed these concepts, the book draws upon river-related experience across the world to show that the Ecological and Open approaches are
more conducive to sustainable development than the Commercial and Cordon
approaches.

It is sometimes thought that the Commercial approach is anthropocentric,
and puts the interests of humans above those of other living species, while the
Ecological approach is non-anthropocentric, and treats the interests of human
beings and of other living species equally. This may not be an accurate presentation of the alternative approaches, because, in the long run, human interests
will also suffer if the ecological base on which human economy and society rests
is undermined. Hence, in the long run, the Ecological approach is more beneficial for human interests than the Commercial approach. At the same time, the
Ecological approach is more friendly to non-human species and life forms and to
the environment in general.

1.6 Overview of the book

This is a wide-ranging book, drawing upon different disciplines and experiences
of different parts of the world. To help the reader navigate through it, a brief
overview of its chapters follows.

1.6.1 Frontal version of the Commercial approach

The initial chapters of the book focus on the frontal version of the Commercial approach. Chapter 2 shows that this approach to rivers is one of the outcomes of the first Industrial Revolution that began in the second half of the eighteenth century and gained full speed in the nineteenth century. While rivers were used for some commercial purposes by pre-industrial societies, the Industrial Revolution made it possible to raise these uses to an entirely new level. It gave human societies both the technological capacity and the desire to use this capacity to dominate rivers. In addition, commercial underpinnings of the Industrial Revolution provided the motivation for full-scale commercial use of rivers. The Commercial approach finds its most clear manifestation in thousands of large dams and barrages that have been constructed on rivers across the world, beginning with developed countries and spreading thereafter in developing countries. The chapter provides an overview of this process.

Chapter 3 discusses the consequences of the Commercial approach. It shows that large dams and barrages built since the Industrial Revolution helped to regulate rivers, produce electricity, expand irrigation, extend navigation, set up industries, and expand urbanization and human settlements. However, the initial view of dams and barrages as unmixed blessings gave way later to doubt and apprehension as their damaging environmental, financial, and human consequences became more evident. Dams and barrages were found to fragment river basins and damage their morphology and ecology; encourage wasteful use of water and create the "More water, more thirst!" syndrome; generate problems of salinity and toxicity in irrigated areas; and diminish the downstream flow of water and sediment, causing harm to floodplains, deltas, estuaries, and the marine environment. In many cases, the Commercial approach led to complete exhaustion of flows, so that rivers became dry before reaching the sea. The Commercial approach, in general, also proved to be socially unfair, because the commercial gains generated by this approach were generally concentrated in the hands of those who either had more property to begin with or were more powerful socially. The Commercial approach also generally led to conflicts among co-riparian communities, states, and countries, because it encouraged each of them to abstract as much of the river flow as they could, leaving less for others.

Chapter 4 illustrates the consequences of the Commercial approach by examining a few concrete examples of the application of this approach. In particular, it reviews the experience of several famous rivers of the world, namely the Colorado River of the United States, the Murray-Darling River system of Australia, the Amu Darya and Syr Darya rivers of Central Asia, the Nile of

Africa, and the Indus River that flows through India and Pakistan. The chapter shows how, in each case, application of the Commercial approach led to the exhaustion of the river and also to substantial damage of river basin ecology.

Chapter 5 illustrates how the Commercial approach gives rise to conflicts by reviewing the experience of the South Asian subcontinent. It shows how application of this approach has led to conflicts between India and its neighboring countries, namely Bangladesh, Bhutan, Nepal, and Pakistan.

1.6.2 Frontal version of the Ecological approach

The Ecological approach arose in response to the adverse effects and various shortcomings of the Commercial approach. Chapters 6–8 are devoted to the frontal version of this approach. Chapter 6 presents the approach, explaining the distinction between its pre- and post-industrial versions and the relationship between the Ecological approach and "basin-wide strategies" for rivers. It shows that the Ecological approach is compatible with both the enlightened anthropocentric as well as the non-anthropocentric views regarding rivers. The chapter ends by showing how the Ecological approach is more conducive to sustainable development.

Chapter 7 provides the broader context in which the Ecological approach emerged. It shows how the Industrial Revolution brought about a trend-break in human history, with a transition from largely horizontal increase in production and consumption to an almost vertical increase, leading to breaches in planetary boundaries, which then required urgent action across the board, aimed at protection of the environment. The emergence of the Ecological approach is a part of that general process. There is therefore considerable synergy between adoption of the Ecological approach to rivers and environmental protection efforts in other areas.

Chapter 8 reviews the progress of the Ecological approach around the world. It notes the promising beginnings in the United States and the more comprehensive effort in Europe, as reflected in European Union's 2000 Directive on Water. Among developing countries, it notes the important role that the Ecological approach is playing in India, providing the basis for various river protection movements (such as the renowned Narmada Bachao Andolon), critique of the river linking project, and growing demand for demolition of the Farakka Barrage. The chapter notes the progress of the Ecological approach in other countries of Asia and in Latin America and Africa. However, it also shows that, despite such progress, the Ecological approach is yet to become the official policy in most developing countries as well as in many developed countries.

1.6.3 Cordon approach—lateral version of the Commercial approach

Chapter 9 presents the Cordon approach, the lateral version of the Commercial approach. It begins by explaining the floodplain nurturing functions—sometimes also called ecological functions—that regular river overflows play. It then shows the adverse consequences that result when, following the Cordon approach, floodplains are sealed off from river channels, depriving them of these nurturing functions, including sedimentation. The ecology of floodplains deteriorates, and riverbeds fill up with sediment that cannot settle on floodplains. Cordons encourage "below-flood-level settlement," which, together with river-bed aggradation, creates the risk of catastrophic flooding. As a result, flood damages end up increasing rather than decreasing. The Cordon approach generally also proves socially unfair, because the commercial benefits of cordons are often concentrated in the hands of large owners of land on the protected floodplains. In addition, common property floodplain resources, from which ordinary people can benefit, come under pressure and are often privatized. In particular, the disappearance of open (capture) fisheries and water transportation often affect adversely the employment and livelihood of low-income and non-propertied groups.

1.6.4 Open approach—lateral version of the Ecological approach

Over time, it became clear that cordons too are not an unmixed blessing: the harm that they cause to floodplains and river channels outweigh, in the long run and in a broader sense, the commercial benefits that they confer. This realization gave rise to the Open approach, which emphasizes the organic unity between a river channel and its floodplains. Chapter 10 explains the Open approach and shows how it helps to maintain and increase floodplain elevation, sustain soil quality, recharge floodplain waterbodies, allow natural gravity-based irrigation, and sustain the flora and fauna. It further shows how the Open approach helps to avoid below-flood-level settlement, river-bed aggradation, the danger of catastrophic flooding, the necessity for pump-based irrigation, and soil salinity and waterlogging. The chapter also explains how the Open approach can be socially fair, by preserving common-property land and water resources, which are usually an important source of livelihood for the less propertied sections of society. The chapter reviews the progress of the Open approach in both developed and developing countries and shows how experiences of catastrophic floods have led many developed countries to rethink and re-establish connections between river channels and floodplains. Similar experiences in developing countries have also generated social pressure for adoption of the Open approach.

1.6.5 Cordon approach in a delta

The contrast between the Cordon and Open approaches in promoting sustainable development is particularly stark in deltas, which comprise mostly floodplains and tidal plains. To illustrate this contrast, the books uses the experience of the Bengal Delta, the world's largest and most active delta, formed by three mighty river systems (the Ganges, Brahmaputra, and Meghna) and spreading across both Bangladesh and India. Chapter 11 presents the experience of the Cordon approach in the context of the Bengal Delta. It first reviews the experience of embankments along rivers in the Indian part of the Bengal Delta and then reviews the experience of the Bangladesh part of the delta, where the Cordon approach received wider application and gave rise to numerous cordons of different types. The chapter shows that, despite some initial success in raising agricultural output, the Cordon approach ended up doing more harm than good. The cordons created the risk of catastrophic flooding. Their irrigation and flood protection roles were undermined by the switch to winter rice as the main crop, which relies on ground-water for irrigation and does not require flood protection. Meanwhile, substitution of the natural, gravity system of drainage by the pump-based system, that the Cordon approach entailed, proved to be totally ill-suited for a country with high and extremely seasonal rainfall. As a result, ubiquitous water-logging has now emerged as the most visible sign of the inappropriateness of the Cordon approach in this delta. The chapter also shows how the Cordon approach has made the Bengal Delta more vulnerable to the adverse consequences of climate change.

1.6.6 Open approach in a delta

In view of the adverse consequences of the Cordon approach, there is now an increasing demand in both parts of the Bengal Delta for a switch to the Open approach. Chapter 12 reviews this process by noting first the views of river scholars of the Indian part of the Bengal Delta who argue for restoration of the normal river overflow on to floodplains. In the Bangladesh part of the Bengal Delta, the social demand for the Open approach was reflected in the strong opposition to the Cordon approach–inspired Flood Action Plan (FAP) that was to be imposed on Bangladesh following the historic flood of 1988 and the pressure from local people to modify several cordon projects. It is also reflected in the adoption and implementation of several irrigation projects that do not require construction of cordons. However, the chapter notes that, despite the progress achieved, a whole-hearted switch to the Open approach still faces many difficulties, in part due to path dependence. However, climate change is making this switch more urgent with each passing day.

1.6.7 Rivers, policies, and interests

The last chapter (chapter 13) of the book deals with the question of why it proves difficult to dislodge the Commercial and Cordon approaches from their dominant position and to switch to the Ecological and Open approaches, despite the greater merit of the latter in promoting sustainable development. In explaining this diffi-culty, the chapter notes that policy choices are not random occurrences and instead depend on the relative strength of social forces that rally behind different policy options. Generally, two factors determine why particular social groups or individ-uals advocate particular policies. One of these is knowledge. Some people may ad-vocate a policy because, according to their knowledge, it is the best policy. Another and often more powerful factor is material interests. Some people may advocate a policy because it is in their material interests. Sometimes, these two forces may work in the same direction. However, it is also possible for them to work in opposite directions. For example, some people may favor a policy that serves their own mate-rial interests even when they know that it is not the best for the broader community.

It so happens that the Ecological and Open approaches generally are at a dis-advantage, as compared to the Commercial and Cordon approaches, on account of both these forces. With respect to knowledge, understanding the merits of the Ecological and Open approaches generally requires more knowledge and infor-mation than is required for the merits of the Commercial and Cordon approaches.

With respect to material interests, the chapter distinguishes between imme-diate and end beneficiaries. The former comprises those whose material interests are served from the very implementation process of the policy, while the latter comprises those whose interests are served by the results of the implementation. In case of a dam project, for example, the contractors, consultants, bureaucrats, politicians, and so on are the immediate beneficiaries, while farmers, expecting to get irrigation water from the dam reservoir, are among the end beneficiaries.

The theory of policy choice shows that it is easier for beneficiaries who are small in number but whose potential per capita benefit is large to get organized and mount pressure for adoption of the policy they favor. By contrast, benefi-ciaries who are large in number but whose per capita benefit is small find it diffi-cult to do so. It so happens that Commercial and Cordon projects more generally involve immediate beneficiaries who are small in number but have large per-capita benefits.

It is the asymmetries with regard to both knowledge and material interests that make the switch to the Ecological and Open approaches more difficult. In developing countries, the difficulty is compounded by foreign influence—which takes the form of both knowledge (technical advice) and material interests (financing)—that generally still favor the Commercial and Cordon approaches.

However, nothing remains static. With time, the flaws of the Commercial and Cordon approaches are becoming more evident. Climate change is forcing upon policymakers and people many conclusions that were difficult to reach before. Hence, the prospects for a wholehearted switch to the Ecological and Open approaches are improving. The world community has already adopted the goal of sustainable development. With more diffusion of appropriate knowledge, it may not be too long to the day when the world community will also adopt the Ecological and Open approaches to rivers. More public knowledge will help to overcome the resistance of material interests. This book aims at contributing to this diffusion of knowledge.

Notes

1. The 2030 Agenda, with its SDGs, succeeded the Millennium Development Goals (MDGs), whose reference period expired in 2015. See UN (2015a) for the Agenda 2030 and UN (2000) for the MDGs.
2. See LeBlanc (2015) for a discussion of the linkages among SDGs and the possibility of identification and use of these nexuses.
3. This commission is popularly known as the Brundtland Commission, named after its Chairman, Gro Brundtland, former prime minister of Norway. Set up in 1983, the commission worked for several years and published its report in 1987.
4. See WCED (1987).
5. See Islam (2014), for a discussion of various interpretations of the definition of sustainable development.
6. See Islam (2014) for discussion of the three dimensions of sustainable development.
7. This is a reductive interpretation of the expression social development. Economy is a part of society, therefore, social development subsumes economic development. However, in conventional usage, social development is often considered to be an addendum to economic growth/development. See Islam (2014) for further discussions of these issues.
8. The thorny issue of substitutability between natural and produced capital finds a clear reflection in the context of rivers. For example, dams and barrages are examples of produced capital, while uninterrupted rivers are an element of natural capital. Are they complementary or substitutes? To the extent that they are substitutes, is the next generation better off with rivers fettered by dams and barrages or it will be better off with healthy rivers running free? These are some of the questions that need to be discussed.

9. The following formulations of the various targets of SDG-6 makes this clear:

   (6.4)   By 2030, substantially increase water-use efficiency across all sectors and ensure sustainable withdrawals and supply freshwater to address water scarcity and substantially reduce the number of people suffering from water scarcity.

   (6.5)   By 2030, implement integrated water resources management at all levels, including through transboundary cooperation as appropriate.

   (6.6)   By 2020, protect and restore water-related ecosystems, including mountains, forests, wetlands, rivers, aquifers, and lakes. (UN 2015a).

10. The targets of SDG-15 include:

    (15.1) By 2020, ensure the conservation, restoration, and sustainable use of terrestrial and inland freshwater ecosystems and their services, in particular forests, wetlands, mountains and drylands, in line with obligations under international agreements.

    (15.3) By 2030, combat desertification, restore degraded land and soil, including land affected by desertification, drought and floods, and strive to achieve a land degradation-neutral world.

    (15.5) Take urgent and significant action to reduce the degradation of natural habitats, halt the loss of bio-diversity, by 2020, protect and prevent extinction of threatened species. (UN 2015a).

11. SDG target 2.3 is formulated as follows: "(2.3) By 2030, ensure sustainable food production systems and implement resilient agricultural practices that increase productivity and production, that help maintain ecosystems, that strengthen capacity for adaptation to climate change, extreme weather, drought, flooding and other disasters and that progressively improve land and soil quality" (UN 2015a).

12. The EKC draws its name and inspiration from the Kuznets' Curve (KC) hypothesis regarding the relationship between level of per capita income and degree of income inequality. According to the KC, inequality will first increase, as the economy grows, and will then decrease after the economy has reached a high level of per capita income. For discussion on the IER, KC, and EKC, see Arrow, Bolin, Costanza, Dasgupta, Folke, Holling, Jansson, Levin, Meller, Perrings, and Pimentel (1995), Griffith (1994), Grossman and Krueger (1995), Islam (1997b, 1999b), Jha (1996), Kaufmann, Davisdotter, and Garnham (1995), Kuznets (1955, 1963), Panayotou (1995), Selden and Song (1994, 1995), Shafiq and Bandyopadhaya (1992), Song (1993), Vincent (1997) and others.

13. See Islam (1997b) for relevant analyses.

2

Commercial approach to rivers

Origin and spread

2.1 Introduction

From the very beginning, the relationship of human societies with rivers had two sides. On the one hand, they found rivers to be an important resource. They drank water from them, caught fish, and used them for transportation, bathing, and watering their crops. On the other hand, they respected and venerated rivers; in some cultures, rivers were worshipped as gods and goddesses.

The Industrial Revolution however changed the situation dramatically. It led to the absolutization of the resource side of rivers and gave rise to the Commercial approach, according to which rivers are primarily an economic resource to be exploited to maximize various commercial gains. The philosophy behind this approach is reflected best by the saying that "Any river water that passes to the sea is a waste!"

The Commercial approach therefore undervalues the basic role of rivers in maintaining the earth's hydrological cycle and in preserving the ecology of river basins, including deltas and estuaries. Sometimes the Commercial approach is characterized as an anthropocentric approach, implying that it upholds the interests of human beings. However, this is not accurate, because human interests are not served well in the long run if the earth's hydrological cycle and river basin ecology are harmed in the process. The Commercial approach may therefore be termed more accurately as a myopic approach, because it focuses on short-term gains while ignoring long-term consequences.

The material basis of the Commercial approach lies in the technological advances that the Industrial Revolution brought about, providing human societies the technological capacity (in particular, the power of machines) to impose their will on nature. Along with the capacity came the desire to do so. This capacity and desire combined with the commercial underpinnings of the Industrial Revolution to give rise to the Commercial approach to rivers.

Implementation of the Commercial approach generally requires construction of structures aimed at altering the direction, volume, breadth, and pace of river flows, in a scale that was impossible during the pre-industrial era. These structures may be broadly classified into two groups, namely frontal and lateral.

Rivers and Sustainable Development. S. Nazrul Islam, Oxford University Press (2020). © Oxford University Press.
DOI: 10.1093/oso/9780190079024.001.0001

Accordingly, I distinguish two versions of the Commercial approach, namely frontal and lateral. The current and the following three chapters focus on the frontal version of the Commercial approach, even though, for the sake of parsimony, we do not always include the term frontal version when referring to it. This chapter discusses the origin of the Commercial approach and its link with the Industrial Revolution. It then follows the spread of this approach across the world, first in developed countries and later in developing countries.

The discussion of this chapter is organized as follows. Section 2.2 examines the origin and purpose of the Commercial approach. Section 2.3 catalogues various types of frontal intervening structures that the Commercial approach deploys to achieve its purpose. The rest of the chapter follows the spread of the frontal version of the Commercial approach across the world. Section 2.4 follows this process for developed (or early industrialized) countries. Section 2.5 does the same for developing countries. Section 2.6 presents a set of tables that portrays the current global picture regarding dams and barrages. Section 2.7 concludes the chapter.

2.2 Commercial approach: origin and purpose

The Commercial approach to rivers is basically an outcome of the Industrial Revolution, the hallmark of which, from the technological viewpoint, was substitution of muscle power with machine power. Availability of the latter resulted in a qualitative leap in human capacity that could then be used to "conquer" nature. Early industrializing societies embarked on this conquest with considerable gusto.

One of nature's objects that became the target of the human industrial might is rivers.[1] Reflecting the commercial underpinnings of the Industrial Revolution, rivers were considered as a resource to be utilized for commercial gains, and this commercial motivation drove both frontal and lateral interventions in rivers. In the case of frontal interventions, the motivation was to impound and divert water for various commercial purposes, especially generation of electricity and expansion of irrigation. Other commercial uses included the supply of river water for expansion of urbanization and setting up of industries.[2] Navigation is another form of commercial use of rivers, though it does not usually require water abstraction. Though many dams and weirs were originally aimed at increasing navigation, this particular use of rivers gradually diminished in importance because of the growth of railway and road transportation.

The commercial motivation behind lateral interventions is also clear. The idea here is that the commercial value of floodplains will be enhanced if these can be made flood-free through embankments, floodwalls, etc., thus allowing more intensive agriculture and development in them.

Rivers were used for some of these commercial purposes by pre-industrial societies too, and these uses sometimes required construction of river-intervening structures. For example, the water of the Euphrates, Indus, Huang He (Yellow), Nile, and Tigris rivers was used for irrigation by the Bronze Age civilizations that developed in their valleys. Dams and irrigation canals were built in other parts of the pre-industrial world and during later ages too. For example, aqueducts were constructed during the time of the Roman Empire to transport water from rivers for municipal use in cities. Similarly, canals were constructed in the Middle Ages connecting the Yangtze River with the Huang He River in China and the Ganges River with the Jamuna River in India.[3]

However, these pre-industrial interventions did not have either the design or the capacity to alter river basin ecosystems. Rarely did they take the form of dams over major rivers. For example, dams during the Bronze Age in the Indus and Mesopotamian valleys were generally removable earthen dams on small rivers or rivulets, built during the lean season and removed before the peak season ensued. In other cases, these were low-height dams (weirs) on small rivers to store water or raise the river level to direct water to irrigation channels or aqueducts. Mostly these were modest, gravity flow diversions from existing river flows, which did not lead to significant changes in the morphology of rivers or the ecology of river basins.

The Industrial Revolution enabled human societies to build river-intervening structures of entirely different magnitude and scale. It made available to them entirely different technologies embodied in steel, reinforced concrete, dynamites, motorized earth-moving equipment, hydraulic hammers, modern engineering knowledge, etc. These technologies made it possible to raise the commercial use of rivers to an entirely new level: the consumption of the entire flow of rivers now became the goal. Consequently, the scale of commercial uses during the industrial age is designed to and/or is capable of having fundamental effects on the earth's hydrological cycle and river ecosystems.

It should be noted that the commercial use of rivers during the industrial era had other important consequences, such as pollution and encroachment,[4] both of which are getting more acute in many developing countries as they industrialize. This book however—to keep its scope manageable—leaves these issues aside focuses mainly on those commercial uses that require impounding and abstraction of river water. At the same time, it should be noted in this context that some of the water used for irrigation, industrial, and municipal purposes generate return flows to rivers through surface run-off, seepage, and groundwater aquifers. These return flows often contain many toxic substances that pollute rivers. Thus, the discussion of pollution cannot be avoided entirely, even if the focus is on impounding and abstraction.

2.3 Frontal intervention structures

Frontal interventions in rivers, inspired by the Commercial approach, take different forms. Most common are dams, barrages, and weirs.

2.3.1 Dams

Dams represent the most radical frontal intervention: they stop the natural flows of rivers altogether and then allow only a regulated amount to pass through sluice gates, spillways, and pipes. Dams can be of many different types and varieties, depending on their purposes and the physical characteristics of the rivers they are built on, these being interrelated. Dams are often multipurpose, serving several goals at the same time. River flow regulation is an inherent outcome of all dams, and dams are also invariably associated with the creation of water reservoirs at their heads, creating a hydraulic differential[5] that can be used to generate electricity. The water of reservoirs is also often transported (through canals) to far-away destinations to be used for irrigation as well as industrial and municipal purposes. Hydroelectricity generation and irrigation are the two most prominent rationales offered for dam construction. While hydropower generation requires impounding river flows, it does not require abstraction of water. Expansion of irrigation however always requires abstraction of river water.

Dams also differ in terms of their construction. The simplest are embankment dams, which are constructed in the same way as embankments. These are suitable for wide rivers and are constructed mainly through earth filling or a combination of rock and earth filling, with a brick or concrete outside layer. These dams may or may not have reinforcements in the form of buttresses. Another type is the gravity dam, whose weight should be sufficient to hold it in place. These are suitable in places where the base-anchoring may not be strong. Some embankment dams can be gravity dams too. Concrete dams, a third type, are built in the form of concrete walls often arched toward reservoirs. These are suitable in rocky gorges or canyons where surrounding hard rocks can provide strong moorings for the concrete walls that serve as the dam. There are many other types and variations of dams, depending on the situation. Table 2.1 and Figure 2.1 show that earth dams are the most numerous, accounting for about 65 percent of all dams, with rockfill and gravity dams next in importance.

Dams are often classified in terms of their size. In particular, certain dams are classified as "large" or "major" dams, though there are no consensus definitions for such characterization. For example, the International Commission on Large Dams (ICOLD)—a trade association representing large dams—defines large

Table 2.1 Dams by type of construction

Type of dam	Number	Percent of total (%)
Earth	38,426	65.4
Rockfill	7,670	13.0
Gravity	7,450	12.7
Buttress	419	0.7
Arch and multiple arch	2,332	4.2
Other	2,361	4.0
Total	58,791	100.0

Source: ICOLD. (as of May 3, 2019).

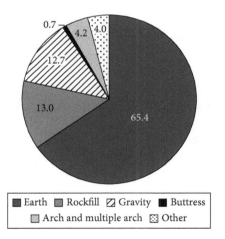

Figure 2.1 Dams by type of construction
Source: Author, based on data in Table 2.1

dams as measuring 15 m or more in height from foundation to crest, which is taller than a four-storied building. On the other hand, a major dam is defined based on either its height (at least 150 m); volume of the dam structure (at least 15 million cubic m);[6] reservoir storage (at least 25 cubic km);[7] or electricity generation capacity (at least 1,000 MW).[8]

2.3.2 Barrages

Barrages are also radical interventions in river flows, though less so than dams, as they do not stop river flows altogether. Instead, they force rivers to flow through sluice gates, which may be used to regulate downstream river flows and to raise upstream river levels. Barrages therefore usually do not form large reservoirs at their heads and cannot produce hydroelectricity. However, raised upstream river levels are generally used to divert water to other destinations, for irrigation or for other purposes. Barrages may also differ with regard to the materials and technology used for construction, desired level of increase in the head level of water, and so on. Barrages are often also called "run-of-river dams."

2.3.3 Weirs

Weirs are generally low-height dams, meant to hold river flow up to a particular height and allow the excess to flow over the top. Barrages and weirs together are often referred to as the run-of-the-river dams, because they do not stop rivers from running, as dams do. Weirs also differ with regard to the material and technology used for their construction, their intended height, and other considerations. They are generally constructed on smaller rivers.

2.4 Spread of Commercial approach in developed countries

The Commercial approach arose in the early industrializing countries and then spread to the developing world. Since dams embody the Commercial approach in its most aggressive form, the progression of this approach can be seen most clearly by following the progression of dam building.[9] Industrial era dam building started in the nineteenth century, and by 1940, about 5,000 large dams had been built, three-fourths of which were in developed countries.[10] However, there was a spree in dam building peaking in the 1970s and 1980s, when at least one large dam was constructed somewhere in the world almost every day. About 5,000 large dams were built between 1970 and 1975. By the end of the 1990s, a total of 45,000 large dams were constructed in 140 countries. In the following, I offer a brief review of the progression of dam building, going by regions.[11]

2.4.1 Europe

Not surprisingly, the first dams built with industrial-era technologies were constructed in the United Kingdom (UK) in the nineteenth century. Being an island country, the UK rivers are not long, with the three longest, the Severn, the Thames, and the Trent Rivers being 220, 215, and 185 km long, respectively. However, they are steep enough in parts to possess hydropower potential, and many dams were constructed to use this potential and to serve other commercial purposes. As of now, the UK has 596 large dams (ICOLD 2019).

Other countries in continental Europe soon followed suit, as they did with regard to industrialization in general. Major rivers of western Europe include the Rhine and the Danube, both of which originate in the Alps but flow in almost opposite directions. The Rhine flows in a north-westerly direction through Switzerland, Germany, and the Netherlands to the Baltic Sea. The Danube, on the other hand, flows east through Germany, Austria, Hungary, Bulgaria, and Romania to reach the Black Sea. Other important rivers of Europe include the Elbe, Order, and Vistula, flowing north to the Baltic Sea through, variously, the Czech Republic, Germany, and Poland. The rivers of France flowing west to the Atlantic Ocean include the Seine, Loire, and Garonne. The Douro, Tagus, and Guadalquivir rivers of the Iberian Peninsula also flow west to the Atlantic Ocean. On the other hand, the Rhone of France, Ebro of Spain, and Tiber of Italy flow to the Mediterranean. Italy's Po flows to the Adriatic Sea.

European countries have built numerous dams and barrages on most of these rivers (see Figure 2.2). In fact, Spain ranks ninth in the world in terms of number of large dams (1,063 of them). Similarly, France and Italy rank twelfth and sixteenth with 712 and 542 large dams, respectively. Other major dam-building European countries include Germany (371), Albania (307), Romania (246), Portugal (217), Bulgaria (181), Austria (171), and Switzerland (167). Rivers in the Scandinavian countries are also not very long, but because of the mountainous terrain (particularly of Norway), they have hydroelectric potential. These countries have also undertaken dam construction on a significant scale, with Norway, Sweden, and Finland having 335, 190, and 56 large dams, respectively.

2.4.2   United States and Canada

Though early industrializing countries of Europe pioneered the new dam era, it was the United States which pushed dam construction to a higher level in the early twentieth century. The river landscape of the United States is dominated by the Mississippi-Missouri system, with its tributaries, such as the Ohio and Tennessee Rivers from the east and Arkansas and Red Rivers from the west. Other major US rivers include the Snake-Columbia River system in the northwest and Colorado River system in the southwest. In the east, important rivers include the Hudson (Vermont, New York, Connecticut), Roanoke, Savanna (Georgia), and Kissimmee (Florida).

Europe: Dams

Global Reservoir and Dam Database, Version 1 (GRanDv1), Revision 01

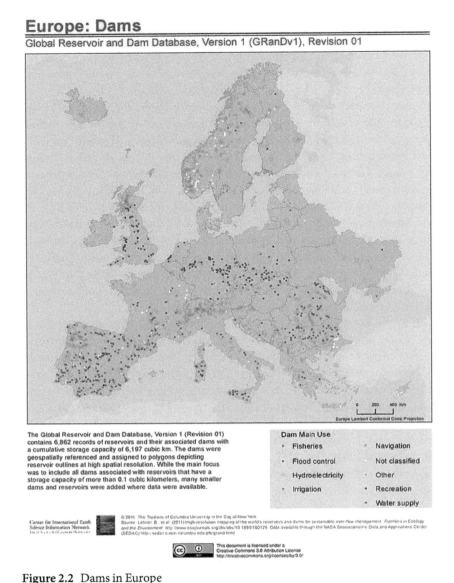

The Global Reservoir and Dam Database, Version 1 (Revision 01) contains 6,862 records of reservoirs and their associated dams with a cumulative storage capacity of 6,197 cubic km. The dams were geospatially referenced and assigned to polygons depicting reservoir outlines at high spatial resolution. While the main focus was to include all dams associated with reservoirs that have a storage capacity of more than 0.1 cubic kilometers, many smaller dams and reservoirs were added where data were available.

Dam Main Use

- Fisheries
- Flood control
- Hydroelectricity
- Irrigation
- Navigation
- Not classified
- Other
- Recreation
- Water supply

Figure 2.2 Dams in Europe
Source: Socioeconomic Data and Applications Center (SEDAC) of NASA

The Bureau of Reclamation (BuRec), officially referred to as USBR and founded in 1902, led the massive dam building campaign in the western part of the United States, building scores of dams on the Colorado and Columbia Rivers and their tributaries as well as on other rivers of the region. These include the Hoover Dam, Grand Coulee Dam, and Glen Canyon Dam. In the east, the Tennessee Valley Authority (TVA) was formed in 1933 to promote generation of electricity and expansion of irrigation using the Tennessee River and its tributaries. Following its mandate, the TVA constructed scores of dams and other river intervening structures in the Tennessee valley. Altogether, the United States now ranks second in the world in terms of the number of large dams, having 9,265 of them. It is also the home of many of the world's major dams (see Figure 2.3).

Major rivers of Canada include the St Lawrence in the east, flowing from the Great Lakes to the North Atlantic, and the Mackenzie River that flows north to the Beaufort Sea of the Arctic Ocean. Several Canadian rivers—such as the Nelson, Severn, and Albany—flow to Hudson Bay. Other important rivers include the North and South Saskatchewan and the Pease. Canada also shares several rivers with the United States, such as the Columbia River, flowing to the south, and the Yukon River in the north, flowing toward Alaska. Many of these rivers originate in the Canadian Rockies and have steep gradients. Canada has built many dams on its rivers and ranks eighth in the world in terms of the number of large dams, with 793.

2.4.3 Australia

The major river system of Australia is the Murray-Darling, located in the southeast of the country. However, there are many other rivers across the continent, flowing out to both the Indian and Pacific Oceans. Australia is also a major builder of dams and ranks fifteenth in terms of the number of large dams, having 486 of them (see Figure 2.4). Many of Australia's dams and barrages are concentrated on the Murray-Darling system, which will be discussed in more detail in chapter 4.

2.4.4 Union of Soviet Socialist Republics (USSR)

The Commercial approach to rivers was not limited to capitalist industrial countries. Countries following the socialist model of industrialization also embraced this approach in their eagerness to outpace the capitalist world with regard to production and industrial accomplishments. In fact, the central planning system of these countries allowed them to concentrate a greater proportion of resources on favored projects and thus to undertake river-intervention projects of even greater scale.

For example, one of the major projects of the Soviet First Five-Year Plan (1928–1933)—the plan of industrialization, as it was called—was the *Dnieprostroi*, the

North America: Dams
Global Reservoir and Dam Database, Version 1 (GRanDv1), Revision 01

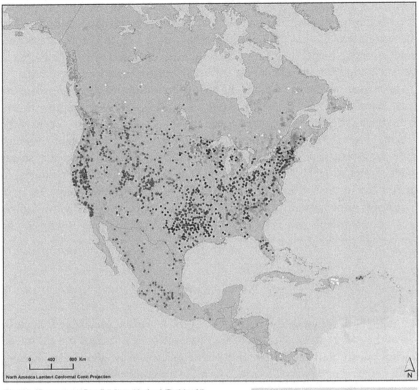

The Global Reservoir and Dam Database, Version 1 (Revision 01) contains 6,862 records of reservoirs and their associated dams with a cumulative storage capacity of 6,197 cubic km. The dams were geospatially referenced and assigned to polygons depicting reservoir outlines at high spatial resolution. While the main focus was to include all dams associated with reservoirs that have a storage capacity of more than 0.1 cubic kilometers, many smaller dams and reservoirs were added where data were available.

Dam Main Use

- Fisheries
- Flood control
- Hydroelectricity
- Irrigation

- Navigation
- Not classified
- Other
- Recreation
- Water supply

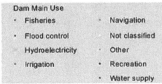
Figure 2.3 Dams in North America

Source: Socioeconomic Data and Applications Center (SEDAC) of NASA

Oceania: Dams

Global Reservoir and Dam Database, Version 1 (GRanDv1), Revision 01

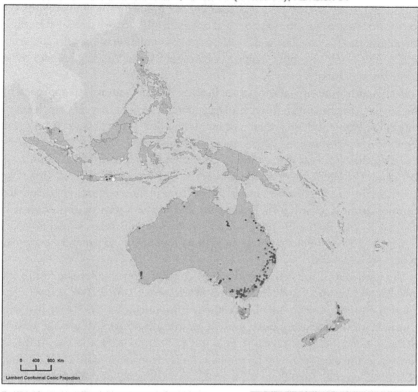

The Global Reservoir and Dam Database, Version 1 (Revision 01) contains 6,862 records of reservoirs and their associated dams with a cumulative storage capacity of 6,197 cubic km. The dams were geospatially referenced and assigned to polygons depicting reservoir outlines at high spatial resolution. While the main focus was to include all dams associated with reservoirs that have a storage capacity of more than 0.1 cubic kilometers, many smaller dams and reservoirs were added where data were available.

Dam Main Use

- Fisheries
- Flood control
- Hydroelectricity
- Irrigation
- Navigation
- Not classified
- Other
- Recreation
- Water supply

Figure 2.4 Dams in Oceania

Source: Socioeconomic Data and Applications Center (SEDAC) of NASA

massive dam on the Dnieper River, one of the major rivers of the former Soviet Union, flowing through both Russia and Ukraine and reaching Black Sea near the city of Odessa. At the 1932 inauguration ceremony for the Dnieprostroi, the world's first major dam, the deputy chief engineer told a crowd of 60,000 workers and dignitaries that the dam was ". . . the mighty foundation of socialist construction."[12] The USSR also built dams and barrages on the Volga, Don, and other major rivers of Russia.

Particularly consequential was application of the Commercial approach by the Soviet government to the Amu Darya and Syr Darya Rivers, which originate in the Pamir and Tian Shan mountains and flow through the central Asian republics of Kirghizstan, Uzbekistan, Turkmenistan, and Tajikistan before reaching the Aral Sea, the second largest inland waterbody in the country at that time. The USSR government dammed and diverted the flows of these two rivers northward to promote cultivation of crops in the steppes (to be discussed in more detail in chapter 4). Toward the end of the Soviet period, there was also a talk of reversing the northward flows of Siberian rivers, namely the Ob, Yenisei, and Lena, in order to use these flows for various commercial purposes in the south and east.

The enthusiasm for dam construction continued even after the fall of the socialist regime and the disintegration of the Soviet Union. For example, the Russian water minister Nikolai Mikheev announced in 1995 that his government was considering once more a mammoth scheme to reverse several of the major Siberian rivers to run south to Central Asia and the Aral Sea. Among future dams planned by the Russian Hydroproject Institute is a 20,000 MW giant dam at Turukhansk, on an eastern tributary of the Siberian River Yenisei. Similarly, Ukraine has completed construction of the Kakhovskaya Dam, which holds back the world's largest capacity reservoir.[13] Since Russia spreads across both Asia and Europe, its dams can be seen in both Figure 2.2 and Figure 2.5.

2.4.5 Japan

Japan, one of the early industrializing countries of Asia, also constructed many dams. Being a mountainous island country, its rivers are short, but they do have steep gradient at certain stretches, thus having considerable hydroelectric potential. Japan built many dams to use this potential and for other commercial purposes. In fact, it ranks fourth in the world in terms of the number of large dams, having 3,113 of these (see Figure 2.5).

Asia: Dams

Global Reservoir and Dam Database, Version 1 (GRanDv1), Revision 01

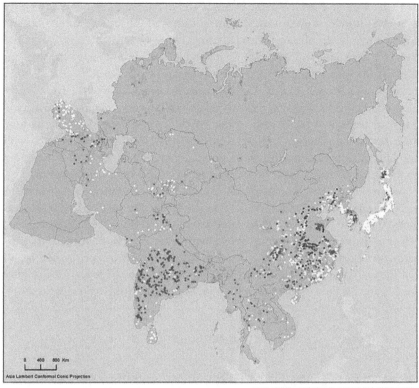

0   400   800 Km

Asia Lambert Conformal Conic Projection

The Global Reservoir and Dam Database, Version 1 (Revision 01) contains 6,862 records of reservoirs and their associated dams with a cumulative storage capacity of 6,197 cubic km. The dams were geospatially referenced and assigned to polygons depicting reservoir outlines at high spatial resolution. While the main focus was to include all dams associated with reservoirs that have a storage capacity of more than 0.1 cubic kilometers, many smaller dams and reservoirs were added where data were available.

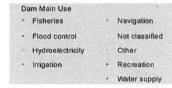

Dam Main Use

- Fisheries
- Flood control
- Hydroelectricity
- Irrigation
- Navigation
- Not classified
- Other
- Recreation
- Water supply

Center for International Earth
Science Information Network

© 2011. The Trustees of Columbia University in the City of New York.
Source: Lehner, B., et al. (2011) High-resolution mapping of the world's reservoirs and dams for sustainable river-flow management. *Frontiers in Ecology and the Environment* http://www.essayreuse.org/doi/abs/10.1890/100125. Data available through the NASA Socioeconomic Data and Applications Center (SEDAC) http://sedac.ciesin.columbia.edu/gis/grand.html

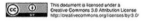
Figure 2.5   Dams in Asia

Source: Socioeconomic Data and Applications Center (SEDAC) of NASA

2.5 Spread of the Commercial approach in developing world

After the Second World War, the main theater of dam and barrage construction shifted to developing countries. Many Asian and African countries became independent and embraced the Commercial approach to rivers. Jawaharlal Nehru, prime minister of newly independent India, famously characterized dams as the "temples, mosques, and cathedrals of our times."[14] Gamal Abder Naser, Arab nationalist leader of Egypt, made construction of the High Aswan Dam one of his main goals. In China, Mao Zedong encouraged dam construction and promoted the dream of the Three Gorges Dam. The International Bank for Reconstruction and Development (IBRD), the newly created development financing institution, more popularly known as the World Bank, took up financing of dam and barrage construction as one of its lending priorities.

As a result of all of this, global dam construction rate witnessed a phenomenal increase during the 1950s and 1960s. The trend peaked in the 1970s, when it is estimated that on average two or three large dams were commissioned each day somewhere in the world. All over the developing world, there was a wave of dam and barrage construction and large-scale abstraction of water from rivers for various commercial purposes. The following offers a brief review of the process, with Figures 2.5, 2.7, and 2.8 showing dams in Asia, Africa, and South America, respectively.

2.5.1 China

Being a mountainous country with many rivers originating from the Himalayan Mountains and the Tibetan Plateau, China has considerable hydroelectric potential. Three main rivers of China are the Huang He River in the north, Yangtze River in the middle, and the Pearl River in the south.

Following the revolution in 1949, China embarked on a major campaign of dam and barrage construction, which received further impetus during Mao's Great Leap Forward program of 1957–1958. China's fervor for dam construction continues to this day, as can be seen in the recent completion of the world's most powerful dam, the Three Gorges Dam on the Yangtze River.[15] China now ranks first in the world in terms of number of large dams, having 23,842 of them (ICOLD 2019). Apart from the huge number of large dams, China is estimated to have around 96,000 small dams (see Figure 2.5).

China's dam-building spree continues. It has plans to build thousands of more dams, including such mega dams on the Yangtze, upstream of Three Gorges, as the 14,400 MW Xiluodo Dam (which would be the second most powerful dam in the world after the Three Gorges) and the 6,000 MW Xiangjiaba Dam.[16]

2.5.1.1 China as the major dam builder in other developing countries

Having gained considerable expertise at home, China's construction companies have emerged as major dam builders abroad, particularly since the end of the 1990s. In fact, the Chinese state-owned Sinohydro Corporation has now emerged as the world's largest hydropower company. The bilateral development assistance and commercial credit offered by China often focus on dam projects. Thus, the China Export-Import Bank (China Exim Bank) has become a major financier of large dams.

As of August 2012, about 308 dam projects underway in 70 countries involved Chinese construction companies and financiers.[17] These are spread across the world, including Europe, where 12 such dams are being built (Table 2.2). However, the largest concentration is in Southeast Asia, followed by Africa, South Asia, and Latin America. Among the Southeast Asian countries, the largest number of dams built with Chinese assistance are in Myanmar, followed by Laos, and Malaysia. Among the South Asian countries, Pakistan has the largest number of dams built with Chinese assistance.[18] Africa has another concentration of dams constructed with Chinese help, with about 85, representing 28 percent of the total. The number of Chinese dam projects in Latin America has grown rapidly since 2008 and accounted for 8 percent.[19]

China's engagement in dam construction overseas received further impetus from the Belt and Road Initiative (BRI), announced by the Chinese leader Xi Jinping in 2013. Since then, BRI has emerged to be a major international program, with more than 65 countries already participating in it and many more

Table 2.2 Dams constructed overseas with Chinese assistance

Region	Number of dams	Country of Southeast Asia	Number of dams
Southeast Asia	131	Myanmar	55
Africa	85	Lao PDR	28
South Asia	36	Malaysia	14
Latin America	23	Cambodia	11
Europe	12	Vietnam	9
East and Central Asia	11	Philippines	4
Middle East	5	Indonesia	3
Pacific	3	Brunei, Papua New Guinea, Thailand	1

Source: International River Network (2012, p. 3).

expected to do so in future. The total investment to be made under this program is estimated to be between $4 and 8 trillion. Though mostly focused on communications projects (such as railroads and ports), dams and barrages are also an important part in the BRI project portfolio, particularly in developing countries. For example, in Laos and Kampuchea alone, about twenty dams have been included in the BRI program.[20]

2.5.2 India

India is a major dam builder, ranking third—after China and the United States—in terms of the number of large dams, having more than 5,000 (with half constructed between 1971 and 1989). As noted earlier, India's nationalist leader and prime minister, Jawaharlal Nehru characterized dams as the "temples, mosques, and cathedrals" of modern times.[21] Accordingly, after gaining independence in 1947, India took up an active program of building dams and barrages, and as a result, dams are now ubiquitous in the country (see Figure 2.5).

India's numerous rivers may be divided into three broad groups. The first comprise those originating from the Himalayan Mountains and their eastern extensions. These rivers may in turn be divided into three subgroups. The first being rivers flowing west and forming part of the Indus River system, including the Ravi, Bias, Sutlez, Jhelum, Chenab, and Indus Rivers. The second subgroup comprises rivers that flow south and serve as tributaries of the Ganges River, including the Jamuna, Gomti, Ghagra, Gandhak, Kosi, and Mechi Rivers. The third subgroup comprise rivers that serve as tributaries of the Brahmaputra and Meghna Rivers, including the Teesta, Sankosh, Manas, and Barak Rivers.

The second group of India's rivers flow out from the Bindhya mountains located in central India. Some of these, such as the Narmada and Tapi Rivers—flow west toward the Arabian Sea, while others—such as the Mahanadi River—flows east to the Bay of Bengal. Yet others—such as the Chambal, Betwa, Ken, and Sone rivers—flow north to feed the Ganges River. The third group of Indian rivers comprises rivers of the Deccan Peninsula and includes such rivers as Godavari, Krishna, Pennar, and Cauvery.

India has built dams on rivers belonging to all these groups and subgroups and has accelerated its pace of dam construction in recent years. Many of these planned dams are on rivers of its seven northeastern states. Among these are the Lower Subansiri and Dibang Dams, with electricity generation capacity of 2,000 MW and 3,000 MW, respectively.

Having gained considerable expertise at home, India is engaging in dam construction abroad, particularly in the neighboring countries of Nepal and Bhutan, as we will see in more detail later.

2.5.2.1 Indian River Linking Project (IRLP)

In a major thrust of the Commercial approach, the Indian government is working on a mammoth project to interlink rivers that flow through India. Under this project, the rivers of the Himalayan Mountains and Deccan Peninsula will be connected among themselves and these two river systems will also be connected with each other (see map in Figure 2.6). The basic premise behind the project is that some river basins are "surplus" in water while others are "deficit," and hence connecting them will enable better utilization of the rivers by allowing transfer of water from the surplus basins to the deficit ones.

The project envisages the transfer of 334 cubic km of water through construction of 30 inter-river links, involving 36 big dams, 94 tunnels, and 10,876 km of canals. Of the 30 canals, 14 are to connect the Himalayan rivers, thus forming the Himalayan part of the project, while the remaining 16 are to connect the Peninsular rivers, thus forming the Peninsular part of the project. The two parts are connected with each other, particularly through Link 10 that connects the Ganges with the Subarnarekha and Link 11 that connects the Subarnarekha with the Mahanadi, which is part of the Peninsular component (see map in Figure 2.6). According to the project proponents, the main surplus river is the Brahmaputra River, and IRLP is geared to transfer water from this river to other rivers of both

Figure 2.6 River Linking Project of India
Source: India Water Portal

the Himalayan and Peninsular components. The preliminary estimated cost of the project is $215 billion.[22]

India's river linking project is the largest river intervention project ever conceived. In terms of its scope, it defeats even the Three Gorges Project of China (TGPC), which, however big, represents intervention in one river only (the Yangtze). By contrast, IRLP proposes simultaneous intervention in several major rivers. Also, while the TGPC concerns a river that lies entirely within the boundary of the country undertaking the project, IRLP involves rivers that are international and hence affects the co-riparian countries, particularly the downstream country of Bangladesh. If implemented, IRLP is likely to change permanently the ecological landscape of these countries. It has therefore emerged as a major source of conflict among these countries, as will be discussed later.

Given its importance, IRLP will appear at several points in the discussion of this book. Chapter 5 discusses its role in generating conflicts between India and Bangladesh, and chapter 6 provides a summary of the main points of critique of the project.

2.5.3 Pakistan

With 163 large dams, Pakistan is another major dam-building country of Asia. In fact, it often seemed that India and Pakistan were competing with each other regarding who would have the taller and bigger dam. The river landscape of Pakistan is dominated by the Indus River and its five tributaries. The combined flow of the Indus moves south through almost the entire stretch of Pakistan, meeting the Arabian Sea near Karachi, the country's largest city. Most of Pakistan's large dams are built on the Indus River and its tributaries, including the Tarbela dam, considered to be the largest embankment dam in the world.[23] Another of Pakistan's famous dams is the Mongla dam on the Chenab River. Apart from electricity generation, Pakistan's dams and their reservoirs play an important role in the expansion of irrigation, particularly in the province of Punjab.

2.5.4 Other countries of South Asia

2.5.4.1 Nepal

The mountainous country of Nepal has many rivers originating in the Himalayas and passing through it, with the major ones being, from west to east, the Mahakali, Karnali, Gandaki, and Kosi. Each of these has its own system of tributaries, flowing mostly from north but some from south too. The Mahakali River forms the western boundary of Nepal with India and continues as the Sarda River in India. It flows in a southeasterly direction and meets the Karnali River, which flows south to India, to become the Ghagra River, a tributary of the Ganges River. The Gandaki River also flows south as another tributary of the Ganges. Finally, the Kosi flows south through the Indian state of Bihar and serves as the largest tributary of the Ganges in terms of volume of water and sediment.

Because of its the mountainous terrain, Nepal has considerable hydro potential, variously estimated to be between 40,000 and 84,000 MW. However, its terrain limits water storage capacity, so that dams with large reservoirs are difficult to build. Nepal's limited financial and technical capacity also act as another constraining factor. However, Nepal has several dams, both run-of-river and with reservoirs, built mostly in collaboration with India, including the Arun I and Arun II on the Arun River, a tributary of the Kosi. Many more dams are planned: the Pancheswar Dam on the Mahakali River; the Arun III on the Arun River; Upper Karnali on the Karnali River; Budhi Gandaki on the Gandaki River; the 700 MW West Seti Dam on the River Seti, a tributary of the Karnali River; and the Soptokoshi Dam on the Kosi River.[24]

2.5.4.2 Bhutan

Bhutan is in a similar situation to Nepal. It is also a mountainous country, with many Himalayan Rivers originating and passing through it. These include— from west to east—Torsa Chhu; Wang Chhu; Sankosh Chhu; Trongsa Chhu, including its tributary Sarangar Chhu; and Drangme Chhu, including its tributaries Lhobrak and Kulong Chhu. Bhutan also has considerable hydro potential, estimated to be around 20,000 MW, but its terrain also limits storage capacity, making run-of-river dams (barrages) technically more suitable than dams with reservoirs. Bhutan also has limited financial and technical capacity, so that most of its dams are built in collaboration with India. The first major hydro project was commissioned on the Wang Chhu River in 1989. Future plans focus mostly on the Wang Chhu cascade, with proposed Chuka I (storage), II (run-of-river), and III (storage) dams.[25]

2.5.4.3 Bangladesh

Bangladesh's lone dam lies in the Chittagong Hill Tracts on the Karnaphuli River. Built during the period 1957 to 1962, its purpose is stabilization of the river, extension of navigation, and production of electricity. It had an original capacity (commissioned in 1962) of 80 MW. Another 50 MW was added in 1982 and a further 100 MW in 1988, for a total current installed capacity of 230 MW. However, a significant part of this capacity has now been lost due to sedimentation of the reservoir, Lake Kaptai. The main function of the dam now therefore lies in river flow management. Bangladesh also constructed the Teesta Barrage, whose purpose however has been frustrated by diversion of water upstream by India through its Gajoldoba barrage, as will be discussed in chapter 3.

2.5.4.4 Sri Lanka

Sri Lanka has a long history of river intervening structures going back to pre-industrial times. The hydroelectric potential of the country is limited, particularly in comparison with other countries of South Asia. However, Sri Lanka too has pursued the Commercial approach, and expansion of irrigation has been an important objective, in addition to electricity generation. Most of Sri Lanka's dams and barrages are concentrated on the Mahaweli River, the main river of the country.[26]

2.5.5 Other countries of Asia

The Commercial approach to rivers spread to many other countries of Asia too (Figure 2.4). The two main rivers of the Mesopotamia region—the Euphrates and the Tigris—originate in Turkey and flow through Syria and Iraq to meet the Persian Gulf near the Iraqi city of Basra. Turkey has built many dams on these rivers and their tributaries. In fact, Turkey ranks tenth in the world in terms of number of large dams, having 972 of them.[27] Turkey has plans to construct some more, particularly in the southeast of the country.[28] Both Syria and Iraq have built dams on the Euphrates and the Tigris Rivers. For example, Syria's Tabqua Dam on the Euphrates has created the Assad Reservoir behind it. In Iraq, there is the Mosul Dam on the Euphrates River near the city of Mosul.

Iran has many rivers, flowing either south to the Persian Gulf or north to the Caspian Sea. Its mountainous terrain offers many sites for dam construction, and Iran has made use of them to build many large dams, making it eleventh in the world in terms of number of such dams. As of 2010, Iran had 588 dams (large and small) built and another 546 planned. Iran's highest dam, Dez, is supposed to irrigate 80,000 hectares.[29]

In East Asia, the Korean Peninsula has many rivers, though they are relatively short in length. However, South Korea is a major dam builder. In fact, it ranks sixth in the world in terms of the number of large dams, with 1,339 of them.

In Southeast Asia, the major rivers are the Irrawaddy and the Mekong, both of which originate on the northern side of the Himalayas and then flow through China before turning south. The Irrawaddy River flows through Myanmar to reach the Bay of Bengal, while the Mekong takes a more southeasterly direction to flow through Laos, Kampuchea, and Vietnam before reaching the sea. Myanmar has built 32 large dams, mostly on the Irrawaddy and its tributaries. Countries of the Mekong Basin have also built many dams, with Vietnam alone accounting for 51 large dams in the Mekong Basin and the basin of the Red River, the other major river of the country. Countries sharing the Mekong River have formed the Mekong River Commission for cooperation regarding the use of this river, as we shall see later. Among southeast Asian countries, other major dam builders are Thailand and Indonesia, with 218 and 132 large dams, respectively.

2.5.6 Africa

The African continent has many major rivers, including the longest river of the world, the Nile. Originating from Lake Tanganyika, the Nile flows north through Uganda, Sudan, and Egypt to reach the Mediterranean Sea. The Blue Nile joins it from Ethiopia. Africa's other major river is the Congo (Zaire) River, which flows through the Democratic Republic of Congo and ultimately heads west to reach the Atlantic Ocean. It is a powerful river, with an estimated average discharge of 41,200 cubic m per second, second only to the discharge of the Amazon River. Several other African rivers also flow out to the Atlantic Ocean. Among these are the Volta and Niger—in the middle of the continent—and the Orange River, in South Africa. On the other hand, the Zambezi River, another important African river, flows east through Mozambique to reach the Indian Ocean. The Limpopo River of South Africa also flows east.

The collapse of colonialism after the Second World War led to the emergence in Africa of a large number of newly independent countries, most of whom embraced the Commercial approach and took to construction of dams and other river intervening structures as a way to development and progress (see Figure 2.7 for dams in Africa.)

For example, Egypt constructed the High Aswan Dam on the Nile during the period 1960 to 1970 with Soviet assistance (this will be discussed in more detail in chapter 4). Ethiopia and Sudan, two large upper riparian countries of the Nile basin, have constructed many dams on the Nile and its tributaries and are planning to build more. For example, Ethiopia is constructing several hydropower dams both to meet domestic electricity demand and export electricity to neighboring countries.[30] Since 2011, Ethiopia is building the Grand Ethiopian Renaissance Dam (GERD) on the Blue Nile near the border with Sudan. With 6,000 MW power generation capacity and a reservoir of 1,680 sq km, it is to be one of the world's largest dams. Similarly, Sudan has built the Merowde Dam, with Chinese assistance. This is one of Africa's largest hydropower projects and has more than doubled Sudan's power generation capacity and created a reservoir with a length of 174 km and a surface area of 476 sq km. Further upstream of the Nile, Kenya has also built dams, such as Kiambere Dam, Masinga Dam, and Turkwell Gorge Dam. The latter, having 106 MW capacity, is one of Kenya's most expensive development projects to date.[31]

Ghana built the Asokombo Dam on the Volta River, creating the Volta Reservoir, one of the world's largest impoundments, and flooding around

Africa: Dams
Global Reservoir and Dam Database, Version 1 (GRanDv1), Revision 01

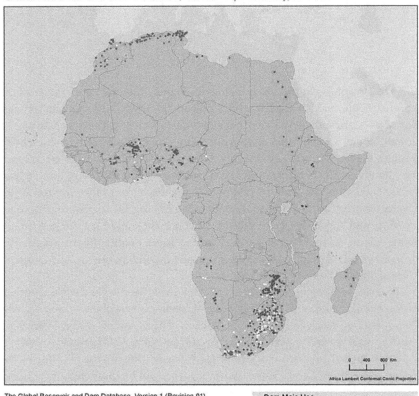

The Global Reservoir and Dam Database, Version 1 (Revision 01) contains 6,862 records of reservoirs and their associated dams with a cumulative storage capacity of 6,197 cubic km. The dams were geospatially referenced and assigned to polygons depicting reservoir outlines at high spatial resolution. While the main focus was to include all dams associated with reservoirs that have a storage capacity of more than 0.1 cubic kilometers, many smaller dams and reservoirs were added where data were available.

Dam Main Use
- Fisheries
- Flood control
- Hydroelectricity
- Irrigation
- Navigation
- Not classified
- Other
- Recreation
- Water supply

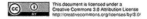
Figure 2.7 Dams in Africa
Source: Socioeconomic Data and Applications Center (SEDAC) of NASA

4 percent of the land area of the country.[32] Nigeria built the Kainji Dam on the Niger River as the "pillar of Nigeria's economic and social development."[33] It has 52 large dams, many of which are on the Niger River and its tributaries, including the Bakolori Dam on the Sokoto, a tributary of the Niger. Senegal built the Diama and Manebali dams on the estuary of the Senegal River. Mali is also a major dam builder, with 112 large dams, including the Manantali Dam. Gabon plans to build, with Chinese assistance, two dams, one of which is the Belinga Dam, near the Kongou Falls in the country's Ivindo National Park.[34]

The Congo River, with its enormous flow, is supposed to possess 13 percent of global hydropower potential, and there is great interest in exploiting it. As of now, there are about forty hydropower plants in the Congo Basin, the largest being the Inga Dams, located about 200 km southwest of Kinshasa, in the Inga Falls part of the river. The plan was to construct five dams, generating a total of 34,500 MW of electricity. So far two of these dams—called Inga I and Inga II— have been constructed, with a combined capacity of 1,776 MW. However, there are plans to construct a giant Inga Dam, with a capacity of 40,000 MW, which could be extended to 120,000 MW. For some perspective, it may be noted that the total electricity generation capacity of the entire African continent in the mid-1990s was about 100,000 MW.[35]

Zimbabwe is a major dam builder, with 254 large dams, many of them on the Zambezi River and its tributaries. Similarly, Zambia has six large dams built at various points of the Zambezi River Basin.

In the south of the continent, South Africa has built many dams on the Orange River and on other rivers of the country. In fact, South Africa ranks eighth in the world in terms of number of large dams, having 1,114 of them, including the Kariba Dam. Lesotho has nine large dams, including the Lesotho Highlands Water Project.

In the north, Algeria, Morocco, and Tunisia have rivers flowing to the Mediterranean Sea and have built many large dams, numbering 154, 150, and 133, respectively. Algeria's Quedd Fodda Dam is one of them.

2.5.7 Latin America

South America is the home of the Amazon River, the largest river of the world in terms of volume of discharge. It originates in the Andes Mountains and flows east across the entire continent to reach the North Atlantic Ocean. The Amazon has numerous tributaries, such as the Putumayo and Negro, flowing from the north, and Maranon, Ucayali, Jurua, Purus, Madeira, Tapajos, and Xingu, flowing from the south. The Parana system is another important river system of South America and includes the Paraguay, Pilcomayo, Salado, and Negro rivers, which combine to flow out to the South Atlantic Ocean between Uruguay and Argentina. Further south, in the Patagonian peninsula—comprising Argentina and Chile—are the rivers Colorado, Salado, and Negro. In the north, several rivers flow out to the Caribbean Sea, among them the Magdalena and Orinoco.

In the upper reaches of the Amazon tributaries, Venezuela, Peru, Colombia, and Bolivia have built many dams, having 77, 66, 62, and 33 large dams, respectively, including the Guavio Dam in Colombia and the Brokopondo Dam in Suriname (see Figure 2.8 for dams built on South American rivers).

In the main part of the Amazon system, Brazil has built many dams on the tributaries of the Amazon River. In fact, Brazil ranks fifth in the world in terms of the number of large dams, having 1,411 of them in total. Among the Brazilian dams are the Balbina, Itaparica, Sobradinho, and Tucurui. The Brazilian government plans to construct at least 60 more new dams in the Amazon basin and is apparently willing to relax environmental and indigenous rights protection laws for this purpose. Some of these planned dams are huge, such as the 11,200 MW Belo Monte Dam, which is the first of a series of planned new dams on the Xingu River in the Parana State of Brazil.[36] Many believe that effective functioning of the Belo Monte will require the building of huge reservoirs upstream, including the Banaquara Reservoir that would flood about 13,500 sq km, more than any of the existing dams in South America.[37] There was a plan to dam the Amazon itself, with a 64 km dam that would create a 190,000 sq km reservoir, larger than the country of Uruguay, and have an electricity generation capacity of 80,000 MW.[38]

Many dams have been built on the Parana River system. The most famous among these is the Itaipu Dam, located at the border of Brazil and Paraguay. Another is the Yacyreta Dam, located at the border of Argentina and Paraguay. Also notable is the Salto Grande Dam of Uruguay.

Both Argentina and Chile have built dams on Patagonian rivers and have 114 and 96 large dams, respectively, with more planned. For example, there is a proposal to construct four dams on the Baker and Pascua Rivers in Chilean Patagonia with a total generation capacity of 2,400 MW. A 1,200 mile transmission line will however be required to transport this electricity to the industrial cities in the north of the country.[39]

2.5.7.1 Dams in Mexico, Central America, and the Caribbean

The countries of Central America, the Caribbean, and Mexico have also built many dams and barrages. The most important river of Mexico is the Rio Grande, which forms the border between it and the United States for a significant stretch. Other Mexican rivers include the Verde, Balsas, Rio Santiago, San Pedro, San Lorenzo, Culiacan, Sinaloa, Fuerte, Mayo, Yacui, Sonora, Concepcion, and Sanoyea, flowing west to the Gulf of Baja and the Pacific Ocean. The Salado, San Juan, San Fernando, Soto la Marina, Panuco, Tecolutla, Jamaca, Papaloapan, and

South America: Dams

Global Reservoir and Dam Database, Version 1 (GRanDv1), Revision 01

The Global Reservoir and Dam Database, Version 1 (Revision 01) contains 6,862 records of reservoirs and their associated dams with a cumulative storage capacity of 6,197 cubic km. The dams were geospatially referenced and assigned to polygons depicting reservoir outlines at high spatial resolution. While the main focus was to include all dams associated with reservoirs that have a storage capacity of more than 0.1 cubic kilometers, many smaller dams and reservoirs were added where data were available.

Dam Main Use

- Fisheries
- Flood control
- Hydroelectricity
- Irrigation
- Navigation
- Not classified
- Other
- Recreation
- Water supply

Figure 2.8 Dams in South America
Source: Socioeconomic Data and Applications Center (SEDAC) of NASA

Coatzacoalcos rivers flow east to the Gulf of Mexico. Mexico has constructed many dams on its rivers, and ranks fourteenth in the world in terms of the number of large dams, with 571 of them.

Central American rivers are short in length, but many have hydropower potential: Panama has 21 large dams; Costa Rica, 13; Honduras, 10; and Guatemala, 6. Notable among these are Aguacapa and Chixoy dams in Guatemala and El Cajon Dam in Honduras.[40] The Dominican Republic in the Caribbean has 17 large dams.

2.5.8   Dam spree continues in developing countries

This brief survey shows that the Commercial approach has spread to all parts of the world, and inspired by it, dams, barrages, and other river intervening structures have been constructed on almost all major rivers. While new dam building has slowed down or even stopped in developed countries, it continues apace in many developing countries. This is in part because developed countries have already utilized most of their "economically feasible" hydro potential, while developing countries are yet to do so. It is estimated that the United States, Canada, and Japan have each exploited around 70 percent of their estimated economically feasible hydro potential, while Europe has utilized about 50 percent. Meanwhile, Africa, Latin America, and China have used about one-tenth of theirs.[41] The dam-building spree can be observed in some transition countries too. For example, it has been reported that across the Balkans (the former republics of Yugoslavia), about 3,000 hydropower plant projects are underway or being planned, some with Chinese assistance.[42]

In the next chapter, I will discuss the consequences of the frontal interventions in rivers, inspired by the Commercial approach. However, to round out this chapter, in the next section I present some summary statistics regarding dams and barrages across the world.

2.6   Global statistics on dams

This section presents some summary tables to portray the global scene regarding dams. Table 2.3 presents the top 20 countries in terms of the number of large dams. As we can see, the top five dam-building countries—China, the United States, India, Japan, and Brazil—account for about 80 percent of large dams worldwide. However, as noted earlier, China dominates the scene, with about 23,841, or about half of the world's large dams. With more than 9,265 large dams, the United States holds the second rank, above India, which has

5,100 large dams. Other than the top five countries, South Korea, Canada, South Africa, and Spain are also large dam builders, with each having more than a thousand. Table 2.4 presents the distribution of large dams across the continents. Clearly, Asia dominates, accounting for about 62 percent, due mostly to China and India. It is followed by North America, due largely to the United States. Europe comes third, while South America is in fourth place, dominated by Brazil.

Tables 2.5–2.8 present the top 20 dams in the world in terms of their height, reservoir capacity, installed electricity generation capacity, and stipulated

Table 2.3 Countries having highest number of large and major dams

Rank	Country	No. of large and major dams
1	China	23,841
2	United States	9,265
3	India	5,100
4	Japan	3,119
5	Brazil	1,364
6	South Korea	1,338
7	Canada	1,169
8	South Africa	1,112
9	Spain	1,063
10	Albania	974
11	Turkey	709
12	France	593
13	United Kingdom	570
14	Mexico	567
15	Australia	541
16	Italy	520
17	Iran	371
18	Germany	335
19	Norway	254
20	Zimbabwe	244

Source: ICOLD (as of May 4, 2019).

Table 2.4 Distribution of large dams by continent

Place	Number of major and large dams	Percent of total (%)
Asia	35,886	61.5
North America	11,004	18.9
Europe	6,735	11.5
Africa	2,167	3.7
South America	1,897	3.3
Australia and New Zealand	663	1.1
Total	58,352	100

Source: Author, based on data from ICOLD (as of May 4, 2019).

command area for irrigation, respectively. We see that all the top 20 tallest dams are in developing countries, except for two in Switzerland. In terms of reservoir capacity, five of the largest dams are in Canada and another six are in Russia. The rest are in developing countries. In terms of installed electricity generation capacity, seven of the top 20 are in China, two each in Canada and Russia, and one in the United States. The rest are in developing countries. In terms of irrigation command area, all the top twenty dams are in developing countries, with Indonesia alone having seven of them, China accounting for three, and Vietnam for two.

As noted earlier, dams can be either single- or multipurpose. Table 2.9 and Figures 2.9 and 2.10 display the distribution of dams in terms of their purpose. We see that, of the single-purpose dams, almost half are focused on irrigation (49.8 percent), with electricity generation (19.8 percent) next in importance, followed by water supply (11.2 percent). The distribution of multipurpose dams is less lopsided. Irrigation still tops the list, with a 23.7 percent share, followed by flood control (19 percent), water supply (17.2 percent), hydropower (15.7 percent), and recreation (11.7 percent).

2.7 Conclusions

The Commercial approach to rivers is one of the outcomes of the Industrial Revolution, which provided human societies the technological capacity to carry out large-scale intervention in rivers. The commercial motive that underpinned

Table 2.5 Tallest dams of the world

Rank	Dam	Country	Dam purpose	Height (m)
1	Rogun (C)	Tajikistan	HI	335
2	Bakhtiyari (C)	Iran	HC	315
3	Jinping 1	China	HC	305
4	Nurek	Tajikistan	IH	300
5	Lianghekou (C)	China	H	295
6	Xiaowan	China	HCIN	394
7	Xiluodu	China	HCN	286
8	Grand Dixence	Switzerland	H	285
9	Baihetan (C)	China	H	277
10	Diamer-Bhasha (C)	Pakistan	HIS	272
11	Inguri	Georgia	HI	272
12	Yusufeki (C)	Turkey	H	270
13	Nuozhadu	China	HCN	262
14	Manuel Moreno Torres (Chicoasen)	Mexico	H	262
15	Tehri	India	IH	260
16	Hacixia	China	H	254
17	Mauvoisin	Switzerland	H	250
18	Laxiwa	China	H	250
19	Deriner	Turkey	H	249
20	Alberto Lleras C.	Colombia	H	243

Note: C = Flood control; F = Fish farming; H = Hydropower; I = Irrigation; N = Navigation; R = Recreation; S = Water supply; T = Tailing; and X = Other.
Source: ICOLD (as of May 4, 2019).

the Industrial Revolution and the desire to dominate the nature, arising from the newly acquired technological capacity, combined to produce the Commercial approach. Dams represent the most radical form of frontal intervention in rivers.

Beginning in the early industrializing countries in the nineteenth century, the Commercial approach spread later to the developing world, particularly during the second half of the twentieth century. As of now, there are about 60,000 large

Table 2.6 Dams with largest reservoir capacity

Rank	Dam	Country	Year completed	Reservoir volume (10^6 m^3)
1	Robert-Bourassa	Canada	1981	460,702
2	Kariba	Zimbabwe/ Zambia	1959	180,600
3	Bratsk	Russia	1964	169,000
4	Aswan High	Egypt	1970	169,000
5	Akosombo	Ghana	1965	150,000
6	Daniel Johnson	Canada	1968	141,851
7	Guri	Venezuela	1986	135,000
8	W.A.C. Bennett	Canada	1967	74,300
9	Hidase (C)	Ethiopia		74,000
10	Krasnoyarsk	Russia	1967	73,300
11	Zeya	Russia	1978	68,400
12	Robert-Bourassa	Canada	1978	61,715
13	La Grande 3	Canada	1981	60,020
14	Ust-Ilimsk	Russia	1977	59,300
15	Boguchany	Russia	1989	58,200
16	Kuibyshev	Russia	1955	58,000
17	Serra da Mesa (Sao Felix)	Brazil	1993	54,400
18	Caniapiscau	Canada	1981	53,790
19	Cahora Bassa	Mozambique	1974	52,000
20	Bukhtarma	Kazakhstan	1960	49,800

Note: (C) = "under construction."
Source: ICOLD (as of May 4, 2019).

dams and barrages constructed all across the world. While further construction of dams and barrages has abated in developed countries, primarily because most potential dam construction sites have already been utilized, it continues in many developing countries.

Asian countries together have more than 60 percent of the large dams, mainly because of China, which alone, with about 24,000 large dams, accounts for more than 40 percent. The United States, with more than 9,000 large dams, ranks

Table 2.7 Dams with highest electricity generation capacity

Rank	Dam	Country	Year	Installed capacity (MW)
1	Sanxia (Three Gorges)	China	2010	22,500
2	Itaipu	Paraguay	1991	14,000
3	Xiluodu	China	2014	13,860
4	Baihetan	China	2019	13,050
5	Belo Monte	Brazil	2019	11,234
6	Guri (Raul Leon)	Venezuela	1986	10,200
7	Tucurui	Brazil	1984	8,370
8	Robert-Bourassa	Canada	1979	7,722
9	Ta Sang	Myanmar	2011	7,100
10	Grand Coulee	United States	1942	6,809
11	Sayano-Shushensk	Russia	1990	6,400
12	Xianjiaba	China	2014	6,400
13	Longtan	China	2009	6,300
14	Hidase	Ethiopia	2018	6,000
15	Krasnoyarsk	Russia	1967	6,000
16	Miytsone	Myanmar	2017	6,000
17	Nuozhadu	China	2014	5,850
18	East Forebay Dyke FF-10A	Canada	1971	5,248
19	Jinping 2	China	2014	4,800
20	Bratsk	Russia	1964	4,500

Note: (C) = under construction.
Source: ICOLD (as of May 4, 2019).

second, while India, with more than 5,000 large dams, ranks third. Other major dam-building nations are Japan, South Korea, Brazil, South Africa, Canada, and Albania, each with more than a thousand large dams.

Reservoirs created by dams across the world have a total storage capacity of about 10,000 cubic km, which is equivalent to five times the (annual) water flow of all the world's rivers. The area submerged by these reservoirs is estimated to be more than 400,000 sq km, roughly the size of California. The reservoirs of the United States

Table 2.8 Dams with largest irrigation command area

Rank	Dam name	Country	Year	Irrigation area (sq km)
1	Shentian	China	1960	315,260
2	Batu Tegi	Indonesia	2002	108,553
3	Ta Trach	Viet Nam	2014	34,782
4	Xiaolangdi	China	2001	26,667
5	Wonogiri/Gjah Mungkur	Indonesia	1982	23,600
6	Gurara	Nigeria	2014	20,000
7	Way Rarem	Indonesia	1984	19,000
8	Ayun Ha	Viet Nam	2002	13,500
9	Zipingpu	China	2006	9,338
10	Ataturk	Turkey	1992	9,325
11	Liujiaxia	China	1974	9,263
12	Telaga Pasir/Sarangan	Indonesia	1941	8,213
13	Tlogo Ngabel	Indonesia	1930	7,164
14	High Aswan Dam	Egypt	1970	7,000
15	Way Jepara	Indonesia	1978	6,651
16	Sempor	Indonesia	1978	6,485
17	Tabqa	Syria	1978	6,400
18	Hancagiz	Turkey	1988	6,250
19	Gunung Rowo	Indonesia	1025	6,052
20	Can Don	Viet Nam	1999	4,800

Source: ICOLD (as of May 4, 2019).

alone have submerged an area equal to the combined area of New Hampshire and Vermont. The river interventions inspired by the Commercial approach have altered the river landscape of the world. As the World Commission on Dams put it,

> Consider: on this blue planet, less than 2.5% of our water is fresh; less than 33% of fresh water is fluid; less than 1.7% of fluid water runs in streams. And we have been stopping even these. We dammed half of our world's rivers at unprecedented rates of one (large dam) per day, and an unprecedented scale of over 45,000 dams more than four stories high (WCD 2000, pp. i–ii) .

Table 2.9 Classification of dams by their purposes

Code	Description of purpose	Number with this as sole purpose	Percent of total (%)	Multipurpose dams with this as a purpose	Percent of the total (%)
C	Flood control	2524	8.6	4778	19.0
F	Fish farming	41	0.1	1395	5.6
H	Hydropower	5786	19.8	3932	15.7
I	Irrigation	14562	49.8	5954	23.7
N	Navigation	97	0.3	580	2.3
R	Recreation	1350	4.6	2942	11.7
S	Water supply	3285	11.2	4330	17.2
T	Tailing	63	0.2	8	0.03
X	Other	1540	5.3	1214	4.8
	Total		100.0		100

Source: ICOLD (as of May 4, 2019).

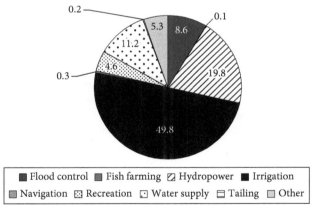

Figure 2.9 Distribution of single purpose dams
Source: Author, based on data presented in Table 2.9

What have been the consequences of this massive scale of intervention in rivers? Have the river intervening structures proved to be as beneficial as their promoters portrayed them? How conducive have they been to sustainable development? These are the questions that I discuss in the next few chapters.

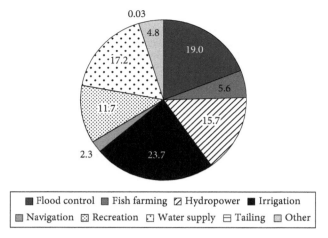

Figure 2.10 Distribution of multi-purpose dams
Source: Author, based on data presented in Table 2.9

Notes

1. Much of information presented in this chapter is drawn from WCD (2000) and McCully (2001). However, to avoid repetition, I do not always present explicit references.
2. Of global water withdrawals from all sources in 1990, about 7 percent went to households and other municipal users. By contrast, industries accounted for about 24 percent of withdrawals. Of the rest, 65 percent went for agriculture, and 4 percent evaporated from reservoirs (McCully 2001, p. 149).
3. The World Commission on Dams took note of the aqueducts built by the Romans to supply drinking water and sewer systems for the towns of Antiquity and of the 2,200-year old Dujiang irrigation project of China, supplying irrigation water to 800,000 hectares (WCD 2000, p. 8).
4. Dumping of industrial and municipal waste, including sewage, is a common practice in many countries. Similarly, rivers have been encroached on for various purposes (such as setting up of industries, infrastructure, dwellings, etc.) For discussion of the impact of Industrial Revolution on pollution of rivers, see Hettige, Lucas, and Wheeler (1992).
5. Hydraulic differential is the difference between the height of the water level in the reservoir and that of the downstream river. See glossary.
6. This would be six times the volume of the Great Pyramid of Cheops. See McCully (2001).
7. This would be enough water to flood the country of Luxemburg to a depth of 1 meter. See McCully (2001).
8. This would be sufficient to power a European city of a million inhabitants. In 1950, ten dams met the criteria for major dam; the number increased to some 305 by 1995. See McCully (2001).

9. The analysis of embankments (lateral interventions) will be provided in chapters 7–10, which are devoted to the discussion of Cordon and Open approaches to rivers.

10. WCD (2000, p.8).

11. As noted earlier, WCD (2000) and McCully (2001) provide excellent accounts of the spread of dam building activities across the world. Instead of repeating that information, this chapter provides an overview, drawing upon these and other sources of information. According to WCD, "Since average construction periods generally range from 5 to 10 years, this indicates a worldwide annual average of some 160 to 320 new large dams per year" (WCD 2000, p. 10).

12. McCully (2001, p. 240).

13. See McCully (2001, p. 22 and p. 125).

14. McCully (1996, 2001, p. lxiii).

15. It should be noted that, though termed as a dam, the Three Gorges Dam has features of a "run-of-the-river" dam, because it mostly raises the upstream river stage to store water, instead of creating a reservoir of large width. This is possible because the upstream stretch of the river passes through three steep gorges (hence its name). As a result, the Three Gorges Dam has proved to be less disruptive for the upstream river basin as otherwise it would have been.

16. China's water ministry planned to increase its hydropower capacity by 55,000 MW (equivalent to three Three Gorges Dams) between 2000 and 2010 and by a similar amount during each decade of this century (McCully 2001, p. lxiii).

17. This represented a 300 percent increase in the number of active hydropower projects over the past four years (IR 2012, p. 3).

18. See IR (2012a, p. 3) for details.

19. See IR (2012a, p. 3).

20. https://asia.nikkei.com/Opinion/Rethink-plans-to-dam-Mekong-after-Laos-disaster (accessed on May 2, 2019).

21. Nehru opened the Nangal canal in Punjab in 1954 and compared the under-construction Bhakra Dam to a temple, mosque, or guruduwara. In his words, "As I walked around the [dam] site, I thought that these days the biggest temple and mosque and gurdwara is the place where man works for the good of the mankind. Which place can be greater than this, this Bhrakra-Nangal, where thousands and lakhs of men have worked, have shed their blood and sweat and laid down their lives as well? What can be a greater and holier place than this, which we can regard as higher?" (McCully 2001, pp. 1–2).

22. Some have traced IRLP to ideas expressed in the past. For example, during British rule, around 1839, Sir Arthur Cotton proposed using rivers as the major means of inland transportation in India rather than constructing railways. More recently, around 1972, Dr. K. L. Rao, the Indian water minister, spoke of a National Water Grid focused on the Ganges-Cauvery Link. Similarly, around 1974, Captain Dastur put forward a proposal for a Garland Canal connecting the Himalayan rivers. However, upon closer scrutiny none of these ideas was found to be sound enough to deserve further attention, and hence were shelved (see Sinha 2004; Iyer 2002).

However, the idea of interlinking rivers was suddenly revived when the Indian president A. P. J. Abdul Kalam made a reference to it in his speech on the eve of India's Independence Day in 2002. Following his speech, Ranjit Kumar, an Indian Supreme Court advocate, filed an application under public interest litigation to the Indian Supreme Court, upon which the court passed a directive to the Indian Government to connect the rivers by 2016. Following the court order, the then BJP-led NDA (National Democratic Alliance) government of India formed a task force to implement the river link, which fleshed out the idea in the form of IRLP and proceeded with its implementation.

The Congress-led UPA (United Progressive Alliance) government that came to power following the defeat of the NDA government in the 2004 general elections, promised to review the IRLP. In January 2005, it replaced the earlier task force with a high-powered inter-ministerial committee, which has now embarked on implementing the project starting with the Ken-Betwa link of the Himalayan part and the Parbati-Kalisindh-Chambal link of the Peninsular part. IRLP received more momentum with the installation of the BJP government led by Narandra Modi in 2014.

For information and discussion regarding IRLP, see the articles in Ahmed, Kholiquzzaman, and Khalequzzaman (2004). In particular, see Alley (2004); Azad, Iqbal, and Sultana (2004); Azad and Alam (2004); Choudhury (2004); Chowdhury and Dutta (2004); Dhungel and Pun (2004); Diwan (2004); Ganguly (2004); Gujja and Das (2004); Haq (2004); Hossain (2004); Hossain and Madina (2004); Hossain and Nasreen (2004); Islam (2006a, 2006b); Jairath (2004); Khalequzzaman(2004); Krishna (2004); Saha (2004); Khalequzzaman, Srivastava, and Faruque (2004), Sultana (2004), Vombatkere (2004), and Vombatkere and Vombatkere (2004). See also Bandyopadhaya and Parveen (2002), Mirza, Ahmed, and Ahmad (2008), and NAPM (2004),

23. The Tarbela Dam of Pakistan is a massive embankment of earth and rock, nearly 3 km long and 143 m tall, at its highest point. However, it is also a problem-stricken major dam, as we shall see (McCully 2001, p. 123).

24. The Pancheshwar Dam on Nepal's far western border with India, which at 315 m in height would be the world's highest dam. See https://india.mongabay.com/2019/06/pancheshwar-dam-vulnerable-to-earthquakes-in-current-form-finds-study/ for information about the current state of affairs regarding this dam. Around 2000, Nepal had less than 300 MW of hydropower capacity (McCully 2001, p. lxiii).

25. Bhutan also plans to install 10,000 MW of hydropower by 2020. It is thought that only 5 percent of the country's hydro potential has been utilized. The country therefore plans to build 74 dams in cascades across the country.

26. Sri Lanka has 27 dams that have a length of more than 100 m or height of 10 m. Among these, 15 were constructed primarily for production of hydroelectricity, and another 10 primarily for irrigation purposes. The 2,000 MW Mahaweli Project, involving a series of dams on the Mahaweli River and supported by several World Bank loans between 1970 and 1998, is the country's largest multipurpose dam project.

27. Among Turkey's prominent dams is the Ataturk Dam.

28. Many dams have been constructed in southeast Turkey under the Anatolia project. There are plans to construct another 22 dams at a cost of some $32 billion. According to McCully (2001, p. 242), these dams have both domestic and international political purposes.

29. However, just as the dam neared completion in the early 1960s, the Shah and his advisers decided that the irrigation water would be better used by foreign agribusiness corporations producing for export (see McCully 2001).

30. According to IRN (2012a, p. 12), the Merowe Dam went ahead because Chinese and Middle Eastern financial institutions agreed to finance it, with China Exim Bank providing up to $520 million, the main part of the finance.

31. The 106 MW Turkwell Gorge Dam, however, threatens Kenya's Lake Turkana, which is the home to the world's largest population of Nile crocodiles, healthy population of hippos, and hundreds of bird and fish species. It has been predicted that the decay of the lake would threaten the livelihood of more than a quarter-million indigenous people who depend on it (see McCully 2001, p. 261).

32. Two hydro dams on the Volta River—the Akosombo and the smaller Kpong Dam downstream—account for about 88 percent of Ghana's total generating capacity of 1,160 MW (McCully 2001, p. 136).

33. See McCully (2001), p. 68. The Kainji Dam was also fitted with a huge lock, served by a 6 km long access canal. The 49 m lift lock, one of the highest in the world, was expected to hold four 5,000 ton barges at a time (ibid).

34. According to observers, the proposed hydroelectric dam is a threat to the famed Kongou Falls in Gabon's Ivindo National Park. The Belinga Dam is one of two proposed dams that would generate power for the Belinga Iron Ore Project, which is located 500 km east of Libreville, Gabon's coastal capital (see IRN 2012, p. 28).

35. See McCully (2001, pp. lxiii–lxiv) and Bosshard (2007, pp. 22–23) for more details on the Inga project. Power from Inga is expected to be transmitted across the entire continent and even to the Middle East, Turkey, and Europe. Apparently, energy officials from South Africa and Egypt have been holding discussions since 1995 about building a high-voltage link between Cape Town and Cairo with the Grand-Inga project as the keystone. The estimated cost of this, the world's largest hydropower project, is about $50–80 billion. Calling it "the mega fantasy to end all hydropower fantasies," McCully (2001) reports that the African Development Bank funded a secret feasibility study of the project, carried out by Électricité de France and German consultants Lahmeyer International. President Thabo Mbeki of South Africa advocated the construction of Grand Inga in his address to the 2000 Assembly of the Organization of African Unity (OAU).

36. See Bosshard (2007, p. 22).

37. See McCully (2001, p. lxiii).

38. See McCully (2001, p. 22).

39. See Bosshard (2007, p. 22).

40. Chixoy is one of a number of very expensive dams built in Central America during the 1970s and 1980s with loans from the World Bank and Inter-American Development Bank (IADB) despite the very high and accelerating rates of erosion in their watersheds.

41. See WCD (2000) and McCully (2001).
42. See https://balkanrivers.net/en/news/new-report-hydropower-tsunami-balkans.
    See also the article by Marc Santora, "Oysters Lead Lives of Excitement and Danger.
    Especially in the Balkans," https://www.nytimes.com/2018/12/31/world/europe/
    croatia-oysters-balkans-bosnia.html.

3

Consequences of the Commercial approach to rivers

3.1 Introduction

The Commercial approach has had a huge impact on the state of rivers and river basins across the globe. The claimed beneficial effects of this approach are well known. Among these are: electricity generation, irrigation expansion, water supply for municipal and industrial purposes, navigation enhancement, and river flow regulation. These benefits have provided the rationale for the spread of the Commercial approach across the world, as noted in chapter 2.

However, over time, many questions have arisen regarding this approach. First, its beneficial effects were often not found to be as pronounced as claimed by its promoters. Second, these beneficial effects were often achieved at a great cost to rivers themselves and the ecology of river basins. These costs are generally less perceptible, because they occur over time and are diffused in nature. Also, the constituencies bearing these costs are generally either politically weak or even not yet in existence (i.e., future generations). However, these costs are real and become more serious in the long run. Hence, they are important from the viewpoint of sustainable development, which has intergenerational equity at its core. Many of the adverse consequences of the Commercial approach are unintended, though that does not lessen their impact. Third, it has been observed that many of the apparent benefits of the Commercial approach could have been obtained through alternative, more sustainable means.

An important adverse consequence of the Commercial approach is conversion of rivers into sources of contention and conflict—instead of being bonds of friendship—among co-riparian countries, regions, and communities. Since river flows are limited, the Commercial approach, based on the philosophy of "abstracting as much water as possible," inevitably leads to a zero-sum game, engendering conflicts among participants.

This chapter reviews the consequences of the Commercial approach. In doing so, it limits its scope to the frontal version of the Commercial approach, leaving the discussion of the lateral version to chapter 6. The goal of the review is to examine the suitability of the Commercial approach for sustainable development.

Rivers and Sustainable Development. S. Nazrul Islam, Oxford University Press (2020). © Oxford University Press.
DOI: 10.1093/oso/9780190079024.001.0001

The discussion of the chapter is organized as follows. Section 3.2 reviews briefly the claimed positive effects of the Commercial approach, noting, in particular, the difficulties of specification and quantification of some of these effects. Section 3.3 discusses the effects of the Commercial approach on river morphology, noting fragmentation of rivers, upstream sedimentation, and downstream disfigurations. It takes note of the "More water, more thirst!" syndrome that the Commercial approach creates, leading to exhaustion of rivers and deleterious effects on deltas, estuaries, and the coastal environment. Section 3.4 discusses broader effects of the Commercial approach on the ecology of river basins, including loss of seasonal pulses and new dangers of catastrophic flooding, waterlogging, salinity, deterioration of soil quality, and damages to the flora and fauna.

Section 3.5 examines the unsuitability of the Commercial approach in the light of climate change, noting GHG emissions from dam reservoirs and increased risks from erratic and greater precipitation. Section 3.6 takes note of the unfair distribution of benefits and costs of Commercial approach, while section 3.7 notes seismic and a few other related risks.

Section 3.8 examines the rates of return issue, noting that financial rates of return of Commercial approach-inspired projects are generally lower than what were originally claimed, and economic rates of return prove to be even lower, if not negative. Section 3.9 explains why the Commercial approach creates conflicts among co-riparian communities. Section 3.10 concludes.

3.2 Beneficial effects of the Commercial approach

The beneficial effects of the Commercial approach are quite well known. They include electricity generation, irrigation expansion, water supply for municipal and industrial purposes, navigation enhancement, and river flow regulation (flood control). However, quantifying and ascertaining their financial and economic value is not as easy at is may appear at the first glance.

Of the benefits just listed, the easiest to quantify is electricity generation. It is estimated that about 20 percent of global electricity is obtained from hydropower.[1] Quantification of irrigation provided by a dam is fraught with problems. Mere acreage does not provide information about how much water is supplied to each of the acres. Also, often the same fields are irrigated through several methods, as can be seen from the widespread presence of tube-wells in the irrigation command areas of dams and barrages. The volume of water flowing out of reservoirs for irrigation purposes is more accurately quantifiable. However, this volume is not a measure of the actual irrigation carried out, as the latter depends on the distance the water has to travel to reach the fields—suffering transfer loss in the process—and how effectively that water is actually utilized when it reaches the fields to be irrigated. Nevertheless, it is claimed that about 30 percent of land irrigated is done so using dams and barrages, though there are considerable disputes regarding this statistic.[2] Even if the volume of water can be quantified, the problem regarding its economic valuation remains, because the relevant prices are either absent or deficient, in particular because they do not capture broader and future effects. Similar issues pertain to the water supplied from dam reservoirs for urbanization and industrialization purposes. Though the water abstracted from a river for these purposes may be quantified, the problems of ascertaining the amount actually utilized and its valuation remain.

So far as navigation is concerned, dams and barrages generally have dual effects. On the positive side, dams and barrages may enhance navigation by increasing flow in the river's upper reaches, thus allowing vessels to travel further upstream than was possible before. Also, regulation of river flow may help to maintain the navigability of the river downstream during the lean season, when it would otherwise not be possible. On the negative side, dams and barrages obstruct free navigation along the river. The lock and lift systems that vessels have to go through following construction of dams and barrages cause long delays, put various restrictions regarding the size and weight of vessels, and can be costly.

As noted, not all dams are intended to have all the beneficial effects listed earlier. Some are meant only for hydropower, some are only for irrigation, and yet others are multipurpose. All dams and barrages can impart some control on river flows, as noted earlier. Such control (or taming) of rivers is generally thought to be beneficial. However, quantification of this benefit is particularly difficult. Moreover, this intended beneficial effect have some unintended consequences too, as we shall see.

In any case, benefits of dams and barrages are well known and advertised. However, these frontal interventions in rivers also have many adverse consequences. In the following sections, I present a brief overview of these consequences, beginning with those on river morphology (i.e., size and shape of the river channels) and the related issues of hydrology (i.e., of volume, speed, etc. of river flows). It may be noted here that the discussion of consequences of the Commercial approach is conducted in this chapter in general terms, instead of providing specific examples, except in certain cases. In other words, the focus here is on systematization rather than on specific examples. This is partly to save space and party because discussions of specific examples are already available in earlier works.[3]

3.3 Adverse consequences for river morphology

River intervening structures, inspired by the Commercial approach, have profound impact on the morphology of rivers. Shape, size, and depth of rivers undergo significant changes, often to the long-term detriment of the rivers themselves. The major changes brought about in river morphology by the Commercial approach are described in the following subsections.

3.3.1 Fragmentation of rivers

Dams, barrages, and other frontal interventions fragment rivers.[4] Fragmentation is more complete in the case of dams, which divide rivers into upstream and downstream parts, radically altering the morphology of both. Upstream, they convert rivers into placid lakes, while downstream, flows of water, sediment, and other nutrients generally diminish (due to abstraction and evaporation of water from reservoirs and collection of sediment in them), causing major impact on floodplains, deltas, estuaries, and the marine environment. Downstream flows become a residual after all commercial needs have been satisfied using reservoirs. Accordingly, the ecological needs of downstream parts of rivers and their basins are often neglected. Rivers with multiple dams become mere staircases of ponds and lose their dynamic nature. This fragmentation affects the aquatic life forms that depend on uninterrupted movements across rivers. It also alters dramatically navigability conditions, as noted earlier.

3.3.2 Upstream sedimentation, filling up of reservoirs, and widening of rivers

Reservoirs trap not only water but also the sediment carried by rivers. As sediment settles on their beds, reservoirs become shallower over time. The rate of sedimentation and filling up of reservoirs with sediment depends on a variety of factors, in particular, on the amount of sediment carried by the river. With decreasing depth, a reservoir's capacity to generate electricity and provide water for other purposes diminishes.[5] In general, the volume of sediment reaching reservoirs increases over time. First, more top soil is exposed due to deforestation and expansion of commercial agriculture. Second, the volume of precipitation in many places is now increasing due to climate change.

Depending on the terrain, the sedimentation process does not remain limited to reservoirs but extends to riverbeds upstream. Upstream sedimentation aggravates the "backwater effect" of dams, increasing lateral pressure on riverbanks, causing bank erosion, and resulting in widening of rivers.

It is true that the dam-building technology has evolved over time, allowing now bed- and mid-level gates for some sediment to pass through. However, most existing dams do not have such gates, and even new dams are not always equipped with them because of technical feasibility and financial considerations.

As a result, sedimentation of both reservoirs and upstream riverbeds, with the accompanying effects, remains an important consequence of dams and barrages.

3.3.3 Downstream effects on river morphology

The morphological effects of dams and barrages on downstream stretches of rivers are important too. The water that comes out of dams and barrages carries less sediment than it would have otherwise. As a result, rivers generally engage in bed scouring immediately below dams and barrages to regain their sediment equilibrium. However, because of the reduced flow, the sediment is then deposited in the riverbed further below. Downstream sedimentation is aggravated when tributaries join the main river, bringing their own sediment, which the main river cannot carry due to its reduced flow.

As mentioned earlier, one of the consequences of dams is stabilization of river flows across seasons, in particular to ensure stable electricity generation rates. However, this stabilization also deprives the downstream section of its high peak season flows that used to flush away sediment and maintain river depth. As a result, rivers often lose depth and flow capacity downstream.[6]

3.3.4 "More water, more thirst!" syndrome and exhaustion of rivers

An important consequence of the Commercial approach is the "More water, more thirst!" syndrome (Figure 3.1). This vicious cycle arises in the following way. Dams, barrages, and reservoirs increase the availability of water for various commercial purposes. The augmented, easy, and "cheap" availability of water leads however to its wasteful use, including settlement in arid areas, cultivation of water-intensive crops, setting up of inappropriate water-intensive industries, promotion of a water-intensive life-style; etc.[7] As a result, soon after more water is made available, demand outpaces supply, and water shortages develop. The perceived water shortage leads to the demand for more river-intervening structures to augment the water supply.[8]

The "More water, more thirst!" cycle aggravates the effects on downstream river morphology and ultimately leads to exhaustion of rivers. In many cases, upstream diversions of water using dams and barrages have been so severe that little river water is left to reach the seas. As the WCD (2000, p. ii) notes, "In some years, our mightiest rivers—Africa's Nile, Asia's Yellow, America's Colorado, Australia's Murray-Darling—do not reach the sea." We will consider these and a few other cases in more detail in chapter 4.

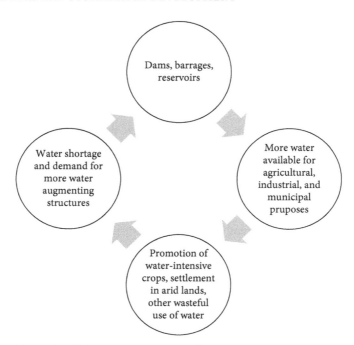

Figure 3.1 "More water, more thirst" syndrome
Source: Author

3.3.5 Degradation of estuaries, deltas, and marine environment

Diminution of river flows, or their outright failure to reach the seas, causes harm to the ecology of floodplains, deltas, estuaries, and the marine environment. The natural floodplain and delta formation process is disrupted, and existing floodplains and deltas face desiccation, subsidence, and erosion. Also, diminished river flows allow tidal flows to move further inland, causing salinity ingress. Increased salinity causes decay of estuarine and delta flora and fauna, particularly mangrove forests. Finally, diminution of the volume of river water reaching the seas has detrimental effects on the marine environment, including fish stocks, which often rely on sediment-laden river flows for various nutrients.[9]

3.4 Adverse consequences for the ecology of river basins

Apart from their effects on river morphology, dams and barrages have broader effects on the ecology of river basins, as briefly noted in this section.

3.4.1 Regular, expected flooding versus unexpected, catastrophic flooding

The regulation of river flows and control of floods are often presented as major benefits of dams. While this is generally true, doing away with regular and expected floods is not always a good thing. Regular river inundation (floods) during high season performs vital ecological functions for floodplains and deltas. These include sedimentation, natural irrigation, recharge of waterbodies, pulses of nutrients needed by flora and fauna, etc. (These functions will be described in more detail in chapter 6.) By stopping regular floods, dams may put an end to these important functions, causing harm to the ecology of river basins downstream.[10]

Meanwhile, dams and barrages can also become a cause of unexpected and damaging flooding. Upstream, dams can cause new flooding, when more water than expected accumulates in reservoirs. Downstream, unexpected floods may be caused by unexpected releases of water from reservoirs, which may result from a variety of factors. First, excessive upstream flow—due to high precipitation or other reasons—may lead to an unexpectedly large volume of water collected in reservoirs. Unexpected releases may then be necessary to avoid dam overtopping. Sedimentation—reducing reservoir capacities—increases the probability of this type of excessive discharge. Second, dam operation failures may cause unexpected floods, for various reasons. For example, decisions regarding discharge are often political and depend on the relative strengths of vested interests of various upstream and downstream quarters. To avoid excessive discharge, it is generally necessary to keep reservoir levels low during pre-peak weeks. However, this may not be possible due to the need to maintain the level of electricity generation demanded by, for example, urban residents, who may not care much for flooding that would affect rural areas. Second, excessive discharge may result from mere errors on the part of dam operators, perhaps because they lack accurate information regarding upstream flows. Sheer negligence may also lead to such errors.[11] Finally, outright dam breaks, due to structural failures or other reasons, provide extreme cases of dam-induced flooding.[12] In short, while mitigating seasonal floods, dams themselves can become a source of unexpected floods.

Furthermore, dam-induced floods often become catastrophic floods because, first, dam-induced floods are unexpected, unlike seasonal floods, which are expected and the people are generally prepared for them. Second, by stopping seasonal floods, dams are generally conducive to "development" (i.e., construction of dwellings, commercial establishments, and infrastructure) in below-flood-level areas downstream (as we shall see in more detail in chapter 6). As a result, when dam-induced flooding occurs, more damage to life and property is caused than seasonal flooding caused previously.

3.4.2 Waterlogging and salinity

Excessive irrigation promoted by "cheap" water from dam reservoirs in turn often leads to waterlogging and salinity. Several factors contribute to this unintended and undesirable outcome. First, river water itself contains dissolved salts that accumulate in the irrigated soil as a residue from evapotranspiration. Second, excessive irrigation causes the groundwater level to rise and reach the top soil through capillary action. The dissolved salt in the groundwater then accumulates in the soil, also as a residue from evapotranspiration. The problem of dam-induced salinity is more severe in desert areas that do not have enough rainwater to wash away some or all of the accumulated salt in the topsoil.[13]

3.4.3 Toxification of return water

The Commercial approach to rivers has generally been accompanied by a switch from pre-industrial organic agriculture to the industrial era's chemical agriculture. Irrigation expansion facilitated by dams led to large increases in the use of chemical fertilizers and pesticides. The surface return flow from irrigated fields to rivers is therefore generally toxic. This toxicity percolates to the groundwater, so that the sub-surface return flow also becomes toxic. In addition to irrigation, abstraction of reservoir water for urban and industrial uses also contributes to toxification, as the return water contains many inorganic and organic pollutants.

3.4.4 Effects on fisheries and aquatic life

The Commercial approach has significant effects on fisheries and other aquatic life in both river channels and floodplains. The fragmentation caused by dams and barrages disrupts the free passage of fish and other aquatic life forms along rivers, as noted earlier. This has a negative impact, particularly on fish species that need to swim upstream to spawn. Pacific salmon and Bay of Bengal hilsa are two prominent examples of such fish species.[14] Attempts to mitigate obstruction to fish-migration by adding fish passages and ladders to dams and barrages have not been that successful.[15] Construction of multiple dams on the same river aggravates the problem. Many aquatic species, such as the Indus dolphin, are on the verge of extinction due to river fragmentation caused by dams, as we shall see in more detail later. It's cousin the Yangtze River dolphin is already extinct. Depending on the circumstances, river temperature and oxygen and other nutrient content in both upstream and downstream reaches of rivers differ from what they used to be in pre-dam conditions. As a result, life forms accustomed to the previous regime can no longer survive.

Dams also adversely affect capture fisheries in floodplains. As noticed earlier, the end of seasonal inundation leads to shrinkage of waterbodies in downstream floodplains, reducing the habitat of capture fisheries. Also, by putting an end to seasonal inundation, dams prevent the recharging of waterbodies necessary for spawning and growth of fish fry. Many other forms of aquatic life that depend on inputs from peak season flows also suffer due to impacts of dams and barrages. Dams may facilitate farm fisheries in downstream floodplains. Also, dam reservoirs themselves may serve as waterbodies for both capture and farm fisheries.[16] However, these effects often fail to compensate for the cumulative loss of fisheries suffered in river channels and floodplains. Moreover, farm and reservoir fisheries generally focus on a few varieties of fish and hence cannot compensate for the loss of fish and other aquatic diversity.

3.4.5 Effects on the flora and fauna of river basins

Thediscussion above makes it clear that dams and barrages have a significant impact on the flora and fauna of river basins. Upstream, reservoirs submerge huge areas, which often contain primeval forests with large amounts of biodiversity. Efforts to collect and introduce elsewhere plant and animal species of areas to be submerged have not generally succeeded. As a result, a great amount of biodiversity have been permanently lost due to dams and barrages.[17] Changes in the seasonality of river flows lead to changes in cropping pattern and the flora and fauna downstream. The species that depend on seasonal flood impulses gradually disappear. Similarly, species that require dry seasons also find it difficult to survive. The deleterious effects of dams and diversionary barrages on the ecology of the deltas, estuaries, and marine shores have already been mentioned above. Overall, as WCD concluded, ecosystem impacts "are more negative than positive, and they have led, in many cases, to irreversible loss of species and ecosystems (WCD 2000, p. 93)."

3.5 Commercial approach and climate change

The Commercial approach to rivers is now facing more controversies in view of climate change, which is altering the parameters within which the dam- and barrage-building calculus is conducted, or has been until now. This section reviews the ways in which the Commercial approach may make it harder to address the climate challenge facing us.

3.5.1   Reservoirs as a source of GHG emissions

Hydropower is often presented as both environment and climate friendly. Accordingly, many countries include expansion of hydropower in their strategy to achieve sustainable development. A closer scrutiny however shows that hydropower is not as climate friendly as has been claimed. In fact, dam reservoirs can be a significant source of GHG emissions for the following reasons. First, the carbon absorbing function of forests submerged by reservoirs is lost forever. Second, submerged trees and other organic matter generate methane, which is a more potent global warming gas than carbon dioxide (CO_2). Third, methane continues to be emitted by algae and other organic materials that either grow in or are brought to the reservoir by the river. The amount of methane generated by reservoirs varies depending on a reservoir's location and is much higher for reservoirs in the tropical region than for those in cold and temperate regions.[18] In tropical reservoirs, the CO_2 equivalent of the methane generated often proves to be greater than for the equivalent amount of electricity generated using fossil fuels. Thus the "clean" attribute accorded to hydropower often proves to be exaggerated or incorrect.[19]

3.5.2   Non-renewability of hydropower

It is also questionable whether hydropower can be termed as renewable. While river flows are renewable (assuming the global hydrological cycle remains, by and large, intact), dams themselves are not renewable. Dams have a finite lifespan and cannot continue to function forever. At some stage, old dams have to be decommissioned and demolished. Whether or not large, old dams can be rebuilt after they have been decommissioned and demolished is difficult to say, because river morphology changes over time due to natural causes and particularly under the impact of dams themselves, as noted earlier.[20]

3.5.3   Consequences of changes in precipitation pattern

Climate change is increasing the risks of dam-induced catastrophic floods. Untimely and excessive precipitation is becoming common, increasing the risk of excessive discharge and dam overtopping. The catastrophic flooding at the Orville Dam in February 2017, when water overflowed across the emergency spillway, is evidence of what may likely result from climate change.[21] At the same time, prolonged droughts in other areas are leading to depletion of reservoirs and their capacity to generate electricity and serve other purposes. Thus, climate change is raising questions regarding dams from both ends.

3.5.4 Adverse consequences for deltas

Earlier, we noted the deleterious effects of dams and barrages on deltas, estuaries, and marine ecology. Climate change is aggravating these effects too. One of the ways it is doing so is through rise in sea level, causing submergence, coastline erosion, and pushing tidal flows further up rivers, thus aggravating salinity ingress. Dimunition of river flows due to dams and barrages aggravates these consequences.

3.6 Adverse social effects of dams and barrages

The Commercial approach generally worsens economic and social inequality.[22] Some ways in which this happens are provided in this section.

3.6.1 Unfair distribution of costs and unfair treatment of indigenous and politically weak peoples

One of the direct social costs of dams and barrages is displacement suffered by people whose lands are acquired for construction, which is extensive in the case of dams that submerge large areas to create reservoirs. According to WCD, the number of people displaced by dams ranged between 40 and 80 million, of whom between 16 and 38 million were in India. It further notes that Yangtze basin dams alone displaced at least 10 million.[23] Often, the displaced people do not get adequate compensation to rebuild their lives. It is generally the indigenous peoples, ethnic minorities, and other politically weak people who bear the direct social costs of dams and barrages. Their land is often confiscated without adequate compensation. Duflo and Pande (2007) note that while scheduled tribes (comprising mostly indigenous peoples) represent only 8 percent of India's population, they constitute 47 percent of the people displaced by dams. Submergence by dams has been a major cause of tribal insurgency in India. The same is true in many other countries.[24]

3.6.2 Unfair distribution of benefits

Meanwhile, benefits produced by dams tend to be concentrated on more well-to-do sections of the society. For example, the electricity generated by dams often goes to urban, relatively well-off residents, bypassing the displaced people as well as rural and poor urban residents. Similarly, the benefits of dam irrigation are generally appropriated disproportionately by large landowners, due both to the size of their holdings and to the influence they can exert to skew decisions regarding the distribution of water in their favor. The landless, by contrast, are completely deprived of any benefits accrued via land owner-ship. Furthermore, dam, barrages, and other river-intervention structures gen-erally constrict common property resources on which the poor often depend for a part of their livelihood. For example, in many countries, capture fisheries are an important resource for the rural poor. However, as noted earlier, dams and barrages reduce capture fisheries in floodplains of the downstream reaches of rivers. Meanwhile, fishing in upstream reaches changes from river to lake fishing, which often proves to be more restricted and capital-intensive. Often, it is women, who in many societies play a more important role in using common property resources, suffer the most.[25]

3.7 Other risks associated with dams and barrages

Apart from the adverse consequences noted in the previous sections, dams and barrages are associated with other risks, some of which are noted in this section.

3.7.1 Exogenous and induced seismic risks

Dams and barrages, particularly the former, are associated with seismic risks, of two sources. First, most dams are located in mountainous areas, which are also generally areas with high seismic activity. These are exogenous seismic shocks. Second, scientists think that large reservoirs created by dams induce seismic ac-tivity; these are endogenous seismic risks. Needless to say, the danger of dam failure, leading to catastrophic flooding, is much higher in areas with seismic activity.[26]

3.7.2 Security risks

Dams and barrages also carry security risks. Many extremist groups threaten to blow up dams in order to cause damage. Similarly, there are worries about terrorist groups blowing up dams in areas they occupy or can infiltrate. For example, there was concern that ISIS would blow up the Mosul Dam on the Euphrates River in Iraq. Dams in other conflict zones are susceptible to this risk too.

3.7.3 Dams and diseases

Dams, or more specifically the reservoirs created by them, are often associated with diseases. Stagnant water in the reservoirs provides favorable conditions for various bacteria and other waterborne vectors to incubate and spread. For example, one-third of the people in the villages downstream of Kpong dam in the Upper Volta were found infected by *Schistosoma mansoni*, which causes intestinal schistosomiasis.[27] Similarly, in Sri Lanka, wider spread of malaria and dengue has been observed around reservoirs.[28]

3.7.4 Loss of lives during construction

Dams and barrages present threats to human lives during their construction; dam construction is a particularly labor-intensive enterprise, with thousands of workers laboring under risky conditions for many years to construct a major dam and fatalities are common (McCully 2001, p. 117).

3.8 Costly benefits of the Commercial approach

The adverse consequences described above raise doubts about the cost-effectiveness of benefits derived from the Commercial approach. For example, dams and barrages are generally presented as cost-effective solutions to the lack of electricity and water for irrigation, urbanization, and industrialization. However, closer analysis reveals that these solutions are not cheap, even in terms of narrow financial considerations. This is more so when the broader economic and environmental effects are taken into consideration.

3.8.1 Cost padding, cost overruns, and delayed completion

The structures necessary for implementation of the Commercial approach are costly, and several reasons combine to increase their costs further. First, river intervention structures are generally carried out by the public sector where accountability is often less stringent than in the private sector. As a result, cost padding, cost inflation, and associated completion delays are quite common. Second, large-scale, public sector infrastructure projects in developing countries are often "donor" financed. For such projects, accountability is even weaker, on the part of both creditors and debtors. Cost and time overruns, generally interrelated, are therefore more the norm than the exception for these projects.[29] WCD found that cost overruns averaged 56 percent for the 81 large dams it surveyed and were found to be worst in South and Central Asia, where they averaged 138 and 108 percent, respectively. As noted earlier, cost overruns are generally associated with time overruns; of of the 99 dams for which WCD had data, only 50 were completed within one year of their stipulated completion date.[30]

3.8.2 Costly hydropower

Production of hydroelectricity has been traditionally a strong argument for construction of dams, and it has been generally claimed that hydropower is cheap. However, a closer scrutiny shows that hydropower is not as cheap as claimed. First of all, construction of dams often ends up costing much more than originally projected, for reasons mentioned earlier. Second, for various reasons, including poor forecasts regarding water availability, the actual amount of electricity generated generally falls short of the installed capacity, raising the per unit cost. Third, even if the dam generates as per installed capacity at the beginning, the generation capacity generally falls with time due to siltation of reservoirs, as noted earlier.[31] As a result, even the financial cost of hydropower proves to be much higher than claimed and sometimes not too different from the cost of electricity generated from other sources.

3.8.3 Costly irrigation

Irrigation provided by dams also proves to be costly. To begin with, dams generally fail to achieve the targeted area of irrigation. According to WCD, 58 irrigation dams it analyzed fell well short of targets in terms of area irrigated and the amount of water applied to the fields in those areas which were irrigated.[32] This is in part because a significant part of the water is lost through evaporation and seepage during travel across long distances (generally through newly dug canals) to reach the fields to be irrigated. Second, many problems arise even after water reaches the fields. The efficient use of large-scale surface water irrigation projects requires cooperation and good management on the part of beneficiaries (irrigators). Cooperation is necessary regarding crop choice, timing of various farming operations, and the amount and ordering of water distribution. Such cooperation however often fails to materialize due, in part, to conflicts of interests among irrigators themselves and the absence of institutions that can effectively mediate and resolve these conflicts. Since dams are generally under public ownership, public officials are appointed for irrigation management, and these officials often do not fit well with local communities and institutions. The result is inefficient use of dam and barrage water. This inefficiency finds reflection in the general inability to recover costs of irrigation from beneficiary farmers. As a result, dam-based irrigation water ends up being distributed either for free or for a very nominal charge, which then encourages excessive and inefficient use of water. For example, dams and barrages have often promoted flood irrigation, which is generally a wasteful method of irrigation. According to some estimates, about 60 percent of water is wasted due to inefficient canal and irrigation systems[33]. This process leads to the "More water, more thirst!" vicious cycle noted earlier. WCD found that the larger the dam, the worse was the record regarding irrigation performance. As a result, dam-based irrigation generally proves to be a financial burden on the government.[34]

3.8.4 Disappointing financial and economic rates of return

As noted earlier, financial rates of return are computed based only on those inputs and outputs that have markets and are evaluated using prices established in these markets. Computation of economic rates of return differs from that of financial rates of return in two respects. First, inputs and outputs that do not have markets are also taken into account. Second, valuation of even marketable inputs and outputs is conducted using not market prices but imputed prices that are supposed to reflect their true social costs and hence are often called social prices.[35] The use of social instead of market prices is particularly important for inputs and outputs that have significant externalities, either positive or negative, and for which markets prices therefore generally do not reflect the true social prices.[36] It is clear that computation of economic rates of return is much more challenging than that of financial rates of return.

WCD found that even information on financial rates of return was hard to obtain, and it therefore relied on project evaluation reports conducted by large financing organizations such as the World Bank, Asian Development Bank (ADB), and African Development Bank (AfDB). As per these reports, nine out of twenty large dams had an Economic Internal Rate of Return (EIRR) less than 10 percent, as compared to 15 percent or higher claimed at project approval. The evidence also showed that multipurpose projects had lower EIRR than single purpose ones.[37]

However, the EIRRs reported by financing organizations are fraught with many issues. The Internal Rate of Return (IRR) equals the discount rate at which the discounted value of the stream of benefits equal the discounted value of costs. It gives a measure of how profitable an investment is.[38] Though the financing agencies describe their IRRs as economic, in reality, these are mostly financial, because inputs and outputs that are not marketed are generally ignored and valuation is generally done using market prices and not social prices, which are difficult to compute.

The difficulty in computing the rates of returns on investment on dams and barrages arises even at the earlier and more crude stage of physical quantification of input and output. For example, the dam industry claims that one-third of the world's food production is due to dam irrigation.[39] However, WCD thought that dam irrigation contributes to at most 12–16 percent of world food production.[40] In case of India, which has one the largest irrigated areas in the world, Thakkar attributes only 12–13 percent of increased food production to dams, and Sengupta thinks this figure to be at most 10 percent.[41]

Duflo and Pande (2007) offer a careful study of rates of return of investments on dams in the context of India. They note that irrigation has been the primary purpose of 96 percent of Indian dams, and the increase in agricultural output has been entirely attributed to dams. Countering this claim, they point out that, first of all, there have been large increases in the use of many other inputs (such as fertilizer, machinery, high-yielding seed, etc.) during the same time. Second, alternative irrigation methods, such as tube wells, either already operate in "command areas" of dams or would have operated there in the absence of these dams. It is therefore not correct to attribute the entire increase in agricultural output to dams. To accurately determine the effect of dams, the authors conduct an econometric study using Indian state-level data on agricultural output and number of dams. However, they use the average surface gradient of a state as an instrument for the number of dams because of the suspected feedback relationship between output increase and number of dams built. Their analysis suggests that large dams contributed to only about 9 percent of the agricultural productivity increase—a number that is close to the Sengupta estimate of 10 percent.[42] Thus the overall contribution of dams to increasing India's agricultural production has been modest. The low realized (i.e., ex-post) financial rates of return of large dam projects find further reflection in the fact that there has been little enthusiasm from the private sector to invest in dams.[43]

Given the low financial rates of return for dams and barrages, the properly computed economic rates of return are certain to be much lower and most likely negative. This is because dam and barrages involve huge negative externalities, in the form of damage done to the environment, lives of displaced people, future generations, etc. As WCD (2000, p. ix) noted, whatever positive contribution dams and barrages made to development, this contribution "has been marred in many cases by significant environmental and social impacts which, when viewed from today's values, are unacceptable." This is more likely to be the case if the unfair distribution, social conflicts, and conflicts among co-riparian communities generated by the Commercial approach are also taken into account.

3.9 Commercial approach and conflicts

One of the pernicious effects of the Commercial approach is the conflicts that it generates among co-riparian communities, regions, and countries. This is not surprising, because the Commercial approach prompts upper riparian countries to abstract as much water as possible before rivers reach lower riparian countries. However, river flows are finite quantities, and hence more abstraction of the flows by one country leaves less for others. The Commercial approach therefore leads to a zero-sum game entailing a "beggar thy neighbor" type of behavior. It pushes upper riparian countries towards wasteful irrigation and other water-diversionary projects in order just to grab as much water as possible. Lower riparian countries, in their turn, often chalk up and implement wasteful irrigation and other water-abstraction projects in order to show to upper riparian countries that they need the water and do not have enough of it. This race to withdraw water often proves to be a race to kill rivers. In the end, what results is known generally as the "tragedy of commons" and, specifically, the exhaustion of rivers.

The Commercial approach generates conflicts even when abstraction is not involved. As is clear from the discussion here, dams constructed for electricity generation, navigation, and flow regulation purposes do not entail abstraction of river water. However, these structures allow the countries constructing them to control river flows, in particular, their timing and volume. Conflicts arise over these control issues, because, in making these decisions, upper riparian countries may be guided more by their own interests and neglect the interests of lower riparian countries. The Commercial approach may generate conflicts in many other ways. In general, it converts rivers into sources of contention and conflicts rather than being bonds of friendship.

Conflicts over river flows also arise among states, provinces, regions, and communities within a country. In such situations, the conflicts are more manageable. However, this is not the case when competing entities are sovereign countries. Conflicts over river flows generated by the Commercial approach can be seen in all parts of the world. Chapter 5 uses the experience of South Asia to illustrate the Commercial approach's tendency to generate conflicts.

3.10 Conclusions

Dams, barrages, and other river intervening structures, inspired by the Commercial approach, led to immediate commercial gains, in the form of electricity, irrigation, and water supply for industrial and municipal purposes. However, the negative consequences of these structures also surfaced with time. It became clear that dams and barrages fragmented river basins; distorted river morphology; damaged the ecology of river basins; encouraged wasteful use of water; created problems of toxic run-off; and gave rise to the "More water, more thirst!" syndrome. Even the apparent benefits of the Commercial approach have darker sides. For example, it was observed that hydroelectricity was not cheap, reservoirs generated GHGs, and irrigation often led to salinity and deterioration of soil quality. Furthermore, the distribution of the benefits of the Commercial approach was often socially not fair, allowing their concentration in the hands of those who had more property and power. Dams and barrages also generated conflicts. The next chapter examines the experience of few particular rivers in order to illustrate the consequences of the Commercial approach discussed in this chapter.

Notes

1. WCD (2000, p. xxix). For benefits of dams, see ICOLD (1997).
2. WCD (2000, p. xxix) states that between 30 and 40 percent of the irrigated land (over 271 million hectares) worldwide depends on dams.
3. For detailed—dam by dam—information about these consequences, see, in particular, WCD (2000) and McCully (2001).
4. According to WCD (2000, p. 93), in the United States, with 19 dams in place, only 70 km of the about 2000 km long Columbia River flows uninterrupted. In Europe, most of the major rivers do not have more than a quarter of their length uninterrupted by dams. Overall, WCD found that dams fragmented 60 percent of the world's rivers. According to McCully (2001, p. xxxii), of the rivers in the continental United States that are longer than 1,000 km, only the Yellowstone River remains undammed. In France, a dam was constructed in 1986 on the only remaining free-flowing stretch of the Rhone River. Other major rivers of Europe, such as the Volga, Weser, Erobo, and Tagus do not have a stretch more than a quarter their lengths that does not have a dam. For a general discussion of river morphology, see Schumm (ed.) (1972).
5. For example, the storage capacity of Nepal's Kulekhani hydro dam was reduced by nearly one-tenth by the sediment scoured off upstream mountainsides during a single 30 hour storm burst in July 1993. There is therefore apprehension that sediments will soon put the 114 m dam out of operation. Yet, Kulekhani dam was predicted to have a design life of 75–100 years when it was completed in 1981 (McCully 2001, p. 109). Similarly, Thailand's 25 largest dams now retain only about half of their combined usable capacity (McCully 2001, p. 105).
6. One subtle way in which dams affect downstream river ecology is by altering the temperature of the downstream flow in undesirable ways.
7. It should be noted that the water supplied by dams and barrage projects appears to be cheap mainly because these projects are generally financed by public money, and private users do not have to pay for it, at least not fully. More on this later. See Reisner (1993) for the experience of American west in this regard.

8. It is important to note in this context that the amount of water deemed necessary may change with perception, attitude, and policies. For example, on the demand side, a water-deficit industrial enterprise can become water-sufficient if it decides to recycle the water that it uses, instead of just discharging it. An otherwise water-sufficient township can become water-deficit if its constituent households all want to have private pools. Similarly, a water-sufficient agricultural region producing wheat may become water-deficit if it wants to grow rice. On the supply side, the local water supply may be augmented by ways other than transferring water from a far-away river. For example, a water-deficit village can become water-sufficient if it implements a successful rainwater harvesting program. Similarly, a water-sufficient agriculture based on groundwater can become water-deficit if it does not care for adequate replenishment of groundwater aquifers.

9. As McCully (2001, p. 36) informs us, Akosombo Dam has virtually put an end to the supply of sediment to the Volta estuary. As a result, the shoreline of neighboring Togo and Benin is undergoing erosion at a rate of 10 to 15 m per year. During a single storm in 1984, the sea advanced by 20 m in places, eroding a large part of the main Ghana-Togo-Benin highway.

10. With regard to river stabilization and flood control, WCD (2000, p. 58) notes that some dams have actually "increased the vulnerability of riverine communities to floods." McCully (2001, p. xix) points to the finding of research in "disturbance ecology" that rivers need floods, just as some types of forests need fires.

11. Suoi Voc Dam of Vietnam provided a recent example of dam-induced catastrophic flooding due to sheer negligence, when, on March 15, 2017, operators discharged excessive water from the reservoir because they were drunk! According to the Vietnam News Agency, about 2 million cubic m of water was released, flooding rural communes and damaging 18 water pumps and a large part of the crop area (http:// www.news.com.au/world/asia/two-vietnamese-communes-flooded-after-drunks-opened-flood-gates/news-story/cadc10a5507ef82276a1ddb6a55b303b).

12. In February 2017, Oroville Dam, the tallest dam in the United States, illustrated the potential danger of catastrophic flooding caused by dams. During heavy rainfall preceeding February 7, damage to the dam's main and emergency spillways had been noticed. As the management closed the spillway to examine the damage, further rainfall raised the water level of the reservoir, ultimately leading water to flow over a concrete weir at the top of the dam's emergency spillway, despite the reopening of the damaged main spillway. The uncontrolled water flow threatened the weir with collapse, which would have sent a 9 m wall of water down the Feather River, and caused devastation downstream. More than 180,000 people had to be evacuated. Fortunately, the collapse did not occur, and the calamity was avoided. But, this shows how dams can be become a source of catastrophic floods.

13. See WCD (2000, p. 66) for evidence on waterlogging and salinization caused by surface water irrigation.

14. For example, the Gulam Mohammed Dam of Pakistan deprived the hilsa, a migratory fish of great commercial importance for South Asia, of about 60 percent of its previous spawning areas in the Indus River (McCully 2001, p. 43).

15. Apparently, World Bank and Thai officials for a long time rejected the claims of independent Mekong fisheries experts and local fishing communities that the fish ladder planned for the highly controversial Pak Mun Dam would be largely useless (McCully 2001, p. 53).

16. The Volta reservoir, the world's largest reservoir by area, apparently exceeded the expectation regarding its fish potential. According to McCully (2001, p. 154), the reservoir produced more than 60,000 tons of fish at the end of the 1960s when the reservoir was first filled. In some subsequent years the income from sales of reservoir fish exceeded the income from the electricity generated by the Akosombo Dam.

17. In case of Thailand's Chiew Larn Dam, for example, only about 5 percent of the animals of the submerged zone could be rescued (McCully 2001, p. 54).

18. According to Eric Duchemin of the University of Quebec at Montreal, mean net emissions from boreal reservoirs are equivalent to 20 to 60 g of CO_2 per KWh generated. By comparison, net emissions from tropical reservoirs range from 200 to 3,000 g of CO_2 per KWh. This shows that emissions from hydropower located in the tropics can be higher than emissions from natural-gas-fired combined cycle power plants—currently the technology of choice of power generators in much of the world—which emit around 430–635g CO_2 equivalent per KWh (McCully 2001, xxxvi).

19. For a detailed discussion of this issue, see McCully (2001, pp. xvii, xxxii–xxxiii, xxxvi), who notes that reservoirs in tropical countries may emit GHG "at levels higher than fossil-fuel-fired power plants (p. xvii). In his view, the United Nations climate panels have not paid enough attention to this issue, and their measurements of methane and CO_2 emissions due to rotting organic matter in reservoirs go back to only 1993 and come from only 30 reservoirs, mostly in Canada and Brazil (p. xxxii). He further notes that though methane is shorter lived than CO_2, its impact on climate change can be stronger. He draws attention to the fact that, according to the IPCC, methane's 100 year Global Warming Potential (GWP) is 21, implying that 100 years after its release, a ton of methane in the atmosphere causes 21 times more warming than CO_2. As per Gaffin's model, this differential is greater—a century later, the cumulative global warming effect of a constant methane emitter is about 39.4 times greater than that of a constant emitter of an equivalent quantity of CO_2 (ibid., p. xxxiii).

20. See McCully (2001) for more on this point.

21. See n. 12 for details on the Orville Dam catastrophe.

22. Regarding the social performance of dams, WCD (2000, p. 129) finds that "the construction and operation of large dams has had serious and lasting effects on the lives, livelihoods, and health of affected communities, and led to the loss of cultural resources and heritage." For social and environmental effects of large dams, see also Goldsmith and Hildyard (1984).

23. See McCully (2001, p. xxxi).

24. McCully (2001, p. 172) points out that the World Bank skipped over the displacement of some 80,000 people in their nine-volume (3 feet thick) feasibility study for Pakistan's Kalabagh Dam. He provides similar evidence regarding submergence of habitat of indigenous peoples by Vietnam's Hoa Binh Dam (p. 79), Thailand's Khao Laem Dam (p. 83), and Sri Lanka's Mahaweli project (p. 241). He also reports

increased poverty in the catchment areas, showing that people affected by dams are not compensated adequately and lack the political clout to demand and ensure adequate compensation.

25. Duflo and Pande (2007, p. 601) note that the distribution of the costs and benefits of large dams across population groups, and, in particular, the extent to which the rural poor have benefited, are issues that remain widely debated.

26. See McCully (2007) for more discussion of dam safety risks.

27. See McCully (2001, p. 89).

28. McCully (2001, p. 7) shows that there is a greater tendency for larvae to breed giving rise to wider spread of malaria and dengue due to stagnant water in reservoirs in Sri Lanka.

29. See McCully (2001) for evidence of cost overruns regarding dams in Colombia (p. 135); Parana River mega dams (pp. 251–251); Itaipu dam (Brazil and Paraguay) (p. 86); Akosombo Dam in Ghana (p.p. 239–420); dams in Malaysia (pp. 139, 262, 265); dams in Sri Lanka (p. 273).

30. See McCully (2001, p. xxviii).

31. WCD (2000) found that of the 63 large dams reviewed, 55 percent generated less power than projected. See also McCully (2001, p. xxxviii).

32. WCD (2000) found that irrigation targets were not achieved for many dams even after 15 years of operation. It also found that water supply dams fared worse than irrigation dams. Only 30 percent of such dams supplied as much water as was stipulated (McCully 2001, p. xxix).

33. See US Foreign Relations Committee (2007, p. 84).

34. As WCD (2000, p. 47) notes, ". . . irrigation dam projects . . . have all too often failed to deliver on promised financial and economic profitability—even when defined narrowly in terms of direct project costs and benefits."

35. In the economic cost-benefit literature, social prices are often called by other names, such as accounting, shadow, and dual prices: dual because these prices are often obtained through a linear programming exercise in which they emerge as the solution to the dual problem; accounting because these prices are result of accounting (computation) and not from observation in actual markets. For further details on these issues, see Little and Mirlees (1969) and Dasgupta, Sen, and Marglin (1972).

36. Needless to say, inputs and outputs that do not have market prices have to be evaluated using social prices, by default.

37. See McCully (2001, p. xxix).

38. As is clear from the discussion here, whether an IRR is economic or financial depends on whether or not non-marketed inputs and outputs are included in the analysis and whether social or market prices are used to evaluate inputs and outputs.

39. See McCully (2001, p. xxx).

40. This was despite the finding that between 30 and 40 percent of the total 270 million hectares of irrigated land received water from dam reservoirs (McCully 2001, p. xxx).

41. See WCD (2000) and Duflo and Pande (2007).

42. Duflo and Pande (2007) note that India's food production increased by four times, from 51 to 200 million tons, between 1951 and 2000. Proponents of dams claim that

at least 25 percent of this increase is due to dam irrigation. The authors note that the number of dams in India increased from fewer than 300 in 1947 to over 4,000 by 2000, and the area irrigated by dams constitutes 35 percent of the total irrigated area in India. However, they point out that it is incorrect to attribute such a large part of the food production increase to dam irrigation, because there was, during the same time period, considerable increase in the use of other inputs such as fertilizer, machinery, etc. much of which would have happened even without the dams. In view of the possible endogeneity of the number of dams, they use average land gradient as the instrumental variable, because, while being correlated with the number of dams, gradient is not likely to have an independent effect on agricultural output of a state. As they explain, "after one accounts for the impact of the overall hilliness of the district and the availability of rivers, the gradient of rivers is unlikely to have a direct impact on changes in agricultural productivity or other district-level outcomes before and after a state builds new dams (p. 622)." This argument provides them the basis to use the variation in dam construction induced by differences in gradient across districts in Indian states to determine the impact of large dams.

43. It has been observed that in cases where private investors had to bear the risks involved, they generally shied away from dam projects. As McCully (2001, p. xvii) put it, "where generous public subsidies and guarantees are offered so that private investors bear little risk, they will indeed gladly fund dams. But where they bear substantial exposure to cost overruns, droughts, accidents, legal challenges, and public opposition, private investors have shown little willingness to risk the huge up-front sums necessary to build big dams."

4

Application of the Commercial approach

A few case studies

4.1 Introduction

The general experience of rivers across the world reveals well the adverse effects of the Commercial approach, as seen in chapter 3. However, these effects become more vivid when examined in the context of specific rivers. This chapter therefore reviews the experience of several renowned rivers from different parts of the world. These are: (i) the Colorado River of North America, (ii) the Murray-Darling river system of Australia, (iii) the Amu Darya and Syr Darya Rivers of Eurasia, (iv) the Nile River of Africa, and (v) the Indus River of Asia. Drawn from different continents of the world, they provide a representative picture of the consequences of the application of the Commercial approach.

The chapter begins with the experience of the Colorado River, which is one of the powerful rivers of the United States, flowing across its southwestern region. Over geological eons, the Colorado River has created canyons, including the famous Grand Canyon, and other unique geological formations and ecosystems. Unfortunately, it has been subjected to a particularly aggressive version of the Commercial approach for many decades so that it now dries up before reaching the Gulf of California. Many unique ecological treasures have been lost, and the river basin is undergoing a slow process of decay.

The chapter next discusses the experience of the Murray-Darling river system, the main river system of Australia. Through extensive interventions and abstractions inspired by the Commercial approach, most of its water was used up, so that little was left to reach the sea. In addition to other adverse effects on the river basin ecology, salinity intrusion from the sea has required construction of several barrages to protect the land.

Next the chapter reviews the experience of the Amu Darya and Syr Darya rivers, the flows of which were diverted to promote cultivation of commercial crops. However, the diversion led to the desiccation of the Aral Sea, once the second-largest inland sea, and the original and natural destination of these two rivers. The destruction of the Aral Sea has been one of the greatest environmental disasters of the world, and it shows that even the socialist ideology didn't prevent

Rivers and Sustainable Development. S. Nazrul Islam, Oxford University Press (2020). © Oxford University Press.
DOI: 10.1093/oso/9780190079024.001.0001

countries from following the Commercial approach aggressively and to the detriment of the river basin ecology.

The chapter then goes on to discuss the experience of the Nile, the longest river of the world and the progenitor of one of the first civilizations of the world. The High Aswan Dam on the Nile greatly enhanced the capacity to abstract river flow for irrigation and other commercial purposes but has also resulted in little water and sediment reaching the Mediterranean Sea and the deterioration of the Nile Delta. Several barrages had to be built to prevent salinity intrusion deep into the delta.

Finally, the chapter presents the experience of the Indus River, the progenitor of another Bronze Age civilization. As with the Nile River, excessive abstraction from the Indus through construction of dams and barrages has left little water to reach the Arabian Sea, subjecting the Indus Delta to erosion, salinity intrusion, and degradation, including the gradual loss of its mangrove forests.

Each of the following five sections (4.2–4.7) is devoted to one of these rivers, while section 4.8 concludes.

4.2 Colorado River

One of the early examples of rivers that have been completely exhausted through the application of the Commercial approach is the fabled Colorado River, located in the southwest of the United States. Originating from the western slopes of the Rocky Mountains, this 2,335 km long river flows in a generally southwesterly direction through or along six states of the United States—Colorado, California, Utah, Nevada, Wyoming, Arizona, and New Mexico—and two states of Mexico—Baja and Senora—before reaching the Gulf of California (Figure 4.1). Along the way, it crosses the Colorado Plateau, where it created Utah's Glenn Canyon and, further south, Arizona's Grand Canyon. Its drainage basin extends over 640,000 sq km, making it the seventh largest river on the American continent. About 97 percent of the watershed lies in the United States, the remaining 3 percent is in Mexico.

Between 1910 and 1970, a series of dams, reservoirs, hundreds of miles of canals and aqueducts, and other structures were built to regulate the Colorado River and divert its flow for various commercial purposes.[1] Two of the largest Colorado dams are the Hoover Dam, near the Arizona and Nevada border, constructed in 1937, and the Glen Canyon Dam, near the city of Page (at the Arizona and Utah border), constructed in 1964. Other major dams include the Imperial (1938), Parker (1938), and Davis (1950). These dams created many reservoirs, including Lake Powell, created by the Glen Canyon Dam, and Lake Meade, created by the Hoover Dam and considered the largest man-made lake in

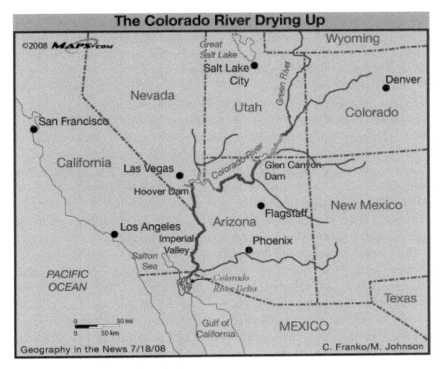

Figure 4.1 Map of the Colorado River
Source: C. Franco and M. Johnson and *maps.com*

the United States. The total hydropower capacity of the Colorado dams is 4,178 MW,[2] and the total area irrigated by the river's water is 1.6 million ha. Diverted Colorado flow meets the municipal needs of about 40 million people both inside and outside its watershed.

The average annual flow of the Colorado River is estimated to be about 20.35 cubic km, amounting to an average flow rate of 637 cubic m per second. The combined storage capacity of the reservoirs created by the Colorado's dams is four times larger than its annual flow. These reservoirs have submerged unique ecological sites, such as the Glen Canyon. Also, these reservoirs themselves have become an important cause of loss of the Colorado flow, as evaporation from these reservoirs now accounts for about 15 percent of the river's flow.

About 90 percent of the Colorado flow is diverted within the United States, the remaining in Mexico. The Imperial Dam, the last dam on the Colorado River in the United States diverts 90 percent of the remaining flow into the All-American Canal to be used for irrigation in California's Imperial Valley. At Morelos Dam in Mexico, almost the entire remaining flow is diverted for irrigation in Mexicali valley. The lower 160 km of the river is now basically dry; the river has not reached the sea consistently since the 1960s. The only exceptions were during the early 1980s, when heavy precipitation due to El Nino effects suddenly augmented the river's flow. Reduced water flow due to upstream diversion is gradually affecting adversely the ecology of the Grand Canyon.

Ironically, even after giving up the last drop of its flow, the Colorado River has failed to quench the commercial thirst, exemplifying the "More water, more thirst!" syndrome. The easy availability of the Colorado River water encouraged the sowing of thirsty crops in the California valley and facilitated the growth of desert cities, such as Las Vegas. It encouraged a settlement pattern and lifestyle, with each household seeming to want its own swimming pool. As a result, there is never enough water. During the summer of 2015, the state of California had to impose restrictions on the use of water to cope with its scarcity.

The Colorado River also illustrates the salinity and toxification effects of the Commercial approach. Extensive use of its water for irrigation, combined with heavy use of chemical fertilizers and pesticides have toxified the runoff back into the river. This is already a problem in the river's upper reaches, such as in the Grand Valley in the state of Colorado. However, it is more serious in the lower reaches where little river flow is left. While the salt content of the lower Colorado used to be about 500 parts per million (ppm) before the dams were built, it had increased to more than 2,000 ppm by 1960.

Not surprisingly, the Commercial approach has made the Colorado River a source of conflicts and disputes among various US states and between the United States and Mexico. These disputes have led to several agreements, the combined set of which is known as the "Law of the River," beginning with the "Colorado River Compact," signed by six US states in 1922 and followed by another nine decisions, federal acts, and agreements, along with a US-Mexico treaty signed in 1944. The Law of the River now stipulates the division of the Colorado flow depicted in Table 4.1. This Table shows how the entire Colorado flow is allocated among US states and Mexico with nothing left for the sea. It thus codifies the Commercial approach's negation of the basic ecological function of the Colorado river, namely carrying the precipitation water of the basin to the sea. Ironically, despite the agreed division shown here, disputes about the actual share continue to be a perennial problem.

Meanwhile, it is not surprising that the Law of River, by leaving no water left for the sea, has, more or less, destroyed the Colorado delta. Before the dams and diversion, the Colorado River used to transport 77–91 million tons (an amount second only to that carried by the Mississippi River) of red-colored sediment—from which the river received its name—to the Gulf of California. The sediment

Table 4.1 Sharing of the Colorado River (Law of the River)

State/country	Amount		Share (%)
	Million acre feet (MAF)	Cubic km	
United States	15.0	18.5	90.9
California	4.40	5.43	26.7
Colorado	3.88	4.79	23.5
Arizona	2.80	3.45	17.0
Utah	1.72	2.12	10.4
Wyoming	1.05	1.30	6.4
New Mexico	0.84	1.04	5.1
Nevada	0.30	0.37	1.8
Mexico	1.50	1.85	9.1
Total	16.5	20.35	100

Source: Author, based on available information, as of 2019

used to nourish the 7,770 sq km that made up the river's delta, once the largest desert estuary in the continent. Now most of this sediment is deposited in Lake Powell and Lake Meade. Termination of both the water and sediment flow has affected adversely the ecology of both the estuary and the Gulf of California. The delta has shrunk in size, as has its species diversity.[3]

The fate of the Colorado River provides a quintessential example of the Commercial approach put to practice. Human beings have conquered the Colorado River by literally consuming it!

4.3 Murray-Darling river system

Australia's Murray-Darling river system provides another example of application of the Commercial approach leading to exhaustion of rivers. This river system spans most of the states of New South Wales, Victoria, and the Australian Capital Territory, and parts of the states of Queensland (lower third) and South Australia (southeastern corner) (see Figure 4.2). The basin covers 1.06 million sq km and comprises about one-seventh of the Australian landmass. The Darling River is 3,376 km long, while the Murray River is 2,528 km long, and the total annual water flow is about 23.44 cubic km.[4]

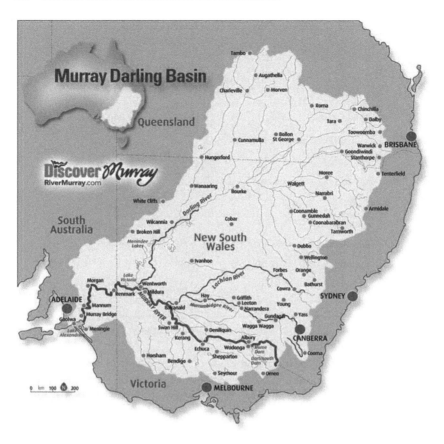

Figure 4.2 Map of the Murray-Darling river system
Source: Discover Murray River

Since the beginning of the twentieth century, Australia has constructed a series of dams, barrages, and reservoirs to divert water from these rivers for irrigation, industrial, and municipal uses. In 1915, the states of New South Wales, Victoria, and South Australia signed the River Murray Agreement, paving the way for construction of reservoirs upstream as well as at Lake Victoria near the South Australian border.[5] Four large reservoirs were built along the Murray River, including Lake Victoria (1920s),[6] Lake Hume near Albury–Wodonga (1936), Lake Mulwala at Yarrawonga (1939),[7] and Lake Dartmouth, which is actually on the Mitta Mitta River upstream of Lake Hume (1979). Along the intervening stretch of the river, a series of locks and weirs were built, aimed at promoting navigation. However, with the improved highway and railway systems, the navigation role of these structures proved largely redundant. The dams and reservoirs also made it possible to divert river water for irrigation. Of the approximately 13 cubic km annual divertible flow, 11.5 cubic km is diverted for irrigation, industrial, and domestic use, with irrigation for agriculture alone accounting for 95 percent of the diverted flow.[8]

The ecology of the river basin changed dramatically due to the impounding and abstractions. The seasonal pattern of the river flows changed, the winter-spring flood and summer-autumn dry pattern giving way to low flows through the winter and higher flows in summer. This altered seasonality disrupted the life cycles of much of the flora and fauna in the river basin ecosystems. Meanwhile, evidently cheap irrigation led to the cultivation of water-intensive crops, such as rice and cotton, and brought about dryland salinity. Increased irrigation also combined with the use of chemical fertilizers and pesticides to produce toxic runoff and increased salinity of the river water. As a result, the health of the river system deteriorated, and much of its aquatic life, including native fish, declined and became endangered.

The diminution and exhaustion of the river flows created the additional problem of salinity ingress for the lower reaches of the Murray-Darling river system. Low gradient at the estuary allowed tidal flows reach deep inside as rivers flows diminished. To prevent salinity ingress, a series of barrages was built near the mouth of the Murray during the period from 1935 to 1940. They include Boundary Creek Barrage located at 243 m; Goolwa Barrage, at 632 m; Mundoo Channel Barrage at 800 m; Ewe Island Barrage at 853 m; and Tauwitchere Barrage at 3.6 km up the river. The experience of Murray-Darling system provides an example of river-intervening structures, inspired by the Commercial approach, not only leading to exhaustion of rivers, but also creating the need for construction of more structures.

4.4 Amu Darya and Syr Darya

Both Amu Darya and Syr Darya originate in the mountains east of the central Asian deserts, through which they flow to the endorheic basin of the Aral Sea (Figure 4.3).[9] The fate of these rivers, and particularly of their destination, the Aral Sea, is a poignant example of the tragedy caused by application of the Commercial approach. It also shows that even the socialist ideology couldn't prevent rivers from becoming tragic victims of this approach.

The Amu Darya River, the more southern of these two rivers, is 2,400 km long, with a drainage basin extending over about 1.4 million sq km, comprising most of Tajikistan, the southwest corner of Kyrgyzstan, the northeast corner of Afghanistan, a long, narrow portion of eastern Turkmenistan, and

Figure 4.3 Map of the Amu Darya and Syr Darya rivers
Source: Kristina Schneider (2001)

about half of Uzbekistan.[10] About 61 percent of the drainage basin lies within Tajikistan, Uzbekistan, and Turkmenistan, while 39 percent is in Afghanistan. The river has an average annual flow of around 97.4 cubic km, almost all of which comes from the high mountains of Pamir and Tian Shan where annual precipitation can be over 100 cm.[11] Even before large-scale abstraction and irrigation began, high summer evaporation meant that not all of this discharge reached the Aral Sea. However, historical records show that the large Pamir glaciers provided enough melt water that the Aral Sea did overflow in the past.[12]

The Syr Darya River, the more northern and eastern of the two rivers, originates in the Tian Shan Mountains in Kyrgyzstan and eastern Uzbekistan and flows for 2,212 km west and northwest through Uzbekistan and southern Kazakhstan to the Aral Sea.[13] It drains an area of over 800,000 sq km, of which less than 200,000 sq km contribute flow to the river, the rest being basically desert.[14] Its average annual flow of 37 cubic km is less than half of its sister river, the Amu Darya.

In the early 1960s, the Soviet government decided to divert the two rivers to irrigate the steppes. This was part of the Soviet plan to grow cotton, or "white gold," for export. The Qaraqum Canal, Karshi Canal, and Bukhara Canal were among the larger canals built to carry the diverted flows of the Amu Darya and Syr Darya. The plan was initially successful and Uzbekistan became the world's largest exporter of cotton in 1988.[15]

However, problems soon surfaced. The first was the wastage of water. Many of these canals were not lined to prevent water leaks. Also, the desert conditions led to high levels of evaporation. It is estimated that 30–75 percent of water flowing through the Qaraqum Canal, the largest, was wasted. Unfortunately, the situation did not improve with time. Only 12 percent of Uzbekistan's irrigation canal length has waterproof lining, and only 28 percent of the 47,750 km of inter-farm irrigation channels in the basin has anti-filtration linings. The percentage is even lower (21 percent) for 268,500 km of on-farm channels.[16] The second problem with the diversion is that excessive irrigation in desert-like conditions led to an increase in soil salinity, so that the productivity of land began to fall over time.

Meanwhile, the diversion of the river water threatened the Aral Sea; once one of the four largest lakes in the world with an area of about 68,000 sq km, it began to shrink. By 1960, between 20 and 60 cubic km of water was diverted to the land instead of the sea. In some years, no water from either of the rivers reached the Aral Sea at all. From 1961 to 1970, the Aral's level fell at an average of 20 cm a year. In the 1970s, this rate roughly tripled to about 50–60 cm per year. By the 1980s, the rate of drop increased further to about 80–90 cm per year. By 1987, as a result of continued shrinkage, the lake split into two separate water bodies, namely the North Aral Sea (the Lesser Sea, or Small Aral Sea) and the South Aral Sea (the Greater Sea, or Large Aral Sea).

The Commercial approach applied to the Amu Darya and Syr Darya rivers led to conflicts, which the strong Soviet state, as long as it existed, managed to contain. It instituted a resource-sharing system under which Kyrgyzstan and Tajikistan shared the Amu Darya and Syr Darya flows with Kazakhstan, Turkmenistan, and Uzbekistan in summer and received in return coal, gas, and electricity from them in winter. Though these former Soviet republics became independent states in the 1990s, the diversion persisted, and, as a result, the decay of the Aral Sea continued.

By 2004, the Aral Sea's surface area had decreased to only 17,160 sq km, 25 percent of its original size. Meanwhile, a nearly five-fold increase in salinity killed most of its natural flora and fauna. By 2007, the Aral Sea surface area declined to only 10 percent of its original size, and it split into four lakes—the North Aral Sea, the eastern and western basins of the once far larger South Aral Sea, and one smaller lake between the North and South Aral Seas. The southeastern lake had disappeared by 2009, and the southwestern lake was converted into a thin strip at the extreme west of the former southern sea. NASA satellite images of August 2014 showed that the eastern basin of the Aral Sea had completely dried up and has become the Aralkum desert (Figure 4.4).[17]

The Aral Sea once supported a flourishing fishing industry, employing about 40,000 people and supplying about one-sixth of the Soviet Union's entire fish

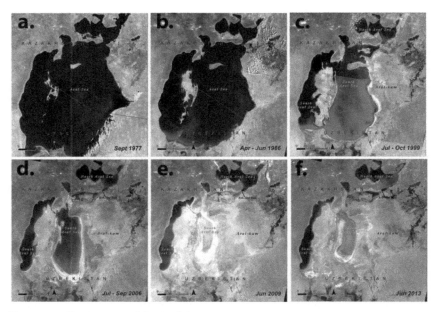

Figure 4.4 Desiccation of the Aral Sea
Source: Landsat of USGS/NASA, visualized by UNEP/GRID-Sioux Falls

catch. With the disappearance of the lake, this entire industry was devastated, creating unemployment and economic hardship among the fishing communities. What used to be fishing towns have now become ship graveyards.[18]

The Aral Sea region is also now heavily polluted, with consequential serious public health problems. Two factors combined to create this alarming situation. One is the increased salinity due to the disappearance and evaporation of the sea. The other is the toxic chemical residue from decades of industrial projects, weapons testing, and runoff laden with chemical fertilizer and pesticides. The salinity of what remains of the South Aral had increased by 2007 to more than 100 grams per liter (g/l), which is about three times higher than the salinity of ordinary seawater (around 35 g/l).[19] Dust storms spread the salt and chemicals to surrounding areas, damaging crops and causing a hazardous situation for public health. Lack of fresh water aggravates the problem.[20] The child mortality rate is 75 in every 1,000 newborns (compared to 3.9 on average in OECD countries in 2017), and maternity death is 12 in every 1,000 women (compared to 0.14 on average in OECD countries in 2015). The drying of the sea has also made regional winters colder and summers hotter. It is no wonder that the destruction of the Aral Sea has been called "one of the planet's worst environmental disasters."[21]

4.5 Nile River

The fabled Nile River of Africa provides an example of how application of the Commercial approach is leading to exhaustion of rivers in the developing world too. Generally regarded as the longest river in the world, the White Nile (which gets its name from the whitish clay suspended in its waters) originates in Lake Victoria and flows north through Uganda, South Sudan, and Sudan. Flowing from Ethiopia, the Blue Nile joins the White Nile at Khartoum, and the combined Nile flows through Sudan and Egypt to reach the Mediterranean Sea (Figure 4.5). Parts of Tanzania, Kenya, Rwanda, Burundi, and Eritrea are also considered part of the Nile basin. About 86 percent of the Nile water originates in Ethiopia.

The Nile is known as the lifeblood of the Egyptian civilization. Its annual summer floods provided the natural irrigation and silt for the floodplains and the Nile delta over thousands of years.[22] The ancient people of the Nile basin often cut small canals to let the river water reach deeper into the floodplain. However, these did not amount to any large-scale intervention in or abstraction of the river flow. The eleventh century Arab ruler of Egypt did attempt to construct a dam to control flooding. However, the expert brought in for this purpose deemed the effort impractical.[23]

The Nile was first subjected to the Commercial approach of the industrial era with the construction of the Old Aswan Dam in 1902 and a few barrages downstream. At the time, this was the largest masonry (gravity-buttress) dam in the world. However, due to its low height (22 m), the amount of water impounded by the Old Aswan Dam was not large.[24] Also, the buttress sections of the dam had numerous gates, which were opened yearly to pass the flood and its nutrient-rich sediments, without retaining any water for carryover to the next year. The Old Aswan Dam therefore did not do away entirely with the seasonal flooding and natural fertilization by the silt of the Nile. The barrages downstream raised the level of river water, which could then be diverted by canals to the interiors of the floodplains. To increase the volume of impounded water, the height of the Old Aswan Dam was raised twice to a final height of 36 m.[25] However, even that did not appear to be enough.

The new government, headed by Gamal Abdel Nasser, that came to power through the Egyptian revolution of 1952, decided to build a higher and stronger dam with a reservoir large enough for inter-year carryover of water, and thus total control on downstream flow. With much larger irrigation and electricity generation capacity, such a dam was supposed to be of pivotal importance for national development. The result was the High Aswan Dam, constructed during 1960–1970, with Soviet assistance, about 6 km upstream from the Old Aswan Dam. With a height of 111 m, length of 3,830 m, and a reservoir (named Lake Nasser) capacity of 132 cubic km, and 2,100 MW electricity generation capacity,

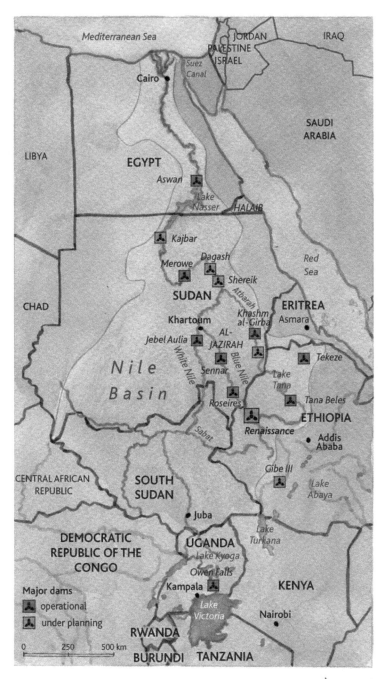

AGNÈS STIENNE

Figure 4.5 Map of the Nile River
Source: Agnes Stienne in Ayeb (2013)

the High Aswan Dam dwarfs the Old Aswan Dam, which now serves as a tail-water regulator of the High Dam.

The major benefits of the Aswan High Dam include increased water availability for Egyptian agriculture; hydropower production; flood control; improved navigation; and the creation of fisheries in Lake Nasser, which is 550 km long, 35 km at its widest, and has a surface area of 5,250 sq. km. [26] With a much larger reservoir, the High Aswan Dam allows controlled irrigation over a much larger area. It releases an average of 55 cubic km of water per year, of which some 46 cubic km are diverted into the irrigation canals. Almost 33,600 sq km of the Nile valley and delta benefit from these waters, with agriculture using up about 80 percent of the available water. Another 8 percent is used for municipal purposes and 6 percent for industries.[27]

However, the negative consequences of the Aswan High Dam became clear with time. First, the dam ended the summer floods, together with their siltation and soil renewal function. The silt now accumulates in Lake Nasser behind the dam, decreasing the capacity of the reservoir over time. Meanwhile, deprived of the natural fertilization provided by flooding, agriculture has become more dependent on chemical fertilizers and pesticides. The result has been increased toxicity of the run-off water and loss of micronutrients.[28]

Second, a major negative consequence of the Aswan High Dam has been increased salinity of both the soil of the Nile valley and the Nile flow itself. Before the High Dam, groundwater levels fluctuated across seasons by about 8–9 m, together with the fluctuation of Nile flow itself. Lowest levels occurred in summer, when evaporation was the highest. As a result, salt contained in the groundwater could not reach the surface through capillary action. With perennial irrigation ensured by the High Dam, the groundwater level is never too low (with a maximum of 1–2 m, depending on the soil conditions and temperature). Groundwater thus reaches the surface through capillary action, creating waterlogging and leaving salt in the soil upon evapotranspiration. Some of this salt that has accumulated in the soil returns back to the Nile flow through run-off. As a result, the salinity of the Nile flow increases from 0.25 kg per cubic m when it comes out of the reservoir to about 2.7 kg per cubic m when it nears the Mediterranean Sea.[29] Faced with this severe problem of waterlogging and salinity (causing crop yield to decrease), Egypt now has to resort to costly soil drainage and desalinization (leaching) projects, including subsurface ones. It is estimated that between 1973 and 2002, Egypt spent about $3.1 billion on agricultural drainage.[30]

Third, largescale impounding and abstraction has now caused the Nile to dry up almost completely before reaching the Mediterranean Sea. As a result, ingress of saline water from the sea into the Nile Delta has become a serious problem, similar to that witnessed in the case of the Murray-Darling estuary discussed earlier. Egypt therefore has constructed a series of delta barrages and now has to pump out a huge amount of saline water across the barrages into the sea, estimated at 12.4 cubic km in 1995–1996. Despite the barrages, seawater seeps into the drainage canals inside the delta. Pumping thus often proves to be a losing battle.[31] Meanwhile, the Nile delta is shrinking, under the combined impact of sea level rise and lack of sedimentation due to the upstream impounding and abstraction of the Nile flow. In some places, the delta is receding by about 100 m a year.[32]

Finally, application of the Commercial approach to the Nile River has given rise to conflicts among the countries of the Nile basin over the amount of water they are entitled to use up. Egypt, being a desert country—with an annual rainfall of 18 mm only—depends on the Nile for 97 percent of its water. As a result, Egypt utilizes 80 percent of the Nile flow, even though Ethiopia contributes about 86 percent of this flow. The 1959 Nile waters treaty between Egypt and Sudan allocated 55.5 cubic km of water per year to Egypt, without specifying any allocation for upstream riparian countries besides Sudan, which was allocated 18.5 cubic km per year. To utilize its hydraulic potential, Sudan has built four dams in the last century with an irrigation command area of about 18,000 sq km, making itself the second most extensive user of Nile water after Egypt. Actual water use by Egypt is widely believed to be in excess of the allocation under the 1959 agreement. Over time, other countries of the Nile basin, such as Ethiopia, have been asserting their rights to the Nile flow, thus giving rise to conflicts.[33]

The Nile Basin Initiative (NBI) partnership was launched in February 1999 to resolve these conflicts, with the participation of nine countries that share the river—Egypt, Sudan, Ethiopia, Uganda, Kenya, Tanzania, Burundi, Rwanda, and the Democratic Republic of Congo—and with Eritrea as an observer. It seeks "to develop the river in a cooperative manner, share substantial socioeconomic benefits, and promote regional peace and security." However, it is proving difficult to reach agreement satisfactory to both the old claimants (Egypt and Sudan) and the new claimants (the other countries of the basin). Thus, Ethiopia, Rwanda, Tanzania, and Uganda signed on May 14, 2010, at Entebbe, a new agreement on sharing the Nile water, despite strong opposition from Egypt and Sudan.

Some of the new claimant countries are moving ahead with their river-intervention projects. For example, Ethiopia, taking advantage of the fact that more than 80 percent of the Nile water originates from its territory, is building new dams. The most prominent of these is the Hidase Dam (also known as the Grand Ethiopian Renaissance Dam), mentioned earlier, built on the Blue Nile, with an electricity generation capacity of 6.45 GW and a reservoir that will take from 5 to 15 years of river flow to fill up. Once completed, it will be the largest dam in Africa and seventh largest in the world. Other countries are also proceeding with their projects. Needless to say, all these projects are likely to compound the problems for lower riparian countries that we have already noted. The Commercial approach is thus leading to more conflicts among the countries of the Nile Basin.[34]

4.6 Indus River

The Indus River provides another example of exhaustion of a river through application of the Commercial approach in a developing country. Originating in the western Tibet, the Indus River flows about 3,180 km through the entire stretch of Pakistan to reach the Arabian Sea near the port city of Karachi (Figure 4.6). Like that of the Nile, the Indus river basin has been the site of an ancient civilization of which Harappa and Mohenjo Daro are the two most important remaining city relics. The total drainage area of the river is about 1.2 million sq km, and its estimated annual flow stands at around 207 cubic km, making it the twenty-first largest river in the world in terms of annual flow. Its main left-hand tributary is the Chenab River, which in turn has four eastern tributaries, namely Jhelum, Ravi, Beas, and Sutlej. (These five rivers together gave rise to the name Punjab.[35])

As with the Nile, the ancient people of the Indus basin relied on annual summer flooding for their agriculture. They often constructed temporary earthen dams on smaller branches of the river and cut canals to let Indus water reach inside floodplains for irrigation.

The British introduced the Commercial approach to the Indus River system during their colonial rule and supervised construction of one of the most extensive and complex irrigation networks in the world at that time. Initially this network was based on barrages, such as the 1,350 m long Guddu Barrage to irrigate Sukkur, Jacobabad, Larkana, and Kalat and the Sukkur Barrage, which has an irrigation command area of more than 20,000 sq km.

After partition and the creation of Pakistan and India as two independent states, sharing of the rivers of the Indus system became a contentious issue, and after considerable negotiations the two countries signed in 1960 an agreement under which India was given rights over the three eastern rivers of Ravi, Sutlej, and Bias, while Pakistan received rights over the Indus, Jhelum, and Chenab Rivers. However, India retained the right of non-diversionary use of these rivers too. The World Bank was given the task of monitoring the treaty's implementation.

While the British had limited commercial interventions to barrages, independent Pakistan raised them up a notch and initiated its Indus Basin Project, which focused on dams. Thus the Mongla Dam was built on the Jhelum River and the Tarbela Dam on the Indus River. With a length of 2,743 m, height of 143 m, and an 80 km long reservoir, the Tarbela Dam is one of the longest

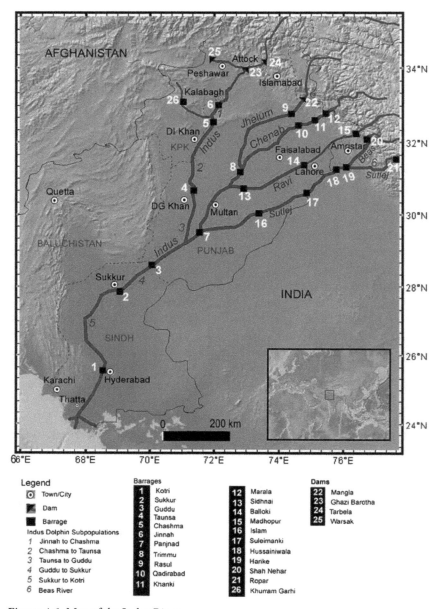

Figure 4.6 Map of the Indus River

Source: Gill T. Braulik, Uzma Noureen, Masood Arshad, and Randall R. Reeves (2015)

embankment dams in the world, as noted in chapter 2. Other dams and barrages constructed as part of the Indus project include the Kotri (also known as the Ghulam Muhammad Barrage),[36] Chasma, Taunsa, and Kalabagh barrages. These barrages and other subsidiary dams and link canals allowed the Indus water to reach areas beyond the Punjab province.[37] They provide the basis for Pakistan's irrigated agriculture, which produces cotton, sugarcane, and wheat and also produce electricity for cities and industries.

However, the dams and barrages on the Indus River have given rise to similar problems as witnessed in the case of the Nile River. First, they fragmented the river, distorting its morphology and restricting the movement of aquatic species. For example, as noted earlier, the Indus dolphins of the downstream reaches are now isolated from those of the upstream. Second, the dams and barrages trap sediment behind them, depriving floodplains of the natural process of soil revitalization through siltation. This has increased dependence on chemical fertilizers and pesticides and the toxicity of runoff water. Third, Pakistan is also facing the problem of waterlogging and salinity caused by perennial irrigation and the consequent rise of the groundwater table, leading to the deterioration of the soil quality.

Finally, impounding and abstraction of water have now left little or no Indus flow to reach the Arabian sea, which is having devastating effects on the Indus Delta. Since the Indus River changed its course to the sea several times before its impoundment, the "total" delta, with a minimum area of 29,524 sq km, is more than six times larger than the "active" delta, which has an area of about 4,762 sq km.[38] Previously, the Indus used to bring 180 cubic km of water and 400 million tons of silt to its delta annually. By 1994, these flows declined to 43 cubic km and 100 million tons, respectively. They have decreased further since 1994. Due to the reduced Indus water flow, seawater now flows more deeply into the delta. Meanwhile, reduced sediment flow is hampering delta formation and sustenance processes and the Indus Delta is shrinking—its length along the axis of the Indus River decreased from about 240 km to about 100 km. The lack of Indus flow is also causing the 17 major creeks of the delta to dry out.

The Indus Delta's condition is aggravated by high wind action, which results in high wave energy levels: it is subjected to the highest wave action of any river delta in the world. In fact, the amount of wave energy received by the Indus Delta in a single day is greater than that received by the Mississippi River Delta in a year. These high waves, caused by high winds from southwest during the summer, lead to submergence of parts of the delta by seawater. As this water retreats, it leaves behind salts in the delta's soil. In the past, large Indus flows helped to counteract the erosional impact of waves.[39] However, with diminished water and sediment flows, there is little left to protect the delta against the wind and wave action.

Diminished Indus water flow reaching the delta is also now highly toxic. Run-off from chemicalized agriculture is an important source of toxicity. Industrial pollution, particularly coming from the port cities of Karachi and Qasim, together with other sources of pollution, is causing eutrophication. The high salinity and toxicity of the water is proving unsuitable for freshwater species. A particular casualty has been the mangrove forests of the Indus Delta that once accounted for 97 percent of the country's mangrove forests. The Coastal Environmental Management Plan for Pakistan estimates that 86 percent of these forests were lost between 1966 and 2003.[40]

Aquatic species are also casualties of these processes. The mangrove forests provided the breeding ground and food for various species, including shrimp. It is estimated that over 1.2 million people lived around the Indus River's mangrove forests in 2002, with many of them depending on fishing. With vanishing forests and fish stocks, many of these communities are now facing a crisis. Other aquatic species are also in crisis, notably the Indus dolphins. Their habitat has been steadily shrinking and fragmenting—as noted earlier—with the increase in the number of dams and barrages constructed. Small groups of dolphins were trapped within parts of the river between two dams or barrages, unable to swim upstream or downstream. According to estimates by World Wildlife Fund (WWF), the Indus dolphin population has declined by 50 percent since 1944 and only 1,100 remain today.[41] Thus the Indus dolphins are likely to face the same fate of extinction as of their cousin dolphins in the Yangtze River.

The Commercial approach has also led to conflicts over the Indus flow. For example in Pakistan, conflicts have arisen between Punjab and other provinces, particularly Sindh, regarding the use of the Indus flow. There has been a general grievance that Punjab was using its upstream advantage (and also its hold over political power and the Pakistan economy) to reap most of the benefits of the Indus flow. As noted earlier, most of large dams, irrigation canals, and irrigated area are located in the province of Punjab. Even though the Indus River is already exhausted by the existing dams and barrages, former Prime Minister Nawaz Sharif (himself from Punjab) approved the construction of two additional massive dams on the Indus River, the 4,500 MW Diamer-Bhasha Dam in the Gilgit-Baltistan area in the north and the 2,160 MW Dasu Hydropower Project in the Khyber Pakhtunkhwa province in the northwest.[42] The conflict between Punjab and Sindh has now found a renewed focus on the issue of the proposed Kalabagh Dam on the Indus. Many people in Sindh are opposed to this dam, while many in Punjab want to press ahead with it. Needless to say, construction of these additional dams will exacerbate the problems noted above and aggravate interprovincial conflicts

4.7 Conclusions

The experience of five major rivers and river systems of the world reviewed in this chapter makes vivid the negative consequences of the Commercial approach. While the specific circumstances differ, in each case, application of the Commercial approach has led to over-extraction of water and damage to the ecology of river basins, estuaries, and the coastal environment. In cases where rivers had deltas, the Commercial approach led to their desiccation and shrinkage. In the particular case of the Amu Darya and Syr Darya rivers, the Commercial approach led to the destruction of the Aral Sea. Finally, in each case, the Commercial approach gave rise to conflicts among countries and among different parts and constituencies of the same country. These conflicts can have serious consequences for the development of individual countries as well as of a region, as we shall see in the next chapter.

Notes

1. About 29 are major dams. It is claimed that each drop of Colorado water is used an average of 17 times in a single year.
2. See for details https://www.crwua.org/colorado-river/uses/power (accessed on May 7, 2019).
3. The vast riparian, freshwater, brackish, and tidal wetlands of the Colorado Delta used to support a large population of plant, bird, aquatic, and terrestrial life. The loss of freshwater flows has reduced delta wetlands to about 5 percent of their original extent. The changed conditions have led to disappearance of native vegetation—such as forests of cottonwood and willow—and allowed nonnative vegetation to grow on sand and mudflats (Briggs and Cornelius 1997).
4. About 6 percent of Australia's total rainfall occurs in this basin. The Murray River also receives water from the Snowy Mountains through a complex dam and pipeline system.
5. The River Murray Commission was established in 1917 under the River Murray Waters Agreement as an advisory body only. For a long time, the commission was concerned only with water quantity. However, with time, salinity became a problem, leading to minor reforms allowing the commission to consider water quality too.
6. Lake Victoria is largely a natural lake (in the Murray floodplain) which has been modified to serve as a storage area for water.
7. Lake Mulwala mostly represents a higher stage of water confined within the river channel, rather than a reservoir with a substantial part outside the channel.
8. See for example, Davidson (1969) for an early dissenting view on water abstraction for irrigation in Australia.

9. In classical antiquity, the Amu Darya and Syr Darya rivers were known as the Öxus and Jaxartes, respectively. See Schneider (2001) for a discussion of the water sharing problem of Amu Darya and Syr Darya Rivers using Game Theory, and also for a source of the Figure 4.3.

10. Part of the Amu Darya's drainage basin in Tajikistan forms that country's border with China (in the east) and Pakistan (to the south).

11. Of the area drained by the Amu Darya, only about 200,000 sq km actively contribute water to the river. This is because many of the river's major tributaries (especially the Zeravshan River) have been diverted, and much of the river's drainage is dominated by outlying desert and steppe, where annual rainfall is about 300 millimetres (12 inches).

12. People began to settle along the lower Amu Darya and the Uzboy River in the fifth century. Around 985 AD, the river was impounded at the bifurcation of the forks by the massive Gurganj Dam, diverting water to the Aral Sea. Genghis Khan's troops destroyed this dam in 1221, and consequently, the Amu Darya flow was divided more or less equally between the main stem and the Uzboy. The river again turned north in the eighteenth century, flowing into the Aral Sea, and the flow along the Uzboy River dwindled.

13. Two headstreams—the Naryn River and the Kara Darya—come together in the Uzbek part of the Fergana Valley to form the Syr Darya River.

14. Indeed, two of the largest rivers in its basin, the Talas and the Chu, dry up before reaching it.

15. The water of the Amu Darya and Syr Darya was used for irrigation in earlier times too and canals were dug for that purpose. However, these earlier canals do not compare with the industrial scale of diversion and use that started during the Soviet rule.

16. Lined channels retain on average 15 percent more water than unlined channels. Only 77 percent of farm intakes have flow gauges.

17. Some groundwater originating in the Pamirs and Tian Shan Mountains finds its way through geological layers to a fracture zone at the bottom of the Aral. At about 4 cubic km (0.96 cu mi) per year, its volume is larger than previously estimated. However, this groundwater inflow did not prove sufficient to prevent the desiccation of the Aral Sea resulting from the diversion of the Amu Darya and Syr Darya flows.

18. Among the communities affected is the town of Moynaq in Uzbekistan, which had a thriving harbor and fishing industry employing about 30,000 people. It is now several miles away from the shore. For several decades fishing boats have lain scattered on the dry bed of the Aral Sea. The muskrat-trapping industry in the deltas of the Amu Darya and Syr Darya, which used to yield as many as 500,000 pelts a year, has also been destroyed.

19. The Dead Sea's salinity varies between 300 and 350 g/l.

20. Rates of certain forms of cancer and lung diseases are high in the area. Respiratory illnesses, including tuberculosis (most of which is drug resistant), cancer, digestive disorders, anemia, and infectious diseases are common ailments in the region. Liver, kidney, and eye problems can also be attributed to the area's toxic dust storms. Health concerns associated with the region are a cause for an unusually high fatality rate among vulnerable parts of the population.

21. There are suggestions that drying up of the Aral Sea was a deliberate decision on the part of the Soviet planners, instead of being an unintended consequence. According to this view, the decision was approved by the council of ministers and the Politburo. Apparently, some Soviet experts considered the Aral to be "nature's error." Such a view of the Aral Sea saga may however be contradicted, because, starting in the 1960s, a large-scale project was proposed to redirect part of the flow of the rivers of the Ob basin to Central Asia over a gigantic canal system, with the main goal of refilling the Aral Sea. Apparently, the project did not go forward due to its huge costs and the opposition of Russians who did not want to see the Ob flow diverted toward Central Asia.

22. According to the Greek historian Herodotus, "Egypt was the gift of the Nile." The silt deposits from the annual overflow of the Nile maintained the fertility of the soil, and the agricultural surplus thus made possible gave birth to Egyptian civilization.

23. It is claimed that Al-Hakim bin Amr, the Fatimid Caliph of Egypt in the eleventh century summoned the Arab polymath and engineer Ibn al-Haytham (known as Alhazen in the West) to Egypt to build a dam to regulate Nile flooding. Finding the task impracticable and to avoid the Caliph's wrath, he feigned madness and was kept under house arrest for 10 years until he died in 1021.

24. The dam is 1,950 m in length.

25. The height of the dam was raised by 5 m during 1907–1912 and by another 9 m during 1929–1933, and this expanded the reservoir capacity to 5.3 million cubic km. An electricity generation capacity of 550 MW was also added.

26. The dam allows a maximum of 11,000 cubic m per second of water to pass through it. In addition, it has emergency spillways for an extra 5,000 cubic m per second. In 1997, the Egyptian government decided to divert Nile water from the Nasser reservoir to the Toshka valley, a part of Egypt's western desert (also known as the Libyan Desert), which is in turn a part of the Sahara Desert. The goal was to convert 2,340 sq km of desert into farmland. This would however require a canal about 312 km in length. The project is still under construction. However, it faces the problem of pre-existing high salinity being further aggravated and spreading to underground aquifers. Successive Egyptian governments are however pressing ahead with this project of additional irrigation using the Naser reservoir.

27. In absolute terms, these amount to 5.3 cubic km of water used for municipal purposes and 4.0 cubic km for industry.

28. Irrigation water that drains back to the river contains chemical fertilizer and pesticide residues along with other toxic organic and inorganic pollutants. Also, about 3.5 cubic km per year of municipal wastewater was discharged into the Nile and the Mediterranean Sea in 2002, of which only 1.6 cubic km per year (about 45 percent) was treated. Similarly, about 1.3 cubic km per year of industrial effluents was discharged, with only small part of it being treated. As a result, east delta drains were found to contain high levels of pathogens, and the people in the northeast of the Nile Delta suffer from higher rates of water-related diseases.

29. Part of this increase is also due to salinity ingress from the Mediterranean Sea.

30. A subsurface drainage system has been put in place for more than 20,000 sq km.

31. The amount of drainage water pumped to the sea in 1995–1996 was about 12.4 cubic km, including about 2.0 cubic km of seawater that seeped into the delta drains. If more dissolved salt is pumped out than enters through Aswan, the salinity level at the delta barrages may decrease. However, this is counteracted by the fact that saline groundwater of marine origin enters the delta through pumping in of brackish water and upwelling in lakes and drains.

32. Rising salinity and sea level is forcing many farmers of the Nile Delta to move out. Others are waging a heroic battle by importing sand to hold back the tide. However, further increase in sea level, as expected from polar ice-sheet and glacier melting, is threating much of the delta, including the coastal city of Alexandria.

33. These conflicts were sharpened by the droughts of the 1980s, when Ethiopia and Sudan faced widespread starvation and famine, while Egypt could rely on the water impounded in Lake Nasser. Egypt has historically threatened war on Ethiopia and Tanzania over the Nile river.

34. Construction of dams on tributaries of the Nile River is reducing the water flow to the Lake Victoria, the world's second-largest lake, which witnessed record low levels in 2006, affecting adversely people in Kenya, Tanzania, and Uganda. See Ayeb (2013) and Bosshard (2007, p. 20).

35. Punjab is a conjugate of two words, *Pancha* meaning five and *Aab* meaning water.

36. The 915 m long Kotri Barrage near Hyderabad is intended to provide an additional supply of water to Karachi.

37. For example, the Chashma-Jhelum link canal was built to extend the water of the Indus system to the regions of Bahawalpur and Multan. Similarly, these canals carried the Indus water to the valley of Peshawar, in the Khyber Pakhtunkhwa province.

38. The arid Indus Delta receives only 250–500 mm of rain in a normal year. See Braulik, Noureen, Arshad, and Reeves (2015) for the effect of dams and Barrages on Indus dolphins.

39. This large amount of wave energy, coupled with lack of silt flowing in from the Indus River, has resulted in the formation of sand beaches.

40. It is estimated that over the past 50 years, about 170,000 hectares of the mangroves have been lost.

41. In view of the multiple threats that Indus Delta faces, it was designated a Ramsar site on November 5, 2002 and WWF is looking into conservation methods to alleviate the shortage of freshwater flowing to the Indus delta (Maria 2015). See UNEP (2015) for further on the Nile River.

42. See Maria (2015).

5

Commercial approach and river-related conflicts in South Asia

5.1 Introduction

One of the undesirable consequences of the Commercial approach is conflicts generated among co-riparian communities, regions, and countries. South Asia provides a good example of how this approach can lead to conflicts, both among and within countries. One reason why inter-country conflicts over rivers are acute in South Asia is that the boundaries of many of the region's countries were drawn on the basis of politics and not physical geography. As a result, the basins of its major rivers were arbitrarily divided among the newly created countries. For example, the Indus River basin was divided between India and Pakistan, while the basins of the Ganges, Brahmaputra, and Meghna rivers were divided among Bangladesh, Bhutan, India, and Nepal. Since India lies at the center of South Asia, having borders and sharing river basins with most other countries of the region, it is also at the center of river-related conflicts in the region.

The Commercial approach aggravated conflicts mainly because it encourages upper riparian countries to divert water from shared rivers, leaving less water for lower riparian countries. However, river-intervention projects inspired by the Commercial approach created conflicts even when there was no diversion.

South Asia also illustrates how the Commercial approach can lead to conflicts over rivers within countries. For example, in peninsular India, various states have been in conflict with each other for a long time over their shares of the Krisha, Cuvery, and Godaveri rivers. In recent years, conflicts over rivers have spread among northern states too. For example, there are disputes among the northwestern Indian states of Haryana and Punjab about sharing the rivers of the Indus system. Similarly, there are now conflicts among Uttar Pradesh, Bihar, and West Bengal about sharing the Ganges River. In Pakistan, as noticed in the previous chapter, there are conflicts between the provinces of Punjab and Sind over sharing the Indus River. However, in reviewing river-related conflicts in South Asia, caused by the Commercial approach, this chapter focuses mainly on inter-country conflicts.

Rivers and Sustainable Development. S. Nazrul Islam, Oxford University Press (2020). © Oxford University Press.
DOI: 10.1093/oso/9780190079024.001.0001

The most extensive and serious inter-country conflicts are between India and Bangladesh, which share more than 50 rivers. Indo-Bangladesh river-related conflicts center mainly on the issue of diversion, carried out by India, using its upstream location and through construction of barrages, such the Farakka Barrage and the Gajoldoba Barrage. In addition, there are conflicts regarding issues of control and timing of water release. Farakka Barrage has become the "Sorrow of Bangladesh," and river-related conflicts have become an obstruction to potentially fruitful cooperation between these two countries in many other areas.

The river-related conflicts between India and Nepal and those between India and Bhutan are similar in character. Because of their limited technical and financial capacity, both Nepal and Bhutan have constructed dams and barrages under joint venture with India. However, these countries are now embroiled in conflicts regarding sharing of costs and benefits. Many people in both Bhutan and Nepal feel that they are bearing more of the environmental and human costs while India derives more of the economic benefits from the joint venture dams and barrages. Several recent incidents have aggravated these conflicts.

It is instructive that river-related conflicts persist between India and Pakistan even though it might have appeared that no such conflict should arise because of the Indus River Treaty that neatly divided up the rivers of the Indus Basin between these two countries. However, the experience of Pakistan and India shows that issues of flow control can also be an important source of conflict even if flow diversion is not involved. The recent conflict over the Baglihar Barrage constructed by India on the Chenab River illustrates this reality. The river-related conflicts between India and Pakistan are more charged because often the structures involved are located in the disputed territory of Kashmir.

The Commercial approach has also pitted the entire South Asia region against China, which reportedly is now planning to divert water away from the upper reaches of the Brahmaputra River, which originates in China and flows east before moving through the Great Bend to enter South Asia through India's northeastern province of Arunachal. Reports indicate that China plans to build several dams to divert Brahmaputra water northward toward the Yangtze and Huang He rivers. China's actions in this regard puts India in a quandary because it is now protesting China's apparent misuse of its upper riparian location when it has been engaging in the same misuse vis-à-vis the lower riparian country of Bangladesh all along. This shows how the Commercial approach leads countries to self-contradictory positions.

The discussion of the chapter is organized as follows. Section 5.2 discusses river-related conflicts between India and Bangladesh. Sections 5.3 and 5.4 examine the river-related conflicts of India with Nepal and Bhutan, respectively. Section 5.5 reviews the Indo-Pakistan conflicts regarding rivers. Section 5.6

notes the burgeoning river-related conflict between China, on the one hand, and the countries of South Asia, on the other. Section 5.7 concludes.

5.2 India-Bangladesh conflicts

The most serious and numerous river related conflicts in South Asia are between India and Bangladesh. These two countries share more than 50 rivers, which enter Bangladesh from India to reach the Bay of Bengal, either by themselves or by merging with other rivers. The most prominent among these are the Ganges, Brahmaputra, Barak, and Teesta rivers. India has built, is building, or plans to build water impounding and diversionary structures on almost all the major shared rivers (Figure 5.1) (Adel 2001, 2017).

As a result, the flow of water and sediment in Bangladesh rivers has declined sharply, affecting adversely the ecology and economy of the country. Following are some of the major conflict-causing structures that India has built on rivers shared with Bangladesh.

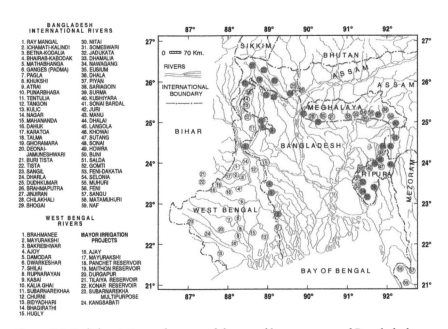

Figure 5.1 India's existing and proposed dams and barrages around Bangladesh
Source: Adel (2001 and 2017).

5.2.1   Farakka Barrage—The "Sorrow of Bangladesh"

The most prominent conflict-causing structure built by India is the Farakka barrage, which diverts the Ganges water away from Bangladesh and into the Bhagirathi-Hoogly channel of the Indian state of West Bengal (Figure 5.2).[1] A canal was constructed for the diversion with a capacity of 40,000 cusec (about 1,133 cumec). As a result of this diversion, the flow of the Ganges in Bangladesh (locally called the Padma River) has declined sharply (Figure 5.3), with drying up of the river (Figure 5.4) and many of its distributaries (Figure 5.5). Some of The latter get completely cut off from the Ganges during the dry season. Prominent among them is the Garai River, which used to be an important river of southwest Bangladesh. But the Modhumati, Kumar, Arial Khan, and many other distributaries of the Ganges in Bangladesh suffer from similar problems. Impounding and diversion of the flows of Ganges and its tributaries further upstream diminish the flow reaching Farakka, aggravating the problem.

   With diminished flow of rivers, saline water from the Bay of Bengal is intruding further inland, affecting the ecology of the southern districts.[2] Aquatic and terrestrial biodiversity in the region has declined, and the changes in the water balance

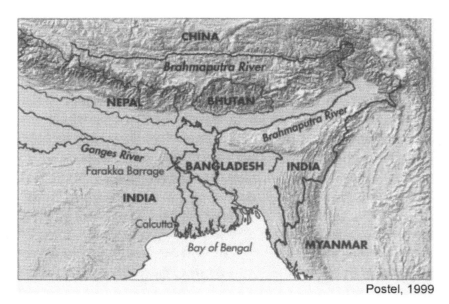

Postel, 1999

Figure 5.2 Map of the Farakka Barrage diverting Ganges water away from Bangladesh
Source: Sandra Postel (1999).

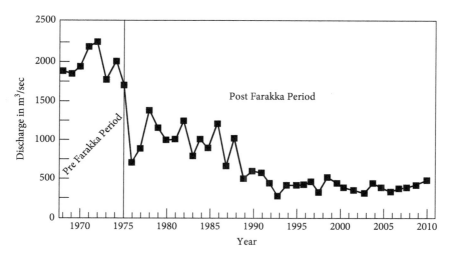

Figure 5.3 Decline in the flow of the Ganges River in Bangladesh due to the Farakka Barrage

Source: Gain and Giupponi (2014).

Figure 5.4 Dried up Ganges River in Bangladesh (under Hardinge Bridge) due to Farakka Barrage

Source: Shamsuddoza Sajen, *The Daily Star* (May 26, 2016).

Figure 5.5 Dried up Garai River, a distributary of the Ganges River in Bangladesh
Source: Amanur Aman, *The Daily Star* (August 29, 2016).

are having harmful effects on the Sundarbans, the largest mangrove forests in the world and a UNESCO-recognized world natural heritage site, as pointed out by a recent UNESCO team that visited the area.[3] The reduced flow of the Ganges has rendered ineffective the Ganges-Kobadak project (to be discussed in more detail in chapter 9), Bangladesh's premier irrigation scheme based on offtakes from the Ganges. According to some scholars, diminished Ganges flow due to the Farakka diversion has led to a fall in the groundwater level and contributed to the arsenic contamination of groundwater in southwestern Bangladesh.

The diminished Ganges flow also affects the northwestern districts of Bangladesh. For example, the Baral River, which originated from the Ganges and flowed in a northeasterly direction through the Chalaan Beel to meet near Baghabari the Jamuna River—as the main channel of the Brahmaputra is called in Bangladesh—is now completely cut off. As a result, this important river that, together with the Chalaan Beel (serving as a vast retention pool), performed the vital function of equilibrating the stages of the Ganges and the Brahmaputra, is dead for the most part, adversely affecting the ecology of the area and ending its equilibrating role.

Treaties and agreements regarding Farakka Barrage have been negotiated between India and Bangladesh, though these could not solve the problem. In 1975, one year after the Farraka Barrage went into operation, a temporary agreement was signed, stipulating Bangladesh's share of the dry season flow to be between 40,500 and 44,500 cusec (about 1,147 and 1,260 cumec), while India was allowed to divert between 11,000 to 16,000 cusec (about 312 to 453 cumec). There was also a guarantee clause, favoring Bangladesh, in case the Ganges flow at Farakka went below the historical minimum.

The 1975 agreement was temporary and was expected to be negotiated further and given a permanent basis. However, the coup of August 15, 1975, the killing of Bangladesh's Father of the Nation, Bangabandhu Sheikh Mujibur Rahman, and the change of political regime stopped that process.

In 1977, following political regime change in India, a five-year treaty was signed, allowing India to divert 40,000 cusec (about 1,133 cumec), that is, the full capacity of the Farakka diversion canal until January 10 annually. Following that date and until May 31 (when the dry season ends), India was supposed to gradually decrease the withdrawal to about 20,000 cusec (566 cumec) to let Bangladesh have around 35,000 cusec (991 cumec) of the flow. This treaty also had a guarantee clause, ensuring Bangladesh a minimum of 27,600 cusec (782 cumec) during the dry season. Following the expiry of this treaty, however, there was no agreement between Bangladesh and India about sharing of the Ganges flow for many years, so that India could withdraw as much as it wished.

In 1996, another treaty was signed, following political change in Bangladesh and installation of the government of Sheikh Hasina, the daughter of Sheikh Mujibur Rahman, specifying how the river would be shared for next 30 years; this is shown in Table 5.1.

One major drawback of this treaty from the viewpoint of Bangladesh is that it does not have a guarantee clause ensuring a minimum flow. This is a serious problem because, following the Commercial approach, India has built and continues to build many more dams and barrages in the upstream reaches of the Ganges and its tributaries, so that the flow of the river reaching the Farakka Barrage is declining over time. As a result, even under equal sharing of the dry season flow, the 1996 treaty has failed to bring about any substantial increase in the flow reaching Bangladesh (Figure 5.3). In fact, in many years, the flow now falls below the environmental flow necessary to sustain the river. Lack of flow causes the riverbed to

Table 5.1 Allocation of the Ganges water at Farakka as per 1996 treaty

Availability at Farakka	Share of India	Share of Bangladesh
70,000 cusec (1,982 cumec) or less	50%	50%
70,000–75,000 cusec (1,982–2,124 cumec)	Balance of the flow	35,000 cusec (991 cumec)
75,000 cusec (2,124 cumec) or more	40,000 cusec (1,133 cumec)	Balance of the flow

Source: Government of Bangladesh (1996).

become silted, reducing its capacity to carry the increased flow during the rainy season, during which the river now expands sideways through riverbank erosion. Thus, the Farakka Barrage is causing the paradox of both drying up and expansion of the river. The Farakka Barrage thus has become the "Sorrow of Bangladesh."[4]

Ironically, while causing much damage to Bangladesh, the Farakka Barrage failed to achieve its stated main goal (for India), which was to desilt Kolkata port to maintain its status as a sea port. This was in fact an impractical goal. Being located almost 100 miles inland from the coast and lying along a narrow, winding river, it was not possible for Kolkata to continue as a major sea port in the age of supertankers and container carriers.[5] Instead, the Farakka Barrage has become a source of problems for India itself, causing upstream siltation and aggravating flooding and riverbank erosion in the Indian state of Bihar and parts of West Bengal. Kapil Bhattacharya, West Bengal's Chief Engineer when the Farakka Barrage was planned, made these predictions in his 1961 report, *Silting of Kolkata Port*. He showed that the siltation of Kolkata port was caused by the dams on the Damodor and Rupnarayan Rivers, the two tributaries of the Hoogli River, and not the low flow of the Bhagirathi River. He pointed out that these dams were constructed without taking into account their impact on tide-borne silt. By decreasing the flows of these tributaries, these dams decreased the flow of the Hoogli, letting more tide-borne silt be deposited in the Hoogli River, near the Kolkata port. Based on his analysis, Bhattacharya predicted that the Farakka Barrage would not be able to flush the sediment. He also predicted that the Farakka Barrage would cause increased flooding upstream in the Maldah and Murshidabad districts of West Bengal and in adjoining districts of Bihar.[6]

Furthermore, being obstructed at the Farakka Barrage, the main channel of the Ganges River is gradually moving northward so that there is the possibility that it may make the adjacent Pagla River its main channel, bypassing the barrage altogether.[7] The barrage will then stand on a dry riverbed as a huge symbol of engineering hubris and misdirected investment. It is not surprising, therefore, that there is now a growing movement inside India for demolition of the Farakka Barrage, as we will see in chapter 6.

5.2.2   Gajoldoba Barrage

Gajoldoba Barrage (Figure 5.6), built by India on the Teesta River, which India shares with Bangladesh, is another important source of intense conflict between the two countries. Originating in the Cho Lhamu Lake of the Himalayas, at an elevation of 5,330 m above sea level, the Teesta River flows through the Indian states of Sikkim and West Bengal before entering Bangladesh and meeting the Jamuna (Brahmaputra) River at Phulchari in Gaibandha District.[8]

The Gajoldoba Barrage is a part of a greater, multiphase project on Teesta River, begun in 1976 by the Indian government. Unlike the Farakka Barrage, which was focused on desilting Kolkata port, Gajoldoba Barrage is focused on irrigation. The goal was to divert Teesta flow to irrigate 922,000 hectares of land in several districts of West Bengal. For this purpose, a 210 km network of distribution canals was constructed. The main canals were to transfer Teesta River flow to the Mahananda, Dauk, Nagar-Tangon, and Jaldhaka rivers. In addition to the main Gajoldoba Barrage, two pick-up barrages were constructed, one across the Mahananda River at Fulbari (in Jalpaiguri District) and the other across the

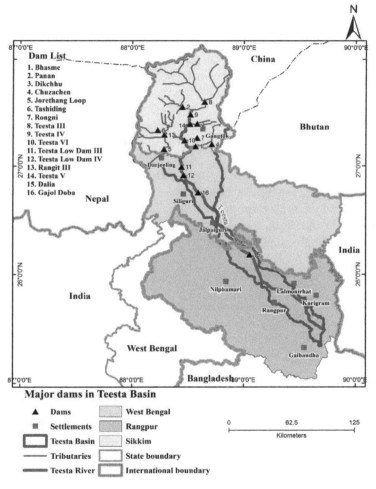

Figure 5.6 Map of Gajoldoba and other dams and barrages on Teesta River in India
Source: Gauri Noolkar-Oak (2017)

Dauk River at Chopra (in Uttar Dinajpur District). In phase 2 of the project, 650 MW of electricity was supposed to be generated through three canal falls on the main Mahananda canal. For this purpose, a reservoir was also to be created.

As with many other projects inspired by the Commercial approach, some of the ambitions of the Gajoldoba project were frustrated by India's own other Commercial approach projects, in this case, by hydropower dams that India's National Hydro Power Company (NHPC) planned to build in the upper reaches of the Teesta River. The planned reservoir and electricity generation components of the project were thus jeopardized. However, the Gajoldoba barrage was completed in 1995–1996, and it started to divert Teesta water away from Bangladesh. In 2008–2009, it was declared a national project of India. The initial irrigation command area of 79,610 hectares was planned to be eventually expanded to 527,000 hectares.[9]

Just as the Farakka Barrage jeopardized Bangladesh's largest surface irrigation project, namely the Ganges-Kobadak project, as noted earlier, India's Gajoldoba Barrage jeopardized Bangladesh's own Teesta Barrage that it had built at Doani in Lalmonirhat District, hoping to use the river water to irrigate about 540,000 hectares of land[10] in several districts of Northwest Bangladesh.[11] Though the project had been approved in 1960, construction of the barrage did not begin until 1979 and of the canal system until 1984–1985. The first phase of the barrage was completed in June 1998.

However, by the time Bangladesh's Teesta Barrage was completed, India's Gajoldoba Barrage was already in place and was diverting Teesta water away, as noted earlier. As a result, the minimum flow—which is important for dry season irrigation—in the Teesta River in Bangladesh fell drastically from about 200 cumec to under 50 cumec (Figure 5.7). In fact, the Teesta River in Bangladesh

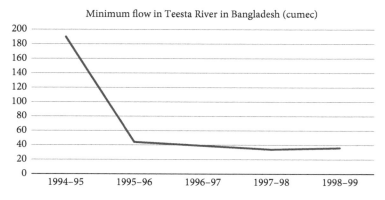

Minimum flow in Teesta River in Bangladesh (cumec)

Figure 5.7 Decline in the Teesta flow in Bangladesh due to India's Gajoldoba Barrage
Source: Author, based on data presented in Mukherjee and Saha (2013).

Figure 5.8 Dried up Teesta River in Bangladesh due to Gajoldoba Barrage
Source: Rashed Shuman in Pinaki Roy (2013)

practically dries up during the winter, when irrigation is most needed. As a result, Bangladesh's Teesta Barrage now mostly sits on a barren riverbed. (Figure 5.8).

Thus, Gajoldoba Barrage has become another sore point in the Bangladesh-India relationship. There have been talks between the two countries about a treaty on sharing the water of the Teesta river. Despite friendly relationships between the government of Bangladesh and the central government of India, such a treaty has remained elusive, apparently due to objections from the government of the state of West Bengal. However, as the experience with the 1996 Ganges water sharing treaty illustrates, even if an agreement is ultimately signed, it is unlikely to restore the adequate natural flow of the Teesta River into Bangladesh.

5.2.3 Tipaimukh Dam

While both Farakka and Gajoldoba Barrages are examples of structures that have already been built, Tipaimukh Dam is one of the intervening structures that India plans to build on rivers that it shares with Bangladesh. This dam is to be built on the Barak River, which is the source for Bangladesh's third largest river, the Meghna River. The dam is to be located 190 km upstream from the border with Bangladesh, in the Indian state of Manipur (bordering Mizoram), 500 m from the point where the Tuivai River meets the Barak River from the south and the latter makes a U-turn

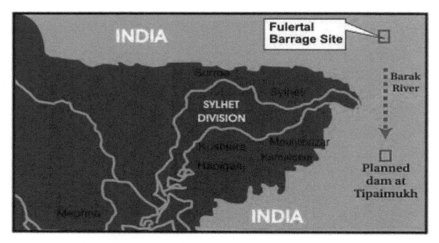

Figure 5.9 Map showing planned Tipaimukh Dam and Fulertal Barrage sites
Source: Subir Bhaumik (2009) and Md. Saidul Islam and Md. Nazrul Islam (2016).

to flow northwest toward the Sylhet district of Bangladesh (Figure 5.9). It is to be a rock filled earthen dam, 390 m long and 162.8 m tall. The dam was originally meant for river flow regulation to mitigate flooding in the Kachar and adjoining districts of the Indian state of Assam. However, most of the area to be submerged by the dam is in the states of Manipur and Mizoram. To make the dam palatable to these states, plans were therefore revised to allow for a hydropower component with a generation capacity of 1500 MW, of which 412 MW was to be utilized in the dry season.[12] Each of the three states—Assam, Manipur, and Mizoram—are to get 12 percent of the electricity generated, and the remaining is to go to the national grid. The area of the reservoir created by the Tipaimukh Dam is estimated to be 2.75 sq km.[13]

Bangladesh has several concerns with the proposed Tipaimukh dam. First, it is likely to have serious impact on the ecology of the *haor* (wetlands) area of the Sylhet division of Bangladesh. After entering Bangladesh, the Barak River bifurcates into the Surma and Kushiara rivers, both of which flow through the *haor* areas, before rejoining to form the Kalni-Meghna River. As will be discussed in more detail in chapters 8 and 9, the *haors* of Bangladesh are environmentally critical areas, which remain under water for most of the year, so that only one crop (*boro* rice) can be grown in winter, when the *haors* are drained, while their deepest parts remain under water perennially. Several of the *haors* are included in the Ramsar list of wetlands of global importance, and they serve as an important waystation for the Siberian birds during their annual migration to and from the warmer south.[14] By changing the river flow pattern, the Tipaimukh Dam is likely to alter fundamentally the ecological features of the area. In particular, by reducing the seasonal variation of river flows, Tipaimukh may lead to early submergence and delayed draining (even though it may reduce the depth of submergence during summer), and thus disrupt the existing cropping pattern of the area.

Second, Bangladesh is worried that soon Tipaimukh Dam will be combined with the proposed Fulertal Barrage, located about 100 km from Bangladesh border, to divert a significant part of the Barak River flow for irrigation in the Kachar district of Assam. Exactly how much water will be diverted is not yet clear. However, the available information suggests that a large part of the lean season flow of the Barak River will get diverted, as has happened with the Ganges and the Teesta Rivers.[15] Reduction of flow will combine with change in seasonality to damage the wetland ecology of the Sylhet Division, causing significant harm. Also, by reducing the Meghna River flow, it will affect the ecology and economy of the downstream districts of eastern Bangladesh. As noted earlier, dams lead to considerable losses of water (through evaporation from the reservoir, etc.) even without any diversion, thus decreasing the potential volume of water that could reach Bangladesh.

Third, Bangladesh is worried that the dam will put control of the river flow in the hands of its Indian operators, who will obviously give preference to Indian interests over those of Bangladesh, causing unexpected floods and droughts in the latter. Given the sensitive ecology of the *haor* area, such arbitrary changes in the river flow can be of serious consequence.

Another worry regarding the Tipaimukh Dam is that it is located in a seismically active area. The probability of the dam being affected by a severe earthquake is therefore not negligible. In case of a dam break, densely populated cities in Bangladesh are likely to be swamped under 8 feet of water in a matter of hours.[16]

It may be noted that Tipaimukh Dam will submerge about 275 sq km, displacing a large number of indigenous families. The United Nations Committee on Elimination of Racial Discrimination has therefore urged the Indian government repeatedly not to construct the dam. For example, it did so in its concluding observation of the Seventieth session (held from February 19 to March 9, 2007) and in its special communications made on August 15, 2008, March 13, 2009, and September 23, 2009. The committee has further urged authorities concerned to follow free, fair, and prior informed consent of the people under the International Labour Organization (ILO) Convention 107.[17]

5.2.4 Indian River Linking Project (IRLP)

River-related conflicts between Bangladesh and India are sure to be aggravated further by the IRLP. As noted in chapter 2, the goal of this mega project is to transfer water from "surplus" basins of the east to "deficit" basins of the west and south. The main target is the Brahmaputra River, which is considered to be the main surplus river. Following the drastic reduction of the Ganges flow, due to the Farakka Barrage, the Brahmaputra River now serves as the source for about 70 percent of the dry season flow in Bangladesh's rivers. Diversion of the Brahmaputra flow through the IRLP will therefore have a devastating impact on the entire river system of Bangladesh. The people of Bangladesh are therefore extremely concerned about IRLP, and this project will only add to the resentment that has already accumulated in Bangladesh against India's policies regarding shared rivers.

5.2.5 "Transit in exchange for rivers"

River-related conflicts are obstructing fruitful cooperation among countries of South Asia in many areas of economic development. A particular example can be seen in the case of the relationship between Bangladesh and India, where river-related conflicts are making it difficult for India to have more direct and easier access to its seven northeastern states—Arunachal, Assam, Manipur, Meghalaya, Mizoram, Nagaland, and Tripura, often referred to collectively as the Seven Sisters. Currently, these states are reached via a long and tortuous route passing through the 35.4 km wide "gooseneck" between Bangladesh and Nepal (see map in Figure 5.10).

This access difficulty is widely regarded as one of the reasons why the northeastern states lag behind in economic development relative to other states. India can have a shorter and direct access to these states through Bangladesh. However, because of the river-related conflicts described above, popular sentiment in Bangladesh is against granting such access. Though the government of Sheikh Hasina has conceded to many of India's requests regarding transit and transshipment, it has to remain wary about popular backlash if it is thought to have given up too much without getting commensurate concessions from India on river-related issues.[18] Under the circumstances above, "Transit in exchange for Rivers" could be a win-win formula of cooperation between Bangladesh and India, as has been pointed out from several quarters in Bangladesh. Under this formula, India would remove the flow intervening structures that it has built on

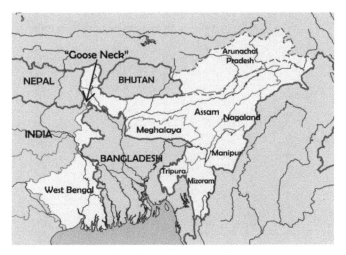

Figure 5.10 Map of Bangladesh and northeastern India
Source: Uwe Dedering at German Wikipedia / CC BY-SA, modified by Nusaybah Islam.

rivers shared with Bangladesh, and in return Bangladesh would provide India access to its seven northeastern countries through land and river routes passing through Bangladesh, subject to the commonly applied tariffs, fees, and other conditions. Though not adopted as an official position yet, this formula remains available to Bangladesh and India for achieving a lasting, mutually beneficial solution to two major problems that these countries face.[19]

5.3 India-Nepal conflicts

The Commercial approach has given rise to numerous river-related conflicts between India and Nepal. The mountainous country of Nepal, as noted earlier, has significant hydroelectric potential, and proponents of the Commercial approach see Indo-Nepal cooperation as the way to utilize it. Several joint projects have therefore been completed and a few more are under construction. Most of these (about 20) are along the Nepal-India border and hence are often called border dams (Figure 5.11), though they are generally run-of-river dams and hence should more properly be called barrages. In the following discussion (regarding Nepal and Bhutan), I use the term dam to include run-of-the-river dams.

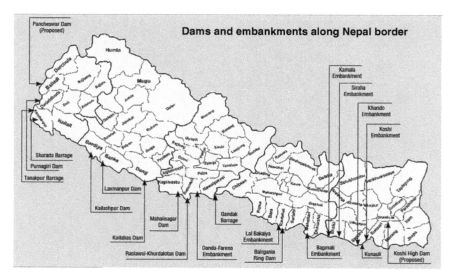

Figure 5.11   Dams in Nepal constructed under joint venture with India
Source: Kalendra Sejuwal (2017).

These dams have become, however, a source of conflict between India and Nepal (Malhotra 2010). The general complaint from Nepal is that these dams have mostly benefited India while Nepal has borne the environmental and human costs, without sharing the benefits in a significant way. As noted in chapter 2, the bulk of the electricity generated by these dams has gone to India, and the water stored by them has also been used for irrigation in India. Meanwhile, many of the dams submerged parts of Nepal's land and forests, displacing many people, often minorities. The management of these dams is generally in the hands of the Indians, who often neglect the consequences of their decisions for Nepal, thus causing unexpected and damaging upstream flooding, in order to avoid downstream flooding in India. For example, in June 2008, the Sarada Barrage on the Mahakali River caused serious flooding in Nepal when, according to Nepali observers, Indian dam operators kept all its 32 gates closed.[20] There are other similar incidents of upstream flooding caused by the border dams. In fact, dams, barrages, embankments, dykes, high roads, etc. built along the border have become a general obstruction to the natural flow of water from the Himalayan foothills to the Gangetic plain, thus causing unnatural, damaging incidents.[21]

Altogether, there is a general feeling in Nepal that the dams built under joint venture with India have not been a fair deal for Nepal (Swain 2018). There are issues regarding the price that India pays to Nepal for electricity generated by these dams. Earnings from export (to India) of electricity generated by the dams have not proved significant. The country instead depends for foreign exchange mainly on remittances sent by Nepalese working abroad. Dams have also not delivered on development for Nepal, and it remains in the UN category of Least Developed Countries (LDC). Dams in Nepal therefore have become associated not only with social deprivation but also with national deprivation (Chintan 2011).

5.4 India-Bhutan conflicts

The nature of India-Bhutan river-related conflicts is similar to that of India-Nepal conflicts. As noted in chapter 2, Bhutan is another mountainous county with significant hydropotential that the proponents of the Commercial approach want to utilize with India's financial and technical help; several joint projects have already been implemented, among them the Kurichu, Basochu, Punatsangchu, and Magdechu hydropower projects. The Sankosh project is the biggest in Bhutan and is the only multipurpose project (Figure 5.12).[22] However, as with Nepal's dams, Bhutan's projects have also been associated with feelings of national deprivation.[23]

The dams in Bhutan, as in Nepal, are generally of run-of-river type—that is, without large reservoirs. As a result, power generation is seasonal, high in the rainy

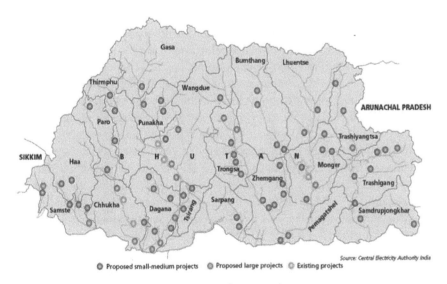

Figure 5.12 Dams in Bhutan—existing and proposed
Source: International Rivers and Central Electricity Authority India

season and much lower during the dry season. Bhutan therefore exports to India much of the electricity generated during the rainy season and imports electricity from India during the dry season. However, it is alleged that the price Bhutan pays for electricity in the dry season is much higher than the price at which India buys electricity in the rainy season.[24] This price differential is an important contributing factor to Bhutan's large negative trade balance with India, from which it imports most of its consumption and investment goods. There are worries that these financial problems will increase in future, as India wants to replace grants with commercial loans and to switch from intergovernmental projects to joint ventures among companies. It is apprehended that these changes in the terms of India's financing may put Bhutan into a debt trap. The government of Bhutan therefore recently expressed interest in opening up the hydropower sector to foreign direct investment (i.e., by countries other than India) — an initiative apparently not seen favorably by India. It is also felt that India is unduly restricting Bhutan's options regarding export of electricity (for example, to Nepal and particularly Bangladesh, which is willing to import electricity from Bhutan).[25]

In addition to the financial concerns just noted, Bhutan is also concerned about the high environmental costs of these hydro projects. The fact that Bhutan dams are generally run -of-river type, not requiring large reservoirs, has lessened the damage to the environment. However, the blasting and tunneling required for construction of these dams have proved no less environmentally destructive, as the experience of the Punatsangchu projects (involving the two largest dams under construction) has shown.[26] Meanwhile, many of the planned dams (such as Sankosh dam) are of the traditional variety, requiring large reservoirs, which are sure to submerge large tracts of agricultural land and forests.

As noted in case of the Indo-Nepal relationship, dam operation issues are also causing conflicts between Bhutan and India, even when dams are of run-of-river type. An example is provided by the Kurichu incident of 2004, when the artificial landslide-dammed Tsatichu River burst, only 34 km upstream of the Kurichu hydroelectricity project in Bhutan, sending water gushing downstream.[27] The Kurichu hydropower corporation opened the reservoir gates to save the dam, but the sudden onrush of water flowed into the Manas and Beki Rivers, tributaries of the Brahmaputra River, with disastrous results for people downstream in the Barpeta and Nalbari districts of the Indian state of Assam. The incident became a source of tension between the two countries, though Bhutan explained that, being a run-of-the-river dam, Kurichu could not have stored so much water as to be responsible for the flooding. It further noted that dams like Tala, Chuka, and Kurichu were normally kept below the full reservoir level to have some cushion against sudden increase in the river flow.[28] Tensions between Bhutan and India surrounding downstream flooding are likely to become more frequent in future in view of climate change and receding glaciers, aggravating seasonality of river flow and electricity generating potential.[29]

5.5 India-Pakistan conflicts

The Commercial approach has made rivers a source of conflict between India and Pakistan too, even though these countries appeared to have removed rivers as a source of conflict through a neat division of the shared rivers under the Indus River Treaty (IRT), reached in 1960. According to this treaty, as noted in chapter 3, India got control of the three southern rivers, the Bias, Ravi, and Sutlej, while Pakistan got control of the upper three rivers, the Chenab, Jhelum, and the Indus itself (Figure 5.13). The World Bank was given the task of resolving any dispute that may arise in implementing this treaty.[30] However, even this neat division of rivers could not eliminate conflicts over rivers between India and Pakistan.

It is not a surprise that the IRT encouraged India to abstract as much water as possible out of the Bias, Ravi, and Sutlej rivers before they entered Pakistan.

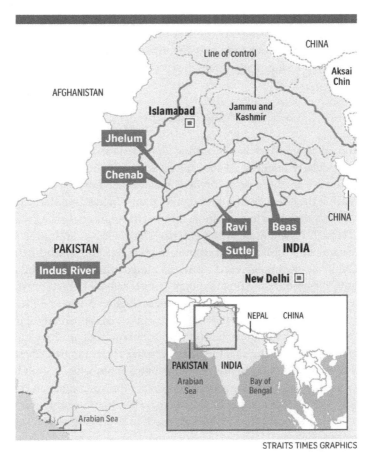

Figure 5.13 Sharing of Indus rivers by India and Pakistan
Source: Nirmala Ganapathy (2016).

Similarly, Pakistan felt unrestrained regarding the use of flows of the Indus, Chenab, and Jhelum within its borders. However, as noted earlier, IRT allowed India non-diversionary use of the latter rivers. In particular, it allowed construction of run-of-the-river power projects with limited reservoir capacity and flow control. Utilizing this option, India constructed several hydro projects on these rivers, with Pakistan objecting to each. In a recent example, Pakistan objected to the Baglihar hydro project that India began to construct on the Chenab River in 1999. According to Pakistan, the height of the dam, reservoir capacity, etc. exceeded IRT limits and put too much control of the river flow in the hands of the Indians. The two countries had to go through a long process of arbitration and adjudication managed by the World Bank in order to resolve the dispute.[31]

As both India and Pakistan pursue the Commercial approach to rivers more aggressively, river-related conflicts are likely to become more common and important, which is a dangerous prospect given that these two countries possess nuclear arms and have already gone into war with each other at least three times since they became independent. The fact that the river intervening structures generating conflict between these two countries are generally located in the disputed territory of Kashmir aggravates the risk and danger.

5.6 Emerging conflict with China

The Commercial approach is now pitting South Asia as a whole against China. This is particularly manifest in the context of the Brahmaputra River, which originates on the northern slopes of the Himalayas and flows east through Tibet until it reaches the Great Bend, where it takes a hairpin turn and starts flowing in a southwesterly direction to enter India through its Arunachal State, from where it flows through Assam State before entering Bangladesh and flowing south to finally reach the Bay of Bengal. According to press reports (Dikshit 2016; Baiyu 2020), China is now implementing projects aimed at transferring the water of the Yarlung Tsangpo (as the Brahmaputra River is called in Tibet) northward to meet the requirements of its northern provinces (Figures 5.14 and 5.15).

India has expressed alarm at these news reports. It has apparently approached Bangladesh with the proposal of putting up a common front against China's plan. Needless to say, there is quite a bit of irony in India's complaint that China is misusing its upstream location to divert Brahmaputra River, when it itself has been doing the same by diverting river water away from the lower-riparian country of Bangladesh. Whether or not the Chinese claim on the Brahmaputra River will persuade India to rethink the Commercial approach to rivers in general is something that remains to be seen. However, these developments show how the Commercial approach has the inherent tendency to lead to more and widening conflicts.[32]

O JIEXU   O ZANGMU   O JIACHA

Figure 5.14 Dams reportedly planned by China on Upper Brahmaputra (Yarlung Tsangpo) River

Source: Sandeep Dikshit (2016).

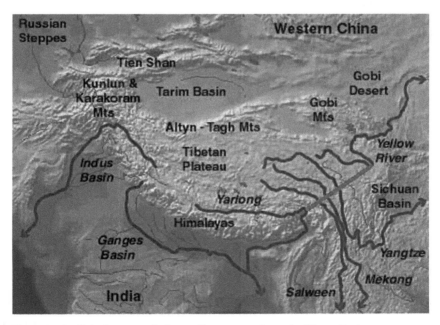

Figure 5.15 China's reported plan to divert water from the Upper Brahmaputra (Yarlung Tsangpo) River

Source: Diganta (2009a) and Ritesh Ghosh (2017)

5.7   River related conflicts as obstacles
to broader cooperation

The experience of South Asia also illustrates how the Commercial approach can hamper fruitful cooperation among countries in many other areas. For the Bangladesh-India relationship, river-related conflicts are a major barrier to India's obtaining direct and shorter access through Bangladesh to its seven northeastern states, as noted earlier. River-related conflicts can also be an obstacle to greater security cooperation between these two countries. Similarly, resentments over perceived unfair deals regarding joint venture dams are making it difficult to expand cooperation in other areas, both between India and Nepal and between India and Bhutan. India and Pakistan have already fought three wars over other issues. They don't need additional problems arising from river sharing to stoke tensions between them any further. Instead, it would be much better if the shared rivers could bolster their realization of the necessity of peaceful cohabitation in the shared habitat provided by the great rivers of the Indus Basin.

The fact that river-related conflicts obstruct fruitful cooperation in economic development among the countries of South Asia is unfortunate in view of the widespread poverty and deprivation suffered by large sections of population in all of these countries. It is therefore important that they earnestly try to remove rivers as a source of conflict and to realize that it is the Commercial approach that generates these conflicts. Only a common position taken by all the South Asian countries against the Commercial approach can provide a comprehensive resolution of river-related conflicts. Climate change is making adoption of such a position more urgent.

5.8   Conclusions

The experience of South Asia illustrates well how the Commercial approach converts rivers from bonds of friendship into sources of conflict among co-riparian countries. The competition that this approach generates in the abstraction of water leads inevitably to conflicts. The Commercial approach creates conflicts even in the absence of water diversion, over issues of control, sharing of costs and benefits of projects, etc. The seeds of conflicts lie in the very nature of the Commercial approach, and they sprout no matter what particular form the Commercial approach takes. It also leads to widening of conflicts—from conflicts among communities and states or provinces within a country to conflicts among countries and then to conflicts among regions comprising countries. No wonder that many have expressed the view that future wars will be fought not over oil but over (river) water.

River-related conflicts are acute in South Asia because the borders of the countries of the region were drawn based mainly on political considerations, ignoring physical geography. India lies at the center of river-related conflicts in the region and has conflicts with almost all other countries of the region. This is not surprising because of its geographically central position, having borders with almost all countries of the region. The nature of these conflicts, however, varies, depending on the type and impact of river-intervening structures. South Asia provides examples of river-related conflicts within countries, between countries, and between groups of countries. The emerging river-related conflict between China and India has exposed the contradictory nature of India's position regarding the conduct of upper riparian countries toward lower riparian ones.

The South Asian experience also shows how river-related conflicts can obstruct fruitful cooperation in other areas of economic development. This is unfortunate in view of South Asia's widespread poverty and deprivation. It is important for the South Asian countries to realize that it is the underlying Commercial approach that gives rise to the river-related conflicts, and hence a common shift away from this approach is needed to resolve these conflicts. Climate change is making such a shift more urgent.

In view of the adverse consequences of the Commercial approach, discussed in previous chapters, and its tendency to generate conflicts, as illustrated in this chapter using the South Asian experience, it is not surprising that there has been a quest for an alternative approach to rivers. This quest has given rise to the Ecological approach, which is discussed in the next few chapters.

Notes

1. The Farakka barrage is about 2,240 m long, and the feeder canal from the barrage to the Bhagirathi-Hoogli River is about 40 km long; it is capable of diverting 1,133 cumec (40,000 cusec) of the river's flow.
2. The distributaries of the Ganges that are dying in Bangladesh due to withdrawal of water at Farakka include the Bhairab, the Garai, the Madhumati, the Arial Khan, the Kumar, and other rivers. Many fish species have already become extinct. Salinity has intruded deeper into the north. Reduced discharge along these distributaries is causing decay of the Sundarbans, the unique mangrove forests along the shores of the Bay of Bengal. For more on the impact of Farakka Barrage on Bangladesh, see Gain and Giupponi (2014), Islam (1997a), Khatun (2004), and Sultana and Azad (2004).
3. In its report of 2016, UNESCO pointed to the necessity of increasing freshwater flow to the Sundarbans in order to protect these from decay (see UNESCO 2016). In 2019 UNESCO's World Heritage Centre declared the Sundarbans as a "World Heritage Site in Danger" (see Gatenby 2019 and *The Daily Star*, June 14, 2019).

4. In view of the conerns noted here, as Adel (2001) notes, Farakka has become in the popular psyche of Bangladesh a symbol of India's unfair treatment of its downstream neighbor. The dry Ganges River in Bangladesh (locally known as the Padma River) shown in figure 5.4 evokes strong sentiment among the people of Bangladesh. They recognize that much of the catchment basin of the Ganges lies in India. However, in their view, it does not give India the right to withdraw its water arbitrarily. India proceeded with the Farakka Barrage more or less unilaterally and presented it to Bangladesh as a *fait accompli*. As we shall see in chapter 6, the 1997 UN convention regarding the use of international rivers forbids a country from undertaking large-scale interventions in rivers without the consent of co-riparian countries. The convention also recognizes the right of co-riparian counties to historic and customary use of river flows. Bangladesh's ecology and economy have for centuries depended on unfettered flow of the Ganges. The construction of the Farakka Barrage thus amounts to disregarding Bangladesh's right to historic and customary use of the Ganges flow, and it remains a serious obstacle to improved relationship between these two countries. See Adel (2001, 2017) for discussions.

5. Eventually, Kolkata had to give up its role of sea port to Haldia, located near the coast. Though irrigation was never presented as an important goal of the Farakka Barrage, it is possible to assume that the barrage promoted and helped to increase agricultural production in the Indian state of West Bengal. However, there is little evidence to that effect. Note in this context that while the shift of the Ganges River from the Bhagirathi to the Padma as the main channel was a natural geological process occurring over many centuries, the diversion of flow by the Farakka Barrage away from the latter to the former is (from the geological viewpoint) a sudden event for which neither Bangladesh nor West Bengal was prepared. In some years, the additional flow in Bhagirathi-Hoogli via the Farakka canal, together with flows from other tributaries, caused riverbank erosion downstream in West Bengal too. See Rudra (2002 and 2003) for further discussions.

6. As Chari (2016, pp. 2–3) put it, "Bhattacharya believed that the proposed (Farakka) Barrage would only increase siltation in the river since in the dry season only half as much water would be available in the dam to divert towards the Hoogly, leading to more siltation. . . . Bhattacharya also said that the dam was designed to discharge too little water at times of floods, which would then lead to devastating floods upstream in Malda and Murshidabad in West Bengal and in several districts of Bihar through which the Ganga flows."

7. See, among others, Rudra (2000). The example of Farakka and of similar dams and barrages elsewhere in the world also shows that there is an important postscript to the human victory over nature mentioned earlier: this victory may be more illusory than real, more ephemeral than permanent. While human beings with their industrial might can apparently subdue rivers, the rivers themselves find ways to subvert human designs and come back with a vengeance. The process of nature taking "revenge" is subtle and may show up only in the long run, but it is a process that proves to be inevitable nevertheless.

8. The Teesta is a rain- and snow-fed river. The permanently snow-covered area of the basin is about 158.40 sq km. The upper catchment receives a total annual

rainfall of 1,328 mm while the middle of the basin receives 2,619 mm. It has been recorded that about 77–84 percent of the annual rainfall is received between June and September.

9. Official information on the Gajoldoba Barrage can be found at: http://www.india-wris.nrsc.gov.in/wrpinfo/index.php?title=Teesta_Barrage,_Phase_-I,_St.I,_Sub_Stage_I_Irrigation_Project_JI02921. For a discussion of impact of Gajoldoba barrage on Teesta River in Bangladesh, see Islam and Higano (1999 and 2002).

10. The objective was to bring 750,000 hectares of land under the irrigation command area with net irrigation area of 540,000 hectares to augment agricultural production.

11. It is spread over 12 upazilas (sub-districts) —Nilphamari, Dimla, Jaldhaka, Kishoreganj, Saidpur, Rangpur, Taraganj, Badarganj, Gangachara, Parbatipur, Chirirbandar, and Khanshama. The project included construction of a barrage, flood embankment, flood bypass, silt trap, main canal, and part of a canal system with improvement of an existing drainage canal.

12. Designed flood discharge is 16,964 cubic m per second. Average annual yield is 12.5 cubic km and the dependable yield (90 percent) is 8.1 cubic km.

13. The Tapaimukh Dam is to be constructed by the North Eastern Electric Power Corporation (NEEPC), a state-owned company of India. For more information, see Rishaduzzaman (2013).

14. According to the Bangladesh Water Development Board (BWDB), the *haors* of north-eastern Bangladesh that are rich in biodiversity include Hakaluki, Muria, Balair, Fultoli, Mojumdari, Dubai, and Lasaitala. Also important in this regard are canals that connect these *haors*, including the Rahimapuri, Moroi, Chagli, Senapoti, Mandi, Teli, Dasher, and Napit canals.

15. According to Diganta (2009b), cultivable area in the Barak valley in Assam is 220,000 hectares. A similar area in Sudan is irrigated by 1.6 cubic km only. Given that India is much wetter, the irrigation water requirement of the Barak Valley should not be more than that. The author notices that India uses 500 cubic km to irrigate a total of 54 million hectares. Since eastern India is wetter, the water requirement would be less. Therefore, the author thinks that 0.5 cubic km should be sufficient to meet the irrigation requirement of the entire Barak valley in Assam. These figures may be compared with the reservoir capacity, which is estimated to be 15.5 cubic km, of which around 10 cubic km would be live. The author therefore concludes that only a small portion will be diverted. This is however a wrong conclusion, based on a wrong comparison. The amount to be diverted by the Fulertal Barrage is a measure of annual flow, whereas the amount of water contained in the Tipaimukh reservoir is a measure of stock. A proper comparison would be between the annual amount diverted with the annual volume of the flow of the river. Also important to note is that the diversion will be made during the lean months, and so the diverted amount has to be compared with the flow during the lean months. Once this is done, it will be clear that a large part of the Barak lean season flow may be diverted, as is the case with the Ganges and Teesta flows, diverted through Farakka and Gajoldoba barrages, respectively.

16. In the last 100 years there have been 16 major earthquakes above magnitude 7 (on Richter scale) in the area where the proposed Tipaimukh Dam is to be located. One

of these occurred as recently as 1959. Another earthquake of magnitude of 6.6 that shook the Manipur-Myanmar border area on August 6, 1988, had its epicenter only 240 km northeast of the Tipaimukh dam site. In 1897, an earthquake measuring 8.7 changed the flow of the Brahmaputra and created the Jamuna River. Thus, there is non-negligible probability of an earthquake of high magnitude that may lead to a break in the 173 m high dam holding 16 cubic km of water, which will rush downstream, inundating Bangladesh's Sylhet city so that it will be under 8 feet deep water in a matter of hours. In short, there is a real seismic risk with the Tipaimukh Dam. See, for further details: https://www.thedailystar.net/news-detail-97382.

17. See ILO (1957). See also Islam S. and Islam M. N. (2016) for further discussion.
18. Bangladesh also has a large negative trade balance with India. It has also been alleged that India has put in place many obstructions to flows of media and cultural services from Bangladesh. The popular demand in Bangladesh regarding reciprocity therefore is not limited to concessions on rivers only but extends to issues of trade and other exchanges.
19. The resolution of river-related conflicts between India and Bangladesh could also help to improve security cooperation between these countries. Both India and Bangladesh have in the past made allegations against each other of allowing their territories to be used as base by insurgency groups. Though cooperation in this regard has improved under the current Sheikh Hasina government, India's concession on river issues is likely to help solidify this cooperation. This shows the multiple directions in which resolution of river-related conflicts can exert positive effect.
20. https://myrepublica.nagariknetwork.com/news/25778/.
21. The two large Indian dams on the Kosi and Gandaki rivers, as well as high roads, embankments and dykes, built parallel to Nepal's 1,751 km border with India, are widely blamed as infrastructure obstructing the natural flow of water. See, for example, Adhikari (2017).
22. See Mahanta (2014) for more information.
23. Bhutan continues to be in the category of Least Developed Countries (LDC), as does Nepal.
24. See Revkin (2013). Currently, earnings from export of hydroelectricity to India account for about 25 percent of GDP and almost half of government revenue. Bhutan wants to raise the latter to 75 percent in the future.
25. See Revkin (2013) for further details.
26. Once completed, these two projects (on the Punakha River) are expected to produce over 2,000 MW in the wet season, but only about 15 percent of that in the dry season. The dam builders have tried to address some of the environmental problems by avoiding construction of reservoirs and instead passing the water through tunnels. However, vital questions, such as whether there will be sufficient water left in the river during the dry season in order to sustain the ecology, including many fish and other aquatic species, have not been addressed. Also, observers are reluctant to rely upon environmental impact studies that are conducted by firms that are related to financiers and builders of the dams. See Jacobs (2015) for details.

27. Rivers in Bhutan flow through steep, narrow valleys of rugged mountains. As noted above, the narrowness of the valleys is one reason why Bhutan dams are more of run-of-river type instead of having large reservoirs. The narrow valleys can also be blocked by relatively small amounts of material (brought down by landslides, for example), damming rivers. When such landslide dams breach and burst, serious hazards follow.

28. For further details on the Kirichu incident, see Mahanta (2013 and 2014).

29. Climate change is posing additional threats to Bhutan's plan to exploit its hydropower potential. Bhutan has already lost 20 percent of its glaciers due to global warming, and the dry season flow is sure to diminish sharply once the glaciers melt away completely. Meanwhile, melting glaciers are likely to cause major floods in the short run. The fact that Bhutan lies in a seismically active region increases the possibility of dam failure and dam breakdown, making the future even more uncertain. See Jacob (2015), Mahanta (2013, 2014), and Revkin (2013) for more discussion.

30. Such a division of a river system, giving exclusive right over particular rivers to particular countries (almost like real estate property), displays the essence of the Commercial approach, which regard rivers as resources over which proprietorship can be established, instead of regarding them as part of the Nature that should be regarded as above proprietorship by humans.

31. For more information regarding the Baglihar dam controversy, see Lafitte (2007).

32. It may be mentioned that, according to press reports (see, for example, Peyton 2020), China also plans to divert water from the upper reaches of the Lancang (Mekong) River. This is likely to put China in conflict with Laos, Kampuchea, and Vietnam.

6

Introduction to the Ecological approach

6.1 Introduction

The Ecological approach to rivers arose as a response to the adverse consequences of the Commercial approach, discussed in chapters 3–5. Instead of thinking that "any river water passing to the sea is a waste," the Ecological approach emphasizes that carrying precipitation water to the sea is the basic function of rivers. By performing this function, rivers sustain the earth's hydrological cycle and unique ecologies of their basins, including the estuaries. The Ecological approach therefore calls for minimization of large-scale interventions in rivers and for preservation of the natural direction and volume of river flows, as much as possible.

The Ecological approach is, in a sense, a return to the practices of pre-industrial societies, although at a higher technological level. While the pre-industrial Ecological approach was a compulsion driven by *low* technology levels, the post-industrial Ecological approach is a choice that industrial and post-industrial societies make to avoid the destructive effects of *high* level of technologies acquired through the Industrial Revolution. It urges for less interventions in rivers.

It is not accurate to characterize the Commercial approach as anthropocentric and by comparison imply that the Ecological approach is non-anthropocentric. By protecting the ecological substratum on which human economy and society rest, the Ecological approach can be more beneficial for human interests in the long run than the Commercial approach. Similarly, the Ecological approach is not synonymous with basin-wide strategies, because the latter can be based on the Commercial approach too.

The Ecological approach does not negate viewing and using rivers as an economic resource. Rather, it promotes the sustainable use of rivers as a resource. It encourages sustainable crop production by avoiding inappropriate crop choice, wasteful irrigation, waterlogging and salinity, excessive chemicalization, and deterioration of soil quality. Similarly, it promotes sustainable fisheries by protecting rivers from fragmentation and the exhaustion and toxification of river water. The Ecological approach promotes sustainable urbanization by encouraging proper location of urban centers; discouraging creation of artificial and desert cities; promoting more economical use of water; and encouraging compact living and shared use of resources. Similarly, the Ecological approach promotes sustainable industrialization by encouraging proper location of industries,

Rivers and Sustainable Development. S. Nazrul Islam, Oxford University Press (2020). © Oxford University Press.
DOI: 10.1093/oso/9780190079024.001.0001

recycling of used water, and treatment of waste water before disposal and by discouraging excessively water-intensive industries. By avoiding construction of unnecessary and often counterproductive river-intervention structures, the Ecological approach can free up resources to be invested on more effective sustainable development programs.

Of particular significance is the Ecological approach's ability to prevent rivers from becoming a source of conflict and thereby to facilitate more comprehensive and deeper cooperation among co-riparian countries. To the extent that river-related conflicts lead to larger defense budgets, the Ecological approach can free up additional resources to be directed toward sustainable development.

Though it discourages interventions in rivers, the Ecological approach is not a passive approach and does not imply inaction. In fact, realization of the benefits of the Ecological approach requires sustained activities, which are more creative, demand more perseverance, and require more participation of people living in river basins.

The discussion of the chapter is organized as follows. Section 6.2 explains the distinction between the pre- and post-industrial Ecological approaches. Section 6.3 discusses the relationship between the Ecological approach and the basin-wide approach to rivers, while section 6.4 explains the relationship between the Ecological approach and non-anthropocentric view of rivers. Section 6.5 shows that, contrary to the general perception, the Ecological approach is an active approach. Section 6.6 explains how the Ecological approach can be more conducive to achieving sustainable development. Section 6.7 concludes.

6.2 Pre- and post-industrial Ecological approach

The modern, post-industrial Ecological approach to rivers arose in response to the negative consequences of the Commercial approach, which had dominated the thinking and practice regarding rivers for several centuries since the first Industrial Revolution. As the detrimental effects of the Commercial approach became clearer with time, there was an urge for an alternative way of treating rivers. This urge led to the Ecological approach, which is based on the basic premise that carrying precipitation water to the sea is the basic function of rivers. By performing this function, rivers help to maintain the earth's hydrological cycle and sustain the unique ecologies and cultures that have developed in their valleys, including their estuaries. According to the Ecological approach, large-scale interventions in the direction and volume of river flow should be kept at a minimum.

The Ecological approach is, in a sense, a return to the practice of pre-industrial societies, which also refrained from large-scale interventions in major rivers.

Their approach may therefore be called the pre-industrial Ecological approach. In this book, I use the expression Ecological approach to refer to the post-industrial Ecological approach, unless otherwise noted. In Hegelian terms, the evolution from pre- to post-industrial Ecological approach can be viewed as a "Negation of Negation."[1] According to this paradigm, the switch from the pre-industrial Ecological approach to the Commercial approach was the first negation, and the switch from the Commercial approach to the post-industrial Ecological approach is the second negation (Figure 6.1).[2]

Despite some similarities, it is important to note that the pre- and post-industrial Ecological approaches are different in many important ways. As just noted, the Ecological approach was a compulsion for pre-industrial societies, because they did not have the technological capacity to carry out large-scale interventions in major rivers. More prosaically, they did not have steel, concrete, dynamite, earth moving machines, hydraulic hammers, etc.; neither did they have the scientific know-how to build such dams as the Hoover Dam or the Three Gorges Dam. They therefore could engage in only limited forms of interventions, such as building seasonal, earthen dams on smaller rivers to hold water for dry season irrigation. More commonly, they dug canals to let river water spread further into floodplains. Interventions in rivers by pre-industrial societies were therefore generally of limited scale and did not radically change major river flows.

Pre-industrial societies were aware of the benefits that rivers provided and tried to make use of them as much as possible. However, they also felt

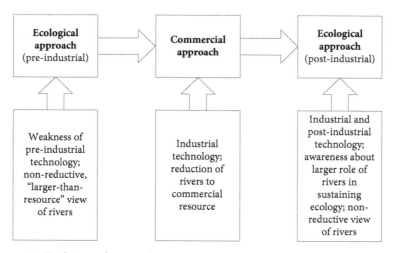

Figure 6.1 Evolution of approaches to rivers
Source: Author

overwhelmed by the rivers' might and power, in the face of which they often were helpless. This combination of feelings of both gratitude and helplessness led pre-industrial societies to revere rivers and impart them supernatural attributes. In many cases, pre-industrial societies treated rivers as gods and goddesses. For example, the name of the river Brahmaputra in South Asia means "son (*putra*) of the *Brahma*," the god of creation, according to the Hindu religion. Rivers were also imparted gender, based either on their purported association with gods or goddesses or on the basis of other considerations. Thus, pre-industrial societies accorded rivers a much larger role in their physical and mental universe. They didn't reduce rivers to just an economic resource. In India, many continue to worship the Ganges River.

While the pre-industrial Ecological approach was due to technological *weakness*, the post-industrial Ecological approach is, in a sense, an outcome of excessive technological *strength*. It is a conscious choice that societies make, even though they have the technologies to carry out large-scale interventions in rivers, to block them, drain them, and exhaust them to death, as we saw in chapters 3 and 4. Instead of human societies being helpless before the might of rivers, it is rivers, which have become helpless, so to speak, before humans. Under the post-industrial Ecological approach, human societies refrain from excessive use of technological strength to intervene in rivers because of their realization that such interventions are not in their best interests in the long run.[3] They are now more aware of rivers' role in maintaining the earth's hydrological cycle and as progenitors and sustainers of ecosystems and cultures in their basins. The Ecological approach is therefore a return to the non-reductive view of rivers. However, unlike that of pre-industrial societies, this view is not laden with superstitions and supernatural interpretations.

The shift of attitude toward rivers is a part of the general shift in attitude toward the intrinsic value of the environment in general and the necessity of protecting it. This intrinsic value, is, in a sense, infinite, because without the necessary ecological substratum, human life on this earth would be in jeopardy. The change in the approach to rivers has been, in part, the result of the negative consequences of river interventions accumulating over time. The change was also facilitated by the fact that, with rivers thoroughly dammed, barraged, and embanked, the desire to prove human ability to dominate and conquer rivers had already been fulfilled.[4]

6.3 Ecological approach and basin-wide strategy

In the discussion on river-related policies, one often encounters references to other approaches and strategies. Among these are the basin-wide approach or basin-wide strategy. The basic idea behind these propositions is that, instead of considering different parts of a river basin separately, they should be considered together. This idea is particularly relevant for international rivers, for which river-related activities of individual countries may conflict with each other and prove to be harmful for rivers as a whole. The concept of a basin-wide approach is however relevant even when rivers are contained in a particular country. This is because any river basin indeed consists of many physically distinct parts. From the longitudinal viewpoint, river basins usually extend from hills and mountains, where they originate, to their estuaries, where they meet the seas. From the cross-sectional viewpoint, river basins comprise not only river channels, but also their floodplains—a point that I will discuss in more detail in later chapters. These different parts of a river basin are interconnected. What happens upstream influences the state of rivers downstream. Similarly, what happens to river channels affects floodplains and vice versa. Also, different parts of river basins are connected not only through surface flows but also through underground flows. Most river basins are underpinned by groundwater tables, which connect different parts of a river basin both longitudinally as well as cross-sectionally. Even in cases where they do not extend deep inside floodplains, groundwater tables near riverbanks serve as stores, whose levels rise during high seasons and decline during lean seasons, when water seeps from these storages into river channels, helping to stabilize river flows across seasons and to maintain environmentally required minimum flows throughout the year.

In view of this interconnectedness, it is necessary that, in planning interventions directed toward any particular part of a river basin, attention is paid to its impact on other parts of the basin. To the extent that basin-wide strategies recognize and draw attention to this interconnectedness, they are better strategies than those which do not.

However, basin-wide strategies can follow either the Commercial approach or the Ecological approach. That is why, the fact that a river strategy is basin-wide does not tell us much about the character of interventions that are to be undertaken. In fact, until recently, interventions under basin-wide strategies mostly followed the Commercial approach. For example, basin-wide strategies were used to argue for construction of multiple dams and barrages at different points of rivers. So far as international rivers are concerned, the main thrust of the conventional basin-wide strategies was on coordination among co-riparian countries on construction of dams, barrages, and reservoirs. Basin-wide strategies under the Commercial approach have therefore often aggravated damage to river basins rather than increasing their protection.[5] It is not surprising, therefore, that many observers viewed the conventional basin-wide approach as "basically an anthropocentric approach" of the myopic variety, because it focuses on the narrow and short-term commercial interests of human beings, ignoring long-run effects.

Actually, the spirit of basin-wide strategies accords more with that of the Ecological approach because of the latter's emphasis on the fundamental role of rivers in preserving the earth's hydrological cycle and the ecology of river basins. Accordingly, it's attention extends from the mountains, where rivers begin, all the way to the seas and oceans, where they end. Similarly, it pays attention to connections between river channels and their floodplains. Furthermore, it takes note of connections between surface and subsurface flows. However, it looks at these interconnections with the goal of preserving the natural flow of rivers, as much as possible, and not with the goal of exhausting their flows, as has been done under the Commercial approach.

6.4 Ecological approach and the non-anthropocentric view

The above arguments for the Ecological approach are largely based on the "enlightened self-interest" of human beings. It is therefore, by and large, anthropocentric, as already noted. However, the Ecological approach does not contradict the non-anthropocentric view, according to which the earth is not just for human beings. Other species and items of the natural environment also have the right to exist and prosper. From this perspective, rivers have the intrinsic right to exist and not be mutilated and killed off (by exhausting their flows through impounding and abstraction) in order to serve human interests. As we shall see later, this non-anthropocentric worldview is gaining ground and providing further impetus to the Ecological approach to rivers. This book however focuses on expounding the anthropocentric arguments for the Ecological approach, while not disagreeing with or diminishing the importance of non-anthropocentric arguments. As its title indicates, the book strives to show that the Ecological approach is more conducive to sustainable development (of human societies). In practical terms, this enlightened anthropocentric view may not conflict with the non-anthropocentric view.

6.5 Ecological approach as an active approach

Sometimes, the Ecological approach is viewed as a passive approach and, by corollary, the Commercial approach as an active approach. This misconception may arise because the latter encourages interventions in rivers, while the former discourages them. However, interventions to alter the volume and direction of river flows are only one type of engagement with rivers. There are many other types of engagement that can make use of rivers as an economic resource, without however changing radically the volume and direction of river flows. The

Ecological approach encourages these other types of engagement, which will be described in more detail later.[6] In fact, as we shall see, activities required by the Ecological approach are more creative, demand more perseverance, and require more participation of people living in river basins. Thus, the Ecological approach is also an active approach, albeit geared to a different set of activities—activities that can make development more sustainable.

6.6 Ecological approach and sustainable development

Sustainable development, as noted earlier, is generally regarded to have three dimensions, namely economic growth, social development, and environmental protection. A close analysis shows that the Ecological approach to rivers is conducive to all of them. That the Ecological approach is conducive to environmental protection is straightforward. Much of this book is geared to proving this point. However, it is important to realize that it is conducive to economic growth and social development too. To begin with, construction of dams, barrages, and other river-intervening structures currently account for a large portion of the budget of many developing countries. The rates of return for investment on these structures are generally low and even negative, as noted in chapter 3. By saving on expenditures on such river-intervening structures, which often prove to be counter-productive and harmful, the Ecological approach can make more resources available for education, health, and technology development, which are more critical for sustainable development.

In the following, I note a few other dimensions along which the Ecological approach can promote economic growth and social development. Many of these issues have been touched upon in the previous discussion of the adverse consequences of the Commercial approach. Here I consider them from a more positive perspective.

6.6.1 Ecological approach and sustainable agriculture

One of the ways in which the Ecological approach can be conducive to economic growth and social development is by promoting sustainable agriculture. It can help sustainable crop production in several ways. First, it promotes appropriate crop choice. As noted in chapter 2, "cheap" dam water often leads to cultivation of water-intensive crops in arid areas where less water-intensive crops are more suitable. The Ecological approach can stop such inappropriate crop choices. Second, the Ecological approach encourages more efficient methods of irrigation (such as drip irrigation) instead of wasteful irrigation methods (such as flooding the entire field). Third, with less irrigation, the Ecological approach can avoid waterlogging and salinity and thus preserve soil quality, particularly in arid regions. Fourth, by preserving the role of silt in soil renewal, the Ecological approach can reduce dependence of agriculture on chemical inputs and thus avoid toxification of run off. Fifth, by striving to avoid toxification of water, the Ecological approach can promote organic agriculture and mixed farming, which are more sustainable. In the case of the latter, the organic waste of one type of crop (or farm activity, in general) can serve as fertilizer for another. Finally, organic and mixed farming—as opposed to chemicalized mono-crop agriculture—are more smallholder friendly. The Ecological approach can therefore be conducive to socially more equitable crop cultivation.

Similarly, the Ecological approach can promote sustainable fisheries. First, by reducing fragmentation of rivers, it can ensure better conditions for fish species to survive and thrive.[7] Second, by preventing the degradation and exhaustion of rivers, the Ecological approach can help to preserve capture fisheries. Third, it can protect fish stock from the damage caused by toxic run off. Fourth, as a result of the above, it can preserve indigenous fish stock diversity, which has its own intrinsic value. Fifth, the Ecological approach can encourage farm fisheries to focus on more diverse fish species, instead of a few specific, often alien, species. Finally, the Ecological approach can promote more socially equitable fisheries by protecting capture fisheries, which play a more important role in the livelihood and consumption of people with less income and capital.

6.6.2 Ecological approach and sustainable urbanization and industrialization

The Ecological approach to rivers can also promote sustainable urbanization by encouraging a less water-intensive settlement pattern; discouraging urban growth in arid regions; encouraging conservation of water use (such as rainwater harvesting and local water storage practices, for example, the *johads* of India)[8]; promoting more shared use of water; helping to strike a better balance between groundwater and surface water use in meeting water requirements; encouraging recycling and reuse of water; and promoting more compact urbanization. It may be noted that the latter also conducive to social development.

The Ecological approach promotes sustainable industries by encouraging their proper location; discouraging excessively water-intensive industries; promoting technologies requiring less water; encouraging reuse and recycling of water; and requiring proper effluent treatment before wastewater is released into neighboring water bodies.

6.6.3 Ecological approach and sustainable power generation

The Ecological approach can promote sustainable power generation in several ways. First, it may dispel the notion that hydroelectricity is a sustainable and climate-friendly method of electricity generation. By doing so, it may help to concentrate efforts on genuinely climate-friendly sources of energy such as solar, wind, geo-thermal, hydrogen, nuclear fusion, etc. Technological progress has already reduced the cost of electricity generated from solar and wind energy to levels that are comparable and, in some cases, even lower than the cost of electricity generated using fossil fuels. Development of hydrogen fuel technology is offering a novel alternative source of power. Further technological breakthroughs (such as in the area of fusion technology) can create unlimited sources of power. New technologies associated with the fourth Industrial Revolution can be conducive to progress in these directions. Thus, it is quite possible to reduce dependence on hydropower either through more intensive use of existing technologies or by achieving further technological breakthroughs in electricity generation. Adoption of the Ecological approach can therefore create more synergy between river restoration activities, on the one hand, and sustainable ways of power generation, on the other.

Second, in situations where use of hydropower is unavoidable, the Ecological approach can promote generation of electricity through run-of-the-river projects instead of dams. These projects—of which there are many types—produce less power, but they are also less destructive to rivers and their basins. Similarly, it can promote the use of energy contained in tidal flows in river estuaries to generate electricity using submersible turbines. It can also promote the use of energy contained in waves.

6.6.4 Ecological approach and peace, fairness, and cooperation

One of the ways in which the Ecological approach can promote sustainable development is by promoting peace and cooperation among co-riparian communities, states, countries, and regions. This role of the Ecological approach accords well with the greater importance given to peace and justice in SDGs, as reflected by the SDG-16.[9]

As the population increases, the level of economic activity rises and the demand for water increases. Meanwhile, freshwater resources are dwindling in many places due to climate change. The drying up of Africa's Lake Chad is a prominent example of this process, but there are other examples around the world. As a result, rivers and other freshwater resources are becoming an important source of conflict. The Ecological approach can reduce the possibility of water-related conflicts, as discussed in chapter 5. The use of rivers under the Ecological approach will prove to be more fair and just for all co-riparian entities, and fairness and justice are preconditions of peace. Just as transition to renewables can undercut the rationale for wars over oil, the transition to the Ecological approach can lower the probability of wars over water. Less tension and fewer conflicts will allow countries to spend less on war and the military and invest more on education, health, and other human capital improvement activities.

6.7 Conclusions

Accumulated evidence regarding negative consequences of the Commercial approach to rivers gave rise to the Ecological approach. In a sense, it is a return to the approach that pre-industrial societies followed toward rivers. However, it is a return at a higher level of technology. The (post-industrial) Ecological approach can be viewed as an enlightened anthropocentric approach, though in practical terms it is compatible with the non-anthropocentric approach. The Ecological approach is inherently a basin-wide approach, though the two are not synonymous, because the latter can also be based on the Commercial approach. The Ecological approach is more conducive to sustainable development, across all three of its dimensions. The Ecological approach is also an active approach, though the activities are directed to preservation of the hydrological and ecological role of rivers and not toward exhaustion of river flows through impounding and abstraction for various short-term commercial gains.

Notes

1. See Hegel (1991) for an original discussion of negation of negation as part of dialectics.
2. The negation of negation is, in a sense, a return to the original position but at a higher level, embodying the progress that has been achieved in the meanwhile. Thus, regarding rivers, the Ecological approach is a return to non-intervention, not because of lack of technological capability, but because of too high level of technology, so to speak. The current movement toward the return to organic agriculture offers a parallel. The agriculture of pre-industrial societies was organic agriculture. With industrialization, there was the switch to chemicalized (inorganic) agriculture, signifying the first negation. The current movement toward organic agriculture is not a simple return to the pre-industrial organic agriculture, instead it embodies the advances in knowledge and technology made during the industrial era. It is a negation of negation.
3. An ideal cost-benefit analysis of interventions in rivers—an analysis that is comprehensive in its lateral coverage and farsighted to include the interests of all future generations (i.e., temporal coverage)—would have revealed this fact. However, such an

analysis is generally not feasible in practice. First, it is not possible to list qualitatively all the different ways in which river interventions affect the current and future generations. Second, even if it were possible to prepare such a list, it is generally not possible to evaluate them quantitatively, because the necessary prices are not available. This is particularly true for future effects, the evaluation of which would require future markets, which generally do not exist. A particular issue regarding evaluation of future effects concerns the discount rate to be applied to compare future with current effects. It is difficult to have agreement on the appropriate value of the discount rate to be applied. These are generic problems of evaluation of natural capital, as discussed in chapter 1. They also apply to rivers, which are an important part of the natural capital. Despite the difficulties involved with this formal analysis, the experience of the Commercial approach has made it clear that rivers have intrinsic ecological value, which cannot be captured by conventional cost-benefit analysis, and rivers therefore should not be treated as an economic resource in the narrow sense.

4. Once conquered, one also often laments the beauty of the vanquished when it was free! By now, the desire to conquer rivers has already been quenched.

5. The Law of River regarding the Colorado River is an example. The basin-wide approach embodied in this law led to the exhaustion of the river. For further discussion on basin-wide management, see Barrow (1998), Jacobs (2017), Pegram, Li, Quesne, Speed, Li, and Shen (2013), Teclaff (1996), Tippett, Searle, Pahl-Wostl, and Rees (2005), and UN (1976).

6. Certain uses of rivers as an economic resource should not pose conflicts with the Ecological approach. For example, uses of rivers for navigation and open-capture fisheries are generally compatible with the Ecological approach. It also does not rule out uses of rivers that require some obstruction and abstraction of water, as long as these do not harm the hydrological and ecological role and integrity of rivers. The latter requires, as we shall see, not only restraint on interventions but also positive actions in many directions. For further discussion of options and alternatives beyond dams, see Brink and McClain (2004) and Brower (2000).

7. As noted earlier, this is particularly important for fish species, such as salmon and hilsa, which swim from the sea upriver to spawn. In the United States, dams on the Elwha River in Washington state led almost to the depletion of the salmon in this river. Following the removal of dams and restoration of the natural flow, the salmon returned.

8. Under the *johad* movement, people construct small-scale, earthen dams to hold rainwater to create small reservoirs from which water can be used for various purposes. These *johads* also help to recharge groundwater aquifers, increasing the availability of water in local wells, which were otherwise drying up. See, for example, Sharma, Everard, and Pandey (2017).

9. Goal 16: Promote peaceful and inclusive societies for sustainable development, provide access to justice for all and build effective, accountable, and inclusive institutions at all levels.

7

Ecological approach as part of greater environmental awareness

7.1 Introduction

The Ecological approach to rivers emerged as a part of the greater, general awareness of the necessity to protect the world's environment. The Industrial Revolution freed human societies from limitations of muscle power, leading to a trend break in human history; the volume of production of goods and the size of the world's population shifted from an almost horizontal to an almost vertical trend of increase. There were similar trend breaks in the volume of waste produced, in particular CO_2 emissions and, consequently, in the global temperature.

For a long time, industrial societies did not pay much attention to the environmental consequences of their rising level of economic activities. This started to change in the 1960s, when unfettered growth in material consumption was first questioned. By the 1970s, a flurry of initiatives drew attention to the earth's limited capacity to provide resources for rising levels of production and consumption and to absorb the wastes generated in the process. These initiatives include the Club of Rome publications in early 1970s pointing to limits to growth; signing of the Ramsar Convention in 1971 for protection of waterbodies; UN Conference on Human Environment (Stockholm Conference) in 1972, putting forward ecological and cultural functions of rivers in addition to their economic functions; UN Environmental Programme (UNEP) set up in 1972, in part to promote protection of waterbodies; formation of the World Commission on Environment and Development (WCED) that put forward the concept of sustainable development in 1987; UN Conference on Environment and Development (UNCED) in 1992 that adopted the wide-ranging *Agenda 21*, calling, in particular, for moving away from unsustainable consumption of rivers and other freshwater resources; the rise of the "Rights of the Nature" movement advocating for inherent rights of nature itself to exist and prosper; adoption of the Aichi Targets by the Commission on Biodiversity in 2010, calling for conservation of 17 percent of terrestrial and inland water and 10 percent of coastal and marine areas; and adoption of SDGs in 2015.

There were also initiatives focused specifically on rivers, such as the 1997 UN Convention on Non-Navigational Uses of International Watercourses,

Rivers and Sustainable Development. S. Nazrul Islam, Oxford University Press (2020). © Oxford University Press.
DOI: 10.1093/oso/9780190079024.001.0001

prohibiting upper riparian countries from doing anything that may harm lower riparian countries without the latter's consent; the formation of the World Commission on Dams and publication of its report, *Dams and Development*, in 2000, cautioning against construction of dams and barrages; and recognition of rivers as living entities having juridical status under law by many countries.

The emergence of the Ecological approach to rivers is part of this broader process of increasing recognition of the necessity to protect the environment and of initiatives to actually do so. There is therefore a synergy between following the Ecological approach and achieving other environmental goals, embodied in the agreements, conventions, resolutions, laws, etc., that resulted from these initiatives.

The discussion of the chapter is organized as follows. Section 7.2 explains the trend break in human history brought about by the Industrial Revolution and its consequence for the earth's environment. Section 7.3 reviews the series of initiatives taken from early 1970s, reflecting rising awareness about the necessity to protect the environment in general. Section 7.4 reviews initiatives that focused specifically on protection of rivers. Section 7.5 concludes.

7.2 Industrial Revolution—a trend break in human history

The Commercial approach to rivers, as noted earlier, was an outcome of the Industrial Revolution, which brought about a dramatic trend break in human history. If in pre-industrial times, human productivity and population size were progressing in an almost horizontal manner, the Industrial Revolution put them on an almost vertical flight. This is evident from the so-called "hockey stick" graphs depicting per capita income (Figure 7.1a), which reflects labor productivity, and the world's population size (Figure 7.1b). This radical trend break was possible because the Industrial Revolution introduced machine power that liberated humans from the limitations of muscle power. With higher productivity there was more sustenance, allowing population size to increase dramatically.[1] The first Industrial Revolution started this process and subsequent industrial revolutions, by and large, strengthened it.

However, greater population and higher levels of production also exerted greater pressure on the earth's ecological resources, of a two-fold nature. On the one hand, greater levels of production required greater levels of natural resources, which the earth has in limited quantities. On the other hand, greater levels of production and consumption also led to higher levels of waste that the earth has a limited capacity to absorb. The latter found its most prominent manifestation in carbon emissions generated by human production and consumption activities. While the earth's atmosphere has a capacity to absorb roughly about 5 gigaton of CO_2 emissions per year, their volume has now increased to more than

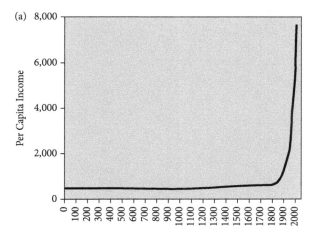

Figure 7.1a Trend break in world per capita GDP growth
Source: Islam (2014)

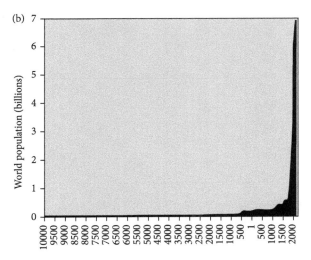

Figure 7.1b Trend break in world population growth
Source: Islam (2014)

40 gigaton. This sharp increase in carbon emissions has caused sharp increase in atmospheric carbon concentration (Figure 7.1c), leading to global temperature rise (Figure 7.1d) and climate change.[2] As we can see, the graphs for carbon emissions and global temperature also have undergone a trend break and display the hockey-stick pattern.

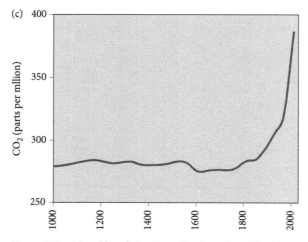

Figure 7.1c Trend break in atmospheric carbon-dioxide (CO_2) concentration increase

Source: Islam (2014)

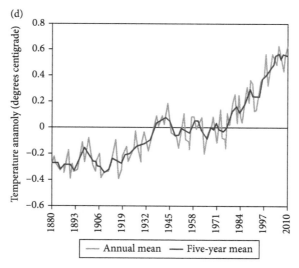

Figure 7.1d Trend break in global temperature increase

Source: Islam (2014)

7.3 Greater awareness of and initiatives regarding environmental protection

For a long time, industrial societies ignored the environmental consequences of their exponentially rising level of economic activities. This started to change in the 1960s, when for the first time the merit of unfettered growth in material consumption was questioned. From the beginning of the 1970s, actual initiatives were taken aimed at protecting the environment in general. The following provides a brief review of these initiatives, which set the background for the emergence of the Ecological approach to rivers.

7.3.1 Club of Rome and "Limits to Growth"

The cultural counter revolution occurring in many developed countries in the 1960s already questioned continuous rise in material consumption as the goal and metric of human development. Soon other, more focused, critiques of unfettered economic growth followed. One of these was reflected in the books published by the Club of Rome in the early 1970s, beginning with *The Limits to Growth* (1972).[3] These books were focused on the problem of "resource limitation," and drew attention to the fact that the earth has limited reserves of minerals (for example, oil) and that, at the prevailing pace of increase in economic activities, human societies would soon hit those limits. Accordingly, club members argued for limits to growth. With time, however, the Club of Rome authors extended their attention to the second problem, namely "waste absorption capacity limitation," and argued for limits to growth on its basis.

7.3.2 UN Conference on Human Environment, 1972

Concerns regarding human impact on the environment found a more comprehensive expression in the UN Conference on Human Environment (popularly known as the Stockholm conference) held in 1972. Proposed (in 1968) and hosted by Sweden, the conference adopted the Stockholm Declaration containing 26 principles, including the principles of safeguarding natural resources and wildlife, maintaining the earth's capacity to produce renewable resources, not exhausting non-renewable resources, and ensuring that pollution did not exceed the earth's capacity to clean itself. In this regard, the declaration paid particular attention to rivers and other waterbodies. Whereas previously rivers were thought to have only one function, namely to serve as a resource, the Stockholm Declaration pointed to two other functions, namely to support ecosystems and to form landscapes.[4] These two together were called ecological functions, as different from the economic function.[5]

7.3.3 UN Environment Programme (UNEP), 1972

An important follow-up step after the Stockholm Declaration was the set-ting up of the UN Environment Programme (UNEP) in 1972, with a concrete mandate regarding water resources, aimed at promoting their environmentally sound management and harmonization of socioeconomic and environmental interests. UNEP strove to maximize all three functions of rivers as identified by the Stockholm Conference, in particular, and called for basin-wide management and an interdisciplinary approach. Clearly, these were initial steps toward the Ecological approach to rivers.

UNEP also undertook many concrete initiatives concerning water resources, among them the UN water conference, held in Mar del Plata in Argentina, March 14–25, 1977. The conference focused on provision of drinking water and sanita-tion facilities and was connected with the UN Conference on Human Settlements (Habitat).[6] In 1985, UNEP launched a new comprehensive water program: the Environmentally Sound Management of Inland Waters (EMINWA) (UNEP 1985).[7] UNEP's projects are spread across the world and are implemented through other UN agencies.[8]

7.3.4 World Commission on Environment and Development (WCED), 1987

An important step in responding to the growing awareness about the environ-mental impact of economic activities was the establishment of the WCED—popularly known as the Brundtland Commission, after its chair, former Norwegian prime minister, Gro Brundtland. In its report, *Our Common Future* (WCED 1987), the commission, as noted earlier, put forward the concept of sustainable development, which now provides the overarching framework for development efforts by the UN and the world community as a whole.[9] As noted in chapter 1, the concept of sustainable development was operationalized as de-velopment that bequeathed the future generation as much natural capital as the current generation received from the previous generation. Clearly, this require-ment enjoins human societies to protect rivers for future generations.

7.3.5 UN Conference on Environment and Development, 1992 and *Agenda 21*

The work of WCED was soon followed by the UN Conference on Environment and Development (UNCED) held in June 1992 in Rio De Janeiro. Popularly known as the Rio Conference, it adopted the wide-ranging *Agenda 21*, which spelled out actions to be undertaken across the board to promote sustainable development.

Agenda 21 included an elaborate discussion of issues pertaining to rivers and other freshwater bodies and put forward sustainable development conforming positions on them. For example, it recognized rivers and freshwater bodies as "an essential component of the Earth's hydrosphere and an indispensable part of all terrestrial ecosystems (18.1)."[10] It expressed alarm at the "gradual destruction and aggravated pollution" of freshwater resources and called for their "holistic management" (18.3), giving due importance to "conservation and wastage minimization" (18.6) and to "demand management" (18.8). It called for a World Water Day to draw attention to the importance of water and its conservation (18.12b) and appealed for international cooperation with regard to transboundary water resources (18.4; 18.10). In particular, *Agenda 21* noted that aquatic ecosystems were adversely affected by "water resource development projects, such as dams, river diversions, water installations, and irrigation schemes" (18.35). It further noted that "creation of reservoirs has, in some cases, resulted in adverse effects on ecosystems" (18.35). It noted that irrigation expansion had taken place without "environmental impact assessments identifying hydrologic consequences within watershed or inter-basin transfers," and had often resulted in waterlogging and salinization (18.65). It further noted that irrigation expansion was carried out without any "assessment of social impacts on peoples in river valleys" (18.65). *Agenda 21* took note of "over-exploitation" of rivers for urbanization and the resulting threat to marine environment" (18.56)[11] and called for the "elimination of unsustainable water consumption patterns" (18.56). Overall, it observed that many of these problems had arisen from a "environmentally destructive" development model (18.35).

In view of its analysis above, *Agenda 21* called for "preserving aquatic ecosystems" and protecting them from any form of "degradation on a drainage basin basis" (18.38). It provided a perceptive account of the destabilizing effects of climate change on rivers[12] and called for "rehabilitation of degraded catchment areas" (18.40) and moving away from the "unsustainable consumption" of river and other freshwater resources.

UNCED also opened for signature several important conventions that are related to rivers, either directly or indirectly. Among these are the Convention on Biological Diversity (CBD),[13] Convention to Combat Desertification (CCD), and United Nations Framework Convention on Climate Change (UNFCCC). Clearly, protection of rivers is necessary for protection of aquatic biodiversity. The relevance of rivers for combating desertification is also clear, because, in many cases, desertification is the result of diversion of rivers. The relevance of rivers for climate change mitigation and adaptation has been discussed earlier and will be discussed further in this chapter and chapter 9.

UNCED led to the formation of the Commission on Sustainable Development (CSD) by the UN General Assembly in December 1992, tasked with overseeing the outcomes of the Rio Conference. CSD played an important role in promoting implementation of *Agenda 21* and served as the preparatory committee for the World Summit on Sustainable Development (WSSD) held in 2002 in Johannesburg. It also played an important role in the preparation for the UN Conference on Sustainable Development (UNCSD) held in Rio De Janeiro in 2012, which marked the twentieth anniversary of the original Rio conference of 1992 and so is often referred to as the Rio + 20 Conference.

7.3.6 Aichi Targets, 2010

An important initiative taken by the Commission on Biodiversity, and having significant bearing on rivers, is the Strategic Plan for Biodiversity 2011–2020, adopted in its 2010 meeting held in Aichi, Japan. The plan includes twenty Biodiversity Targets (known as the Aichi Targets), classified into four groups, and offers guidance about how to achieve them. In particular, the Aichi Targets call for conservation by 2020 of 17 percent of terrestrial and inland water and 10 percent of coastal and marine areas, "especially areas of particular importance for biodiversity and ecosystem services" (Target 11). Needless to say, achieving this target requires protection of rivers, including their estuaries. In addition, Aichi Targets call for sustainable management and harvesting of all fish and invertebrate stocks and aquatic plants, so as to avoid overfishing (Target 6); sustainable agriculture, aquaculture, and forestry (Target 7); reduction of pollution, including from excess nutrients, to nondetrimental levels (Target 8); reduction of the danger of invasive alien species (Target 9); and reduction of anthropogenic pressures on coral reefs (Target 10). The plan includes a target for halting or reducing by half the rate of loss of natural habitats and significant reduction of their degradation and fragmentation by 2020 (Target 5). Needless to say, meeting Aichi targets requires adoption of the Ecological approach to rivers.

7.3.7 Sustainable Development Goals (SDGs), 2015

The effort to achieve sustainable development received a big push from the 2012 UNCSD which recommended adoption of a set of SDGs to replace the MDGs, whose reference period was expiring in 2015. After several years of open, transparent, and inclusive negotiations, UN member countries adopted SDGs in 2015 (in the form of 17 broad goals and 169 targets).[14] The quest for sustainable development, that had attained a strong conceptual advance through the WECD, now found a concrete expression in the form a set of global goals, recognized by all UN member states. As noted in chapter 1, achievement of many SDGs depends directly on protection of rivers and achievement of many other SDGs depends indirectly on their protection.

7.3.8 Movement for "Rights of Nature" and Happiness

The growing awareness about the necessity to protect the environment has progressed further to the Rights of Nature movement, which argues that nature has inherent rights to exist and prosper and these rights should be given explicit recognition and legal standing. The idea is to change the status of ecosystems from being regarded as property to being recognized as rights-bearing entities, having legal status. In other words, the goal is to convert nature from an "object" to a "subject," albeit with certain limitations.

As noted earlier, the idea that nature has rights can be a reflection of the non-anthropocentric worldview, which argues that nature deserves protection for its own sake, and not just because it is beneficial for human beings, directly or indirectly, now or in the future. However, as noted earlier, this idea can also be compatible with the "enlightened self-interest-based anthropocentric view" which emphasizes that the protection of nature is in the long-run interest of human beings themselves.[15] No matter which of these two perspectives one proceeds from in conferring rights on nature, it needs human agency to uphold these rights, so that, in the end, it devolves upon human beings to decide how these rights will be exercised. Hence the practical significance of the difference between a non-anthropocentric view and an enlightened self-interest-based anthropocentric view may not be that great, as noted earlier.

Ecuador was the first country to enshrine the rights of nature in its new constitution adopted in 2008.[16] Article 10 of this constitution confers on nature a status similar to that of persons, communities, peoples, and nations, and recognizes it as the subject and bearer of rights conferred to it in the constitution.[17] Article 71 confers on nature the "right to integral respect for its existence and of the maintenance and regeneration of its life cycles, structure, functions, and evolutionary processes." It allows all persons, communities, peoples and nations "to call upon public authorities to enforce the rights of Nature." In enjoins the state to "give incentives to natural persons and legal entities and to communities to protect nature and to promote respect for all the elements comprising an ecosystem."[18] Article 72 recognizes nature's right to be restored—a right that is particularly important in view of the damage it has already suffered. Furthermore, it makes it an obligation on the part of "the State and natural persons or legal entities to compensate individuals and communities that depend on affected natural systems."

In case of severe and irreparable damages, the article enjoins the state to "establish the most effective mechanisms to achieve the restoration."[19] Article 73 enjoins the state to take "preventive and restrictive measures" to stop "extinction of species, destruction of ecosystems, and permanent alteration of natural cycles." It also forbids the "introduction of organisms and organic and inorganic material that might definitively alter the nation's genetic assets."[20] Article 74 recognizes the right of persons, communities, peoples, and nations "to benefit from the environment and the natural wealth" that enables them "to enjoy the good way of living." However, it forbids private appropriation of environmental services, and puts in the hands of the state the responsibility of their "production, delivery, and development."[21]

The rights of nature movement is closely associated with the *Buen Vivir* (Good Living) idea that has spread across many Latin American countries in recent years.[22] Rooted in the ideas and beliefs of the indigenous, pre-Columbian societies, *Buen Vivir* represents a philosophy of life that emphasizes harmony with nature and within communities. Clearly, the rights of nature initiative reflects, in part, the desire to live in harmony with nature.

Bhutan's idea of Gross National Happiness (GNH) is also an attempt to move away from using material consumption as the index of well-being. The idea was proposed by Jigme Singye Wangchuk, former king of Bhutan, in 1972, as a substitute for, or at least complement to, the conventional Gross National Product (GNP), in measuring collective well-being. As is the case with *Buen Vivir*, GNH also emphasizes harmony with nature as one of its four pillars.[23] The idea of GNH has spread to other countries too. In 2011, the UN General Assembly passed the resolution, "Happiness: Towards a Holistic Approach to Development," declaring happiness a "fundamental human goal," and urging member states to follow the example of Bhutan and measure happiness and well-being.

It is clear that rights of nature, *Buen Vivir*, GNH, and similar other initiatives, all represent a negation of the commercial approach to nature. Instead, they place the emphasis on the intrinsic role of nature as the basis of life and source of well-being and happiness.

7.4 Initiatives focused on rivers and waterbodies

The initiatives described in section 7.3 provide a general background for the emergence of the Ecological approach to rivers. However, there were also important initiatives focused specifically on protection of rivers and waterbodies that can be considered as direct antecedents of the Ecological approach. Some of these are reviewed below.

7.4.1 Ramsar Convention for protection of wetlands, 1971

A practical step reflecting the environmental concerns focused on water-bodies was the signing in 1971 of the Convention on Wetlands, aimed at conservation and sustainable use of wetlands. Popularly known as the Ramsar Convention—after Iran's city of Ramsar where the treaty was signed—it now has 169 contracting parties. One of its most important initiatives has been the preparation of the List of Wetlands of International Importance, which includes 2,186 sites, commonly known as Ramsar sites, covering a total area of about 209 million hectares.[24] Needless to say, many of these waterbodies are either part of or intricately connected with neighboring river systems. The Ramsar Convention obliges its contracting partners to maintain in good health the Ramsar sites under their jurisdiction.[25]

7.4.2 UN Convention regarding international rivers, 1997

There have also been several developments focused on rivers more directly. One of these was the adoption in 1997 of the convention on international rivers, formally titled the Convention on Non-Navigational Uses of International Watercourses (UN 1997), which went into effect on August 17, 2014, following accession of the stipulated minimum number of countries.[26]

The 1997 convention is the successor to the earlier Helsinki Rules on the Uses of the Waters of International Rivers, adopted by the International Law Association at its fifty-second conference held in Helsinki in August 1966 in an early attempt to set certain rules regarding the use of waters of international rivers, beyond what the co-riparian countries might have voluntarily agreed among themselves.

Over time, however, it was felt that the Helsinki Rules were not comprehensive enough. The 1997 UN Convention regarding international rivers strives to overcome this deficiency. Divided into 37 articles, it provides guidance for the amicable use of rivers by co-riparian countries. It urges signatory states to utilize international watercourses in a sustainable manner, ensuring "adequate protection of the watercourse" (Article 5.1) and urges co-riparian states to use watercourses in an "equitable and reasonable manner," making it a duty for them "to cooperate in the protection" of the watercourse (Article 5.2). In making equitable and reasonable use of the watercourse, co-riparian states are obliged to take note of "geographic, hydrographic, hydrological, climatic, ecological, and other factors of a natural character"; "conservation, protection, development, and economy of use of the water resources" of the river; "the cost of measures" taken for those purposes; and "availability of alternatives, of comparative value" (Article 6.1).

In particular, the 1997 convention forbids a signatory state from using the international watercourse in a way that causes "significant harm" to any other co-riparian state (Article 7). In this regard, it enjoins the signatory states to take note of "existing" uses of the watercourses (Article 6). It also obliges the signatory states to cooperate in protecting the ecosystems of the international watercourses (Article 20) and emphasizes the need to protect and preserve the "marine environment, including estuaries" (Article 23). The convention further obliges signatory states to prevent, reduce, and control pollution of rivers (Article 21) and draws attention to the danger of alien invasive species, obliging the signatory states to protect ecosystems from damages that such species may cause (Article 22).[27]

It is clear that the Commercial approach would be in conflict with the convention's emphasis on the protection of river basin ecosystems, including estuaries, and the marine environment. Upholding the 1997 convention therefore requires the Ecological approach.

7.4.3   World Commission on Dams, 2000

Alongside the inter-governmental processes just described, vigorous initiatives were taken in the 1990s at the non-governmental and civil society levels toward the Ecological approach to rivers. The California-based International River Network (IRN) and India's Narmada Bachao Andolon—campaigning against the Sardar Sarovor Dam on the Narmada River—played an important role in this regard.

The 50th anniversary of the World Bank (in July 1994) was an important occasion for river movements across the world, giving them the opportunity to galvanize their campaign for the Ecological approach to rivers. Until then the World Bank had been the most important financier of river-intervention projects inspired by the Commercial approach. It was also one of the original major financiers of the Sardar Sarovar Dam. On the eve of the World Bank anniversary, the IRN and Narmada Bachao Andolon, on behalf of river movements across the world, issued a declaration against large-scale river-intervention projects. The Manibeli Declaration, signed in June 1994 by 326 activist groups and NGO coalitions from 44 countries, called for a moratorium on World Bank funded large dams until a comprehensive and independent review of dams funded by the bank was completed.[28]

Initially the World Bank responded to the call by conducting an evaluation through its own Operations and Evaluation Department (OED) and organizing a meeting with International Union for Conservation of Nature (IUCN) in April 1997 in Gland, Switzerland to discuss the evaluation report.[29] However, critics of dams were not satisfied with the OED evaluation and instead demanded an independent evaluation. Their position was supported by the First International Meeting of Peoples Affected by Dams, which had been held at Curitiba in March 1997. Faced with this broadly based demand, the World Bank conceded, but wanted the evaluation to extend to all dams, and not just the ones it had financed. The result was the Gland Agreement and setting up of the WCD, comprising independent experts and overseen by a Reference Group consisting of participants of the Gland meeting.[30]

WCD did commendable work, over two and a half years, holding nine meetings in different cities around the world, with four regional consultations, and receiving about 1,000 submissions. It conducted in-depth studies of eight large dams as well as two country studies (China and India) and 17 thematic reviews, covering economic, technical, social, environmental, and technical issues. It also did a cross-check survey based on brief performance audits of 125 dams.

The resulting WCD report, *Dams and Development*, is a serious indictment of dams, showing convincingly that claims regarding their benefits were exaggerated, while harmful consequences were underestimated or outright ignored, as noted in chapter 2. WCD did not rule out the role of dams in development altogether. However, it put a big question mark on them and suggested that water development projects were to be viewed in the light of the UN Declaration of Human Rights (1948), the UN Declaration on the Right to Development (1986), and the Rio Principles agreed to at the UN Conference on Environment and Development (1992). It further suggested that development (for which water projects are intended) should have five objectives: equity, sustainability, transparency, efficiency, and accountability (WCD 2000, p. ix). Proceeding from this general context, it recommended that decisions regarding water projects should be guided by seven principles: (i) gaining public acceptance; (ii) comprehensive options assessment; (iii) addressing existing dams; (iv) sustaining rivers and livelihoods; (v) recognizing entitlements and sharing benefits; (vi) ensuring compliance; and (vii) sharing rivers for peace, development, and security (WCD, pp. xxxii–xxxv).

Not all dam critics were happy with the WCD report; Medha Patkar, leader of Narmada Banchao Andolon, for example, wrote a separate Comment (WCD 2000, pp. 321-322). However, by documenting and establishing convincingly the exaggeration of dam benefits and underestimation of the multidimensional harm caused by dams, WCD and its report provided a big boost to the Ecological approach to rivers.

7.4.4 Recognition of rivers as "living entities"

While Ecuador's rights of nature initiative is comprehensive, several countries have moved to offer similar rights to rivers. For example, New Zealand passed a law on March 16, 2017, granting its Te Awa Tupua River the status of a person having the right to bring legal action to protect its interests. The decision was taken in response to the demands of the North Island Maori tribe of Whanganui, which had been fighting for such recognition for 140 years. The right will be exercised through two guardians, one to be appointed by the government and the other by the Whanganui iwi (tribe). The law reflects the worldview of the Maori tribes who regard themselves as part of the universe—including mountains, rivers and seas[31]—and were aghast at the treatment of the river as a property and a commercial resource.

Similar steps granting "living entity" (or "subject") status have now been taken in several other countries. In Bangladesh, for example, the high court in a ruling of July 2019 declared the country's rivers to be "living entities" having juridical rights. Appointing the country's National River Conservation Commission as the "legal guardian" of rivers, it further directed other state agencies to fully assist the commission in carrying out its stipulated role.[32]

In the United States, the issue of recognizing rivers and other elements of nature as subjects endowed with rights came up prominently during the landmark environmental law case of *Sierra Club v. Morton*, 405 US 727 (1972). In his dissenting opinion in this case, Justice William O. Douglas argued that "inanimate objects" should have the legal standing to sue in court (so that cases could be litigated in their name and not in the name of their owners). Singling out rivers, Justice Douglas noted, "The river, for example, is the living symbol of all the life it sustains or nourishes—fish, aquatic insects, water ouzels, otter, fisher, deer, elk, bear, and other animals, including man, who are dependent on it or who enjoy it for its sight, its sound, or its life. The river as plaintiff speaks for the ecological unit of life that is part of it" (*Sierra Club v. Morton*, 405 U.S. 727, 741–43; USSC 1972).

In treating rivers—along with other components of nature—as subjects and not just objects or property, we can see a reflection of the pre-industrial Ecological approach, wherein rivers were revered and often regarded as gods and goddesses. This is explicit in the case of the Whanganui River of New Zealand, noted earlier, and in case of the Ganges and Brahmaputra Rivers of India. Indeed, pre-industrial views of the world, as expressed through their worldviews brought to attention by the indigenous populations of Ecuador (comprising about 40 percent of the population of the country), played an important role in formulation of the country's 2008 constitution. However, current steps toward granting subject status to rivers are not necessarily tied to myths and superstitions and instead are based on profound understanding of the role of nature in sustaining life and civilization in the long run, illustrating the process of negation of negation discussed in chapter 6.

7.5 Conclusions

The rise of the Ecological approach to rivers is not an isolated phenomenon. Instead, it is an integral part of the general increase in awareness of the need for and initiatives toward environmental protection. The Industrial Revolution was a watershed event in human history: while before it, humans were helpless vis-à-vis nature, after it, nature (including rivers) itself became in a sense helpless before human beings, equipped as they were with the power of machines conferred on them by the revolution. From being subjects, rivers became objects for exploitation. However, it soon became clear that mistreating nature is not in the long-term interests of human beings themselves. Initiatives for protecting the environment began in the early 1970s, with many initiatives focused specifically on rivers and wetlands. The 1971 Ramsar Convention aimed at protecting wetlands of international significance; the 1997 UN convention forbade river intervening structures that can harm co-riparian countries. WCD documented the adverse consequences of dams and urged caution against taking up new dam projects. The process has gone further, and many countries have declared rivers to be living entities with legal rights protected by courts. This is a return of rivers as subjects— not due to the technological helplessness of humans but due to the realization that human life has to be in harmony with nature. The Ecological approach is a continuation of this progression of human relationship with rivers.

Notes

1. The Industrial Revolution also made possible many breakthroughs in medicine, helping to bring about a sharp reduction in the mortality rates—another factor that led to the dramatic increase in world population.
2. For basic information regarding climate change, its consequences, and necessary mitigation and adaptation efforts, see IPCC (2007, 2014), Stern (2007, 2009), UN (2015c), UNDP (2007), and UNFCCC (1992).
3. For publications by the Club of Rome, see Meadows, Meadows, Randers, and Behrens (1972), and Meadows, Randers, and Meadows (2002).
4. See David (1986, p. 391).
5. The 26 principles of the Stockholm Declaration are as follows:
   1. Human rights must be asserted, apartheid and colonialism condemned
   2. Natural resources must be safeguarded
   3. The Earth's capacity to produce renewable resources must be maintained

   4. Wildlife must be safeguarded
   5. Non-renewable resources must be shared and not exhausted
   6. Pollution must not exceed the environment's capacity to clean itself
   7. Damaging oceanic pollution must be prevented
   8. Development is needed to improve the environment
   9. Developing countries therefore need assistance
   10. Developing countries need reasonable prices for exports to carry out environmental management
   11. Environment policy must not hamper development
   12. Developing countries need money to develop environmental safeguards
   13. Integrated development planning is needed
   14. Rational planning should resolve conflicts between environment and development
   15. Human settlements must be planned to eliminate environmental problems
   16. Governments should plan their own appropriate population policies
   17. National institutions must plan development of states' natural resources
   18. Science and technology must be used to improve the environment
   19. Environmental education is essential
   20. Environmental research must be promoted, particularly in developing countries
   21. States may exploit their resources as they wish but must not endanger others
   22. Compensation is due to states thus endangered
   23. Each nation must establish its own standards
   24. There must be cooperation on international issues
   25. International organizations should help to improve the environment
   26. Weapons of mass destruction must be eliminated

The conference thus established a link between development and the environment by adopting the principle that "development is needed to improve the environment" (Principle No. 8). In particular, at the urging of Indian Prime Minister Indira Gandhi, the conference recognized the need to alleviate poverty in order to protect the environment: the link between ecological management and poverty alleviation was highlighted. After the conference, the European Community formulated its first Environmental Action Program and created the Environmental and Consumer Protection Directorate.

6. The resolution adopted by the 1977 UN water conference urged all "national governments to provide all people with water of safe quality and adequate quantity and basic sanitary facilities by 1990, according priority to the poor and less privileged and to water scarce areas; and ensure larger allocation to this sector from the total resources available for general economic and social development" (United Nations 1977). To achieve this goal, the conference asked for greater international cooperation and greater financial and technical assistance by developed countries to developing countries. The conference did not deal with conservation or environmental protection issues. However, it shows the UN's long-standing interest in drinking water and sanitation issues, which were included later in both MDGs and SDGs.

7. This program is aimed at assisting governments to integrate environmental considerations into the management and development of inland waterbodies, both national and international, with priority given to the latter. As David (1986, p. 400) noted, the program would "contribute to the sustainable development in river basins." UNEP's other programs and activities, such as the Global Environment Monitoring System (GEMS), combatting desertification, human settlements, and regional seas also have important water components (ibid, p. 399). UNEP also helped prepare the International Drinking-Water Supply and Sanitation Decade (IDWSSD).

8. UNEP is not an executing agency.

9. As noted in chapter 1, the emergence of the concept of sustainable development signified a paradigm shift and eventually became the overarching framework for all development efforts of the international community, as symbolized by the adoption of the SDGs by 192 nations at the UN in 2015.

10. In addition, *Agenda 21* notes that global climate change and atmospheric pollution are having adverse impact on freshwater resources, for example, through sea level rise. It observes that demand for water was increasing rapidly, with 70–80 per cent required for irrigation, less than 20 per cent for industry, and a mere 6 percent for domestic consumption (18.6).

11. *Agenda 21* notes that "rapid urban population growth and industrialization are putting severe strains on the water resources" (18.56). It observes that "a high proportion of large urban agglomerations are located around estuaries and in coastal zones. Such an arrangement leads to pollution from municipal and industrial discharges combined with overexploitation of available water resources and threatens the marine environment and the supply of freshwater resources" (18.56).

12. For example, *Agenda 21* notes that "higher temperature and decreased precipitation would lead to decreased water supplies and increased water demands; they might cause deterioration in the quality of freshwater bodies, putting strains on the already fragile balance between supply and demand in many countries. Even where precipitation might increase, there is no guarantee that it would occur at the time of year when it could be used; in addition, there might be a likelihood of increased flooding. Any rise in sea level will often cause the intrusion of salt water into estuaries, small islands and coastal aquifers and the flooding of low-lying coastal areas; this puts low lying countries at great risk" (18.82).

13. CBD has three main goals, namely (i) conservation of biological diversity (or biodiversity); (ii) sustainable use of its components; and (iii) fair and equitable sharing of benefits arising from genetic resources. By June 1993, it had 168 signatures. Following signature and ratification by sufficient number of countries, CBD went into force in December 1993.

14. It also proposed the formation of a High Level Political Forum (HLPF), to be working under the auspices of both the General Assembly and the ECOSOC, with the task of monitoring and promoting the implementation of the SDGs. The CSD expired following the formation of HLPF.

15. According to this view, protection of nature is ultimately in the interests of human beings themselves, because, first, they are part of nature too, and second (if human beings are contraposed to the rest of nature), nature provides the basis on which human life rests, and hence degrading nature in the end cannot but adversely affect human life itself.

16. However, there were some prior initiatives in this direction. For example, in 2000 Switzerland's constitution recognized the right to dignity of animals, plants, and other organisms; but the implications of these rights were not made clear. Also, as noted already, the step to incorporate the rights of nature in the Constitution in Ecuador was a part of the greater move toward the philosophy of *Buen Vivir* (Good Living), which is an alternative concept of development, relying on the idea that a good life is possible only within a community that includes nature. A clear implication is that development has to be in harmony with other people and with nature. This alternative concept of development is gaining support in many countries of South America and elsewhere in the world.

17. "Article 10. Persons, communities, peoples, nations and communities are bearers of rights and shall enjoy the rights guaranteed to them in the Constitution and in international instruments. Nature shall be the subject of those rights that the Constitution recognizes for it (GoE 2008)."

18. "Article 71. Nature, or *Pachamama*, where life is reproduced and occurs, has the right to integral respect for its existence and for the maintenance and regeneration of its life cycles, structure, functions and evolutionary processes. All persons, communities, peoples and nations can call upon public authorities to enforce the rights of nature. To enforce and interpret these rights, the principles set forth in the Constitution shall be observed, as appropriate. The State shall give incentives to natural persons and legal entities and to communities to protect nature and to promote respect for all the elements comprising an ecosystem (GoE 2008)."

19. "Article 72. Nature has the right to be restored. This restoration shall be apart from the obligation of the State and natural persons or legal entities to compensate individuals and communities that depend on affected natural systems. In those cases of severe or permanent environmental impact, including those caused by the exploitation of nonrenewable natural resources, the State shall establish the most effective mechanisms to achieve the restoration and shall adopt adequate measures to eliminate or mitigate harmful environmental consequences (GoE 2008)."

20. "Article 73. The State shall apply preventive and restrictive measures on activities that might lead to the extinction of species, the destruction of ecosystems and the permanent alteration of natural cycles. The introduction of organisms and organic and inorganic material that might definitively alter the nation's genetic assets is forbidden (GoE 2008)."

21. "Article 74. Persons, communities, peoples, and nations shall have the right to benefit from the environment and the natural wealth enabling them to enjoy the good way of living. Environmental services shall not be subject to appropriation; their production, delivery, use and development shall be regulated by the State (GoE 2008)."

22. https://www.theguardian.com/sustainable-business/blog/buen-vivir-philosophy-south-america-eduardo-gudynas.

23. The four pillars of GNH are: (i) sustainable and equitable socio-economic development; (ii) environmental conservation; (iii) preservation and promotion of culture;
and (iv) good governance. It also has nine domains: psychological well-being, health,
time use, education, cultural diversity and resilience, good governance, community
vitality, ecological diversity and resilience, and living standards. Actual measurement
of GNH is based on subjective (based on surveys) and objective indicators and was
included in Bhutan's 2008 constitution as one of the country's goals.

24. See https://www.ramsar.org/sites/default/files/documents/library/current_con-
vention_text_e.pdf for Ramsar Convention and see https://www.ramsar.org/sites-
countries/wetlands-of-international-importance for a list of Ramsar sites. As of May
12, 2019, Ramsar sites covered a total of 2.1 million sq km. The United Kingdom
had the highest number (170) of Ramsar sites, while Bolivia had the greatest area
(more than 140,000 sq km) covered by Ramsar sites. Eighteen Ramsar sites are
transboundary.

25. Apart from national governments, the Ramsar Convention works in cooperation with
six other International Organization Partners: Birdlife International, International
Union for Conservation of Nature (IUCN), International Water Management
Institute (IWMI), Wetlands International, WWF International, and Wildlife and
Wetlands Trust (WWT). It has also instituted February 2 as World Wetlands Day.

26. https://treaties.un.org/Pages/ViewDetails.aspx?src=TREATY&mtdsg_
no=XXVII-12&chapter=27&lang=en.

27. The 1997 UN Convention on Non-navigational Uses of International Watercourses
states, in part:

Article 7 (Obligation not to cause significant harm)

1. Watercourse States shall, in utilizing an international watercourse in their territories, take all appropriate measures to prevent the causing of significant harm
to other watercourse States.

2. Where significant harm nevertheless is caused to another watercourse State,
the States whose use causes such harm shall, in the absence of agreement to
such use, take all appropriate measures, having due regard for the provisions of
articles 5 and 6, in consultation with the affected State, to eliminate or mitigate
such harm and, where appropriate, to discuss the question of compensation.

Article 20 (Protection and preservation of ecosystems)

"Watercourse States shall, individually and, where appropriate, jointly, protect
and preserve the ecosystems of international watercourses."

Article 21 (Prevention, reduction, and control of pollution)

1. For the purpose of this article, "pollution of an international watercourse"
means any detrimental alteration in the composition or quality of the waters of
an international watercourse which results directly or indirectly from human
conduct.

2. Watercourse States shall, individually and, where appropriate, jointly, prevent,
reduce and control the pollution of an international watercourse that may cause

significant harm to other watercourse States or to their environment, including harm to human health or safety, to the use of the waters for any beneficial purpose or to the living resources of the watercourse. Watercourse States shall take steps to harmonize their policies in this connection.

3. Watercourse States shall, at the request of any of them, consult with a view to arriving at mutually agreeable measures and methods to prevent, reduce and control pollution of an international watercourse, such as:

    (a) Setting joint water quality objectives and criteria;

    (b) Establishing techniques and practices to address pollution from point and non-point sources;

    (c) Establishing lists of substances the introduction of which into the waters of an international watercourse is to be prohibited, limited, investigated or monitored.

Article 22 (Introduction of alien or new species)

Watercourse States shall take all measures necessary to prevent the introduction of species, alien or new, into an international watercourse which may have effects detrimental to the ecosystem of the watercourse resulting in significant harm to other watercourse States.

Article 23 (Protection and preservation of the marine environment)

"Watercourse States shall, individually and, where appropriate, in cooperation with other States, take all measures with respect to an international watercourse that are necessary to protect and preserve the marine environment, including estuaries, taking into account generally accepted international rules and standards."

28. See McCully (2001, p. xix).

29. The OED found that 37 (74 percent) of the large dams reviewed are acceptable or potentially acceptable and hence concluded that, overall, most large dams were justified (McCully 2001, p. xx). With this report in hand, OED arranged with IUCN to hold a workshop, to be held in Gland, Switzerland, in April 1997, to discuss the findings of the report. To influence the Gland meeting, IRN got hold of a leaked copy of the OED report and wrote a critique. It then wrote a letter, supported by 44 organizations, and sent it, along with the critique, to World Bank president Wolfensohn just before the workshop. The letter called for a "comprehensive, unbiased, and authoritative review of past World Bank lending for large dams" (McCully 2001, p. xx).

30. River activists viewed formation of the WCD as a big victory. McCully (2001), for example, draws attention to the expansion of the review to all dams. He thinks that this was because World Bank could not defend its OED review and that a focus just on World Bank financed dams would be an embarrassment. As to why dam builders themselves agreed to the review, McCully thinks that they were facing resistance in any case. Also, they hoped that despite some criticisms, overall (on balance) the report would favor dams and thus help them continue their business. McCully further thinks that the dam industry might have also expected the review to support hydropower dams because of climate change. In his view, it was not a foregone

conclusion which way the report would go. The budget for WCD, estimated to be $10 million, was financed by contributions from more than 50 governments, international agencies, private corporations (including many from the dam industry), foundations, and NGOs (McCully 2001, p. xxii).

31. In fact, the Maori tribes trace their genealogy to—and consider themselves children of—rivers, mountains, and seas.

32. https://www.ndtv.com/world-news/bangladesh-high-court-declares-countrys-rivers-legal-persons-2063061.

In India, the high court of the state of Uttarkhand, in a ruling on March 20, 2017, granted the Ganges and Jamuna Rivers "living entity" status. However, the decision was overruled by the country's supreme court in July of the same year upon appeal by the state government of Uttarkhand. See https://www.bbc.com/news/world-asia-india-40537701.

8

Spread of the Ecological approach across the world

8.1 Introduction

Not surprisingly, the post-industrial Ecological approach arose initially in developed countries, which had already pursued the Commercial approach for a long time and experienced its consequences. However, it is now spreading to developing countries, many of which were following the pre-industrial Ecological approach until recently. In many cases, the strong and recent heritage of the pre-industrial Ecological approach is making it possible for developing countries to take more advanced steps toward the post-industrial Ecological approach. As at the world level, at the level of individual countries too, the Ecological approach has progressed in tandem with a general movement for environmental protection.

One reflection of the Ecological approach gaining ground is the deceleration in new dam construction. Globally, the number of large dams commissioned per decade decreased from more than 5,400 (at its peak) in the 1970s to around 2,000 in the 1990s.[1] Part of this decrease is due to the exhaustion of potential dam building sites in developed countries. However, the decrease has also been the result of advocacy for the Ecological approach and opposition to dams.

Another reflection of the progress of the Ecological approach can be seen in the decommissioning and demolition of existing dams. As of now, these measures have been concentrated mainly in developed countries, where many dams have outrun their stipulated life. The removal of dams has, in many cases, gone together with efforts at restoration of rivers, which is another dimension along which the Ecological approach has progressed.

The progress of the Ecological approach can also be seen in burgeoning river (protection) movements. This is particularly the case in developing countries, where people are no longer ready to accept, unquestioned, proposals for dam construction. Instead, they are putting up stiff resistance toward such proposals, particularly where the inequity of the distribution of costs and benefits or potential destruction of environment or cultural treasures are evident.

In following the progression of the Ecological approach, it is important to keep in mind that people can arrive at this approach (or positions conforming to

Rivers and Sustainable Development. S. Nazrul Islam, Oxford University Press (2020). © Oxford University Press.
DOI: 10.1093/oso/9780190079024.001.0001

it) via different routes. Some may arrive at it through deep and comprehensive realization of its merit in sustaining the earth's hydrological cycle and ecologies of river basins, while others may arrive at it because of the unequal distribution of costs and benefits of the river-intervention projects inspired by the Commercial approach. People who are directly affected by the adverse consequences of the river-intervention projects can reach the Ecological approach conforming positions because of their own material interests. It is also possible for people to arrive at the Ecological approach because of real or perceived national deprivation caused by river-intervention projects. No matter which particular route a person or group follows in reaching the Ecological approach, they add to the gathering force behind this approach and can play a useful role in making it the dominant approach in their respective countries.

The discussion in this chapter proceeds as follows. Section 8.2 reviews the progress of the Ecological approach in the United States, where both dam removal and river restoration activities have made some headway. Section 8.3 does the same for Europe, where this progress has assumed a more comprehensive character, as evidenced by the European Union's Directive on Water Policy of 2000. Section 8.4 reviews the progress of the Ecological approach in other developed countries, such as Australia and Japan. Turning to the developing world, section 8.5 reviews progress in India, where the Ecological approach is providing the basis for a host of activities. Section 8.6 reviews the progress of the Ecological approach in other counties of Asia, including Bangladesh, Bhutan, China, Laos, Myanmar, Nepal, and South Korea. Section 8.7 reviews the progress of the Ecological approach in Latin America and section 8.8 does so for Africa. Section 8.9 concludes.

8.2 Ecological approach in the United States

As at the global level, the shift in the attitude toward rivers in the United States also occurred in the context of a general increase in awareness of the need for protection of environment. Reflecting this awareness, a number of environmental protection laws were enacted in the United States at the end of the 1960s and beginning of the 1970s, including the Environmental Protection Act (1969), Clean Air Act (1970), and Endangered Species Act (1974). The plight of rivers in the United States as a result of industrial pollution was brought to the fore quite vividly when the heavily polluted Cuyahoga River of Ohio caught fire in 1969. Partly in response to this incident, the Clean Water Act was enacted in 1972 and amended in 1977.[2] Under this Act, a massive operation began to clean polluted rivers and prevent further pollution.[3]

Alongside efforts toward mitigation of river pollution, a movement against dams and for the restoration of rivers developed. As a result, about 500 dams have been decommissioned in the United States (Figure 8.1), and since 1998 the rate of decommissioning has exceeded the rate of construction of large dams.[4] The removal of dams has in many places helped to restore rivers, in many cases bringing back fish species that were prevented by dams from swimming up river to spawn. A prominent example is provided by the Columbia River basin, where salmon has returned.

It is true that the dams removed so far have been mostly of relatively small size, and their lifespan was already over or nearly over, so that it was politically easier to remove them. However, the spread of the Ecological approach has prevented the construction of large, potentially viable dams (based on the usual yardsticks). In fact, some dams were successfully opposed in the 1950s. For example, conservationists were able to resist plans to construct the 175 m high Echo Park Dam on a tributary of the Colorado River in the 1950s. In the following decade, they were able to thwart plans for construction of two more dams on the main Colorado River in the Grand Canyon. The Sierra Club, a prominent environmental organization, based in the western part of the United States, played an important role in this regard.[5]

In an encouraging move, there are now initiatives to revive the Colorado River Delta, which, as noted in chapter 4, was desiccated through application of the Commercial approach. In June 1993, the United Nations Educational, Scientific and Cultural Organization (UNESCO) designated over 12,000 sq km of the Upper Gulf of California and the Colorado River Delta a Biosphere Nature Reserve.[6] More than 4,000 sq km closest to the delta have been designated the reserve core area, while the remaining 8,000 sq km of open water and shoreline are designated the buffer area. In addition, 2,500 sq km within the Colorado River Delta have been designated a Ramsar Wetland.

Figure 8.1 Dams removed in the United States
Source: American Rivers.

There are now efforts to restore some of the Colorado River flow to its delta and the sea. To this end, the United States and Mexico formulated in November 2012 Minute 319—an agreement that permits Mexico to store its share of the Colorado flow in US reservoirs during wet years.[7] The plan is to use this water to provide both an annual base flow and a spring pulse flow, to mimic the original seasonal pattern of the river flow, based on the original snowmelt-driven regime. Accordingly, the first pulse flow, an eight-week flow of 130 million cubic m, was released beginning on March 21, 2014, with the aim of revitalizing 950 hectares of wetland. It reached the sea on May 16, 2014, marking the first time in 16 years that any water from the Colorado River flowed into the ocean. The event was praised as an experiment of historic political and ecological significance. It was also recognized as a landmark in US–Mexican cooperation in conservation. The pulse flow was to be followed by restoration of a base flow of 64 million cubic m during the following three years. These are commendable efforts, conforming with the Ecological approach. However, they are not adequate for restoring the health of the Colorado River and its delta, because the planned base flow is only a small fraction of the river's original flow.

Restoration efforts are also underway for the San Joaquin River, the second longest river in California, with an agreement among the co-riparian entities reached in 2006 to restore some water to its dry stretch. This agreement was reached only after an 18 year legal battle over how much water should be permitted to flow from the Friant Dam to allow the return of salmon. If properly implemented, the agreement will lead to restoration of 247 km of the San Joaqin River—making it one of the biggest restoration projects in the United States.[8] Similar river restoration efforts are also underway for the Kissimmee River in Florida, following the realization that too much of the Florida swamps has been drained to promote agriculture and settlement (including urbanization). All these are modest beginnings, and it may be hoped that the process of appreciating and adopting the Ecological approach to rivers will gain further momentum in the coming years in the United States.

8.3 Ecological approach in Europe

A more comprehensive effort to move toward the Ecological approach to rivers can be seen in Europe. Following the Stockholm conference (1972), these countries, both collectively—through the European Union (EU)—and individually, took initiatives toward mitigating the adverse effects of economic activities on the environment. Many of these initiatives were directed toward protection of rivers and other surface waterbodies.

8.3.1   EU Directive on Water Policy, 2000

An important step toward restoration of rivers and waterbodies was the Water Framework Directive (WFD), adopted by the European Union in 2000. It begins with the observation that "water is not a *commercial* product like any other, but rather a heritage which must be protected, defended, and treated as such" (EU 2000, p. 1, italics added). The WFD aims primarily at quality of water but notes that ensuring quantity is important for preserving quality. The directive also notes that waters in the European Union are under increasing pressure from continuous growth in demand for their use for different purposes. The WFD goes on to note the important role of conserving river flow for protection of fish populations and for the health of the coastal and marine environment (p. 2).

Emphasizing sustainable use of water based on long-term protection of available water resources, the WFD asks for classification of water bodies into three categories, namely high, good and moderate status. From the conditions it sets for a waterbody to be classified as high status, such as "continuity of river," it is clear that it advocates minimal anthropogenic intervention in the hydrology and morphology of water bodies (pp. 34–35).[9] To ensure sustainability of water use, the WFD calls for laying down "overall principles" for control on abstraction and impoundment of water systems (p. 4).

Regarding rivers, the WFD advocates a basin-wide approach and calls for formulation of River Basin Management Plans (RBMPs), extending, if necessary, across countries. It asks member countries to identify individual "river basin districts," defined as "the area of land and sea, made up of one or more neighboring river basins together with their associated groundwaters and coastal waters" (p. 6). It also asks them to designate river basins covering more than one member-country as an "international river district" (p. 8). The WFD called on each member state to conduct for each river basin or portion of an international river district falling in its territory (i) an analysis of its characteristics; (ii) a review of the impact of human activity on the status of surface waters and on groundwater; and (iii) an economic analysis of water use (p. 11). It calls on member states to establish for each river basin district a "program of measures," consisting of two parts, basic measures and supplementary measures. Basic measures comprise the "minimum requirements," such as (i) legislation for protection of water; (ii) measures to promote an efficient and sustainable use of water; (iii) controls over the abstraction of fresh surface water and groundwater, and impoundment of fresh surface water, including a register or registers of water abstractions and a requirement for prior authorization for abstraction and impoundment; (iv) establishing control, including requirement for prior authorization of artificial recharge or augmentation of groundwater bodies; (v) ensuring that the hydromorphological conditions of waterbodies are consistent with the achievement of the required ecological status or good ecological potential for bodies of water designated as artificial or heavily modified; and (vi) control of pollution (p. 14). The WFD also asks "river districts" to prepare a detailed inventory of water resources and the amounts of abstraction and impoundment in order to monitor the sustainability of those water resources.[10] It pays attention to groundwater and the connection between surface and groundwater to form a whole river basin. It aims at controlling and ending pollution of both groundwater and surface water. In this regard, it advocates application of the "polluter-pays" principle (p. 12).[11] The WFD asks member states to prepare similar RBMPs for their parts of international river districts in cooperation with other co-riparian member states (and countries that are outside the EU).[12]

This summary shows that the EU's WFD reflects, to a considerable degree, the Ecological approach to rivers.[13] Its renunciation of river water as a commercial product; its refrain against impoundment and abstraction; its emphasis on preservation of waterbodies, avoidance of wastage, application of cost-recovery and polluter-pays principles, etc. are all in agreement with the Ecological approach to rivers. Many individual European countries have taken concrete steps in accordance with the WFD.

In tandem with the WFD, a movement developed in Europe for removal of dams and restoration of rivers. As a result, many dams have been removed (Figure 8.2). It is true that—as in the United States—most of these dams are old

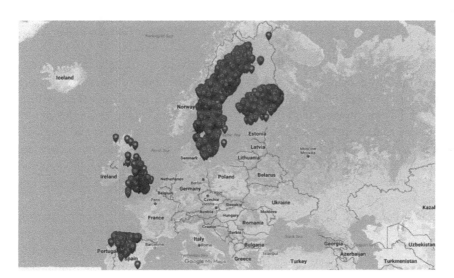

Figure 8.2 Dams removed in Europe
Source: Ajit Bajaj and downtoearth.org.in.

and had either already reached or were close to reaching the end of their stipu-lated lifetime. However, the movement is also aiming at larger and more recently constructed dams, and it is certainly putting up resistance to plans for construc-tion of new dams.

8.4 Ecological approach in other developed countries

8.4.1 Australia

The influence of the Ecological approach can also be seen in recent river-related policies in Australia. The most prominent in this regard are efforts to restore the Murray-Darling river system, which, as we saw in chapter 4, be-came a victim of the Commercial approach. It has been recognized that the river system has been overexploited and diversion of the river water needs to be reduced in order to restore the ecological health of the river system. The Murray-Darling Basin Authority (MDBA) was given the task of formulating and overseeing a legally enforceable Murray-Darling Basin Plan (generally known as the Basin Plan), which is to set "environmentally sustainable limits" on the quantities of water that may be abstracted.[14] It is also required to set basin-wide environmental, water quality, and salinity objectives; to develop efficient water trading regimes; to set requirements for state water resource plans; and to improve water security for users of the basin. MDBA also intends to minimize social and economic impacts while achieving the Basin Plan's environmental goals.

In October 2010, the MDBA released the Guide to the Proposed Murray–Darling Basin Plan.[15] According to this Guide, water allocation from the river system was to be reduced by 2.75 cubic km so as to raise the environ-mental flow, mitigate the salinity intrusion problem, and improve the quality of water. The proposal however drew opposition from farmers and other af-fected groups, so that the plan was revised through negotiations with state governments and other stakeholders, keeping intact, however, the planned amount by which diversions would be reduced. The Basin Plan became law in November 2012.

The Basin Plan for the Murray-Darling River System is certainly a step in the right direction and conforms with the Ecological approach. However, the planned reduction of diversion is inadequate to restore the system. The Environmental Water Requirement (GoA 2015) estimates that a minimum re-duction of abstraction by 7.60 cubic km is required to restore the health of the river system.

8.4.2 Japan

The movement for ensuring and restoring the natural flow of rivers has spread to developed countries in other parts of the world too. In Japan, for example, the opposition to the Nagara River Estuary Dam in the 1980s sparked a nationwide movement against construction of dams. In January 2000, opposition to the Yoshino River Estuary Dam led to the first ever peoples' referendum, which led to suspension of the project.[16] The newly elected governor of the Nagano Prefecture issued a declaration opposing new dam construction in the prefecture, which also stopped construction of the Asakawa Dam, already underway with half the budget spent. Similarly, plans for dams on the Yodogawa River system of Osaka Prefecture were abandoned even though construction had already begun. Japan has also embarked on dismantling existing dams. For example, the Arase Dam on the Kumagawa River was to be dismantled during 2012-2018 (Tanabe 2014).

These examples show that the Ecological approach has gained good ground in many developed countries. Other instances of this progress include popular movements leading to the postponement or cancellation of such large, prestigious dams as the Franklin Dam in Australia, Sierra de la Fare Dam in France, Nagymaros Dam in Hungary, and Katun Dam in Russia.[17] However, more efforts are necessary, particularly in ensuring implementation of the various relevant resolutions, decisions, and directives that have already been adopted.

8.5 Ecological approach in India

It is encouraging that the Ecological approach is also spreading in developing countries. As noted earlier, many developing countries have until recently been practicing the pre-industrial Ecological approach to rivers. The Commercial approach is therefore relatively new for them. However, their own national experiences and the experience of developed countries are convincing many of them of the adverse effects of the Commercial approach and prompting them to move toward the post-industrial Ecological approach.

India, as noted in chapter 2, is the third largest builder of dams and barrages after China and the United States. Moreover, the government of India plans to build many more dams and other river intervening structures in the coming years. Its IRLP is currently the most ambitious and perilous river-intervention plan in the world, as noted earlier. Therefore, there is no doubt that the Commercial approach dominates India's government policies. However, there is also now a strong civic movement in India advocating for the Ecological approach. This movement finds expression in various efforts focused on saving specific rivers; in various all-India organizations working for the protection of rivers in general; and in certain quarters of the Indian political establishment coming out against some iconic river-intervention structures that were built in the past. The following subsections review some of these Ecological approach-inspired movements of India.

8.5.1 Narmada Bachao Andolon

Among the river movements of India, the most prominent is NBA (Save the Narmada River Movement), the impact of which was felt worldwide. As noted earlier, together with the IRN, NBA played a leading role in initiating the process that ultimately led to the formation of the WCD.

The Narmada, as noted in chapter 2, is one of India's major rivers, originating in the Bhindya Mountains and passing through the Madhya Pradesh, along the border of Maharashtra, and then through Gujarat before flowing into the Arabian Sea (Figure 8.3). Following the independence in 1947, this river became a target of dam construction, with the Narmada Valley Development Authority (NVDA) established under the Madhya Pradesh government. Since the river flows across several states, placing NVDA under a particular state government

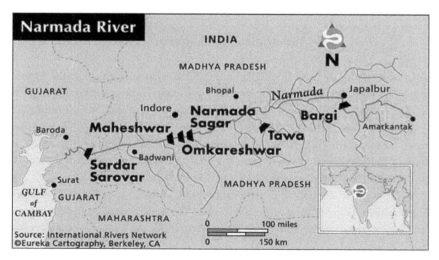

Figure 8.3 Narmada River and the Sardar Sarovar Dam
Source: International Rivers Network.

obviously created problems. In 1969, the central government therefore formed the Narmada Tribunal to adjudicate disputes regarding sharing of the river by different states and intra-state entities and also to propose plans for use of this river, including conditions regarding the resettlement and rehabilitation of the people who would be displaced by the proposed plans. The tribunal submitted its report in 1979, approving construction of 30 major, 135 medium, and 3,000 small dams, of which Sardar Sarovar Dam (SSD) is the largest project. The Narmada Control Authority (NCA) was set up to implement this plan.

The idea of the SSD goes back further into the past. In fact, Jawaharlal Nehru, the first Indian prime minister, laid the foundation of this dam on April 5, 1961. The dam proposal became more concrete over time, and through successive recommendations from the government and approval by the Indian Supreme Court, its height increased from the 80 m first proposed to the 163 m in the final design. The dam's total electricity generation capacity is 1,450 MW, and its reservoir has a capacity of 9.5 cubic km. It is supposed to irrigate about 18,000 sq km, mostly in the drought-prone areas of Gujarat and Rajasthan states (of Kutch and Saurashtra regions). It is also supposed to provide flood protection to 30,000 hectares, covering 210 villages, Bharich city, and a population of 400,000 in Gujarat.[18] It is to be the second largest gravity dam by volume after the Grand Coulee Dam in the United States, and will have the world's third largest spillway.

There was a strong reaction against the dam proposal. Some civic groups were opposed to the dam altogether. Others were more focused on ameliorating the adverse effects of the dam on the people—particularly the indigenous peoples— whose habitats would be submerged by the dam. These various groups coalesced under NBA, led by Medha Patkar.[19]

The movement against SSD intensified after construction began in April 1987. Medha Patkar organized a 36 day solidarity march along the Narmada River, from Madhya Pradesh to the project site in Gujarat. She undertook a 22 day fast that almost took her life. Along with other protesters, she testified in Washington DC in 1989 on the World Bank's role in the dam's construction, pointing out in particular that, in approving the project, the bank had ignored its social and environmental consequences.[20] Her actions prompted the World Bank to form the Morse Commission in 1991 for an independent review of the project. The commission's 357 page report noted the lack of environmental assessment by both the Indian government and the World Bank. Nevertheless, in an internal referendum, the bank voted very closely for the continuation of the SSD. In 1993, Medha Patkar undertook another fast and resisted evacuation from the dam site. In 1994, the World Bank withdrew itself from the project. However, seeing this negative decision coming, the Indian government itself canceled the World Bank loan on March 31, 1993, and proceeded with the project through alternative financing. In 1994, after the NBA office was ransacked, Medha Patkar and her NBA colleagues staged another fast. They were arrested and force fed intra-venously after 20 days. Many victims of evacuation allowed themselves to be sub-merged rather than be evicted (Figure 8.4). Many eminent Indian personalities came forward in solidarity with NBA, and several documentary movies were made in support of it.[21]

NBA could not, however, stop the dam construction. Although it obtained a court injunction in 1993 stopping work on the project, the court's final decision, given in 2000 after seven years of deliberation, allowed the construction to pro-ceed.[22] Construction resumed and was completed in 2006.

Despite its defeat, NBA had a huge impact. First, it contributed hugely to getting more satisfactory compensation and resettlement of the affected people: the Indian Supreme Court, while allowing the project to go ahead, made it contingent on ensuring proper compensation and resettlement. For this purpose, the court required the setting up of Grievance Redressal Authorities (GRAs) in each party state to monitor the progress of resettlement in tandem with raising the dam height. Second, before NBA's efforts, dams and barrages were accepted as more or less unquestioned instruments of development in India. NBA prompted the entire nation to question that premise. It put forward the idea that the Commercial approach to rivers is not the only way to develop-ment and made people aware that there was an alternative, Ecological approach,

Figure 8.4 Women protest against Sardar Sarovar Dam
Source: Avantika Tewari (2017).

to rivers and development as well.[23] Third, NBA had an important impact on the global movement for protection of rivers. As noted earlier, together with the IRN, it set off a process that led to the establishment of WCD and publication of its report, *Dams and Development* (WCD 2000).

8.5.2 Critique of the IRLP

The critique of and opposition to the IRLP that have developed inside India also reflect the spread of the Ecological approach in the country. Though the Indian government presented IRLP as vital for national interests, many Indian river experts and activists are opposing it. Despite the paucity of information furnished by the government on this project, river experts and activists have taken initiatives to gather relevant information and offer their assessments. The criticisms they have offered can be summarized into the following eight points.

First, the very premise, according to which certain river basins are surplus while others are deficit is questionable, because each river is unique, with its unique hydrology and morphology, giving rise to a unique ecosystem and culture in its basin. Therefore, specific to its ecosystem and culture, a particular river's water is neither surplus nor deficit. Any large-scale transfer of water from one basin to another is therefore bound to harm the ecosystems of both basins. Different ecosystems are part of nature; they need not be human-engineered so as to be uniform. Diversity is a part and parcel of nature and perhaps its main attraction.[24]

Second, the concepts of surplus and deficit are relative and not absolute. In particular, they depend on demand and supply conditions, which may change endogenously. As noted earlier, due the "More water, more thirst!" phenomenon, augmentation of river water may even increase the perceived deficit. Thus "perceived" surplus and deficit are not the right guide for making decisions on interbasin water transfer issues. Also important is the fact that climate change is altering the objective reality of water availability across different states. Pointing to the relative and changing nature of the perceptions and reality regarding surplus and deficit basins, Dinesh Mishra, the noted river scholar of India, observes the following:

> Floods have disappeared since 2007 from Bihar, the state that—along with Assam— is supposed to have surplus water and is expected to feed the nation by its waters. The flood of 2008 was an accident. Drought relief has become a regular feature in the state. Will someone take cognizance of the situation? The states where Bihar's water is planned to be sent are facing floods now, whereas Bihar reels under drought. What will the Inter Linking of Rivers project result in such a case is the question that will have to be answered someday. Think of the newly developed flood prone areas before asking of water from states like Bihar or Assam (Mishra 2015, p. 2).

Third, even if certain river basins are thought to be surplus, and it is agreed in principle to transfer part of its water to another basin, the actual transfer may not be a technically sound proposition. This seems particularly to be the case with respect to transferring water from the Brahmaputra River, which according to IRLP proponents is the main surplus basin. Transferring Brahmaputra water to south and west India will require canals passing through the "goose neck"[25] and cutting across such major rivers as the Teesta, Dharla, Mahananda, and Atrai. To reach peninsular India, the water will also have to cross the Bindhya mountain range. Such a trajectory will entail "engineering at cross-purpose," a problem that is also endemic with most other proposed IRLP link canals. Capital construction carried out under IRLP will annihilate a lot of capital already in place in the form of water structures and other infrastructures. The project will also entail the problem of "engineering begetting engineering," which refers to the phenomenon of one engineering project requiring many more to be successful. Thus, IRLP has serious problems of technical feasibility.

Fourth, IRLP's flood mitigation effects are doubtful. Flood mitigation through inter-basin transfer in South Asia is made difficult by the fact that the months of heavy precipitation (from south-westerly monsoon winds) are basically the same for almost all parts of the region, except the Eastern Ghats of Deccan which receive some rainfall during the winter. So, during the dry months, when water is required in west and south India, river flows are lean in eastern India too. Similarly, when rivers are full in the east and floods occur, providing the basis for them being characterized as surplus, rivers in western and southern India also have their peak season. Thus, contemporaneous transfer of water cannot be an effective flood mitigation strategy in South Asia. Meaningful flood mitigation would instead require inter-temporal transfer, entailing storage of monsoon water in the east on a scale that would be difficult to achieve. Furthermore, there is much misconception regarding flooding. As we shall see in chapters 7 and 9, overflow of river water onto the river's floodplain is a normal, natural phenomenon, necessary for the economy and ecology of the floodplain. Ironically, much of the flood aggravation in recent period has been caused by ill-conceived structures built to prevent such overflow. Thus, transferring water through IRLP may not be the best way to mitigate flooding, as we shall see in following chapters.

Fifth, a mammoth river-intervention project such as IRLP cannot but have many unintended, negative consequences. For example, many of the project's dams and reservoirs are to be built in seismically active regions, engendering significant risks of catastrophic floods in the event of dam failure or collapse. More importantly, as noted above, climate change may soon make IRLP's calculations obsolete. Hundreds of billion dollars of investment, which could have been directed to more worthy causes, will then become redundant, as pointed out by Mishra, noted above.

Sixth, IRLP is almost certain to generate conflicts both within India and between India and its neighboring countries. Particularly intractable will be conflicts with the lower riparian country of Bangladesh, with which India already has serious conflicts over diversion of river water, as noted in chapter 5. Additional transfer of water from the Ganges and the Brahmaputra, as envisioned by IRLP, will surely aggravate these conflicts.

Seventh, even if the necessary engineering might and sophistication are marshalled and the proposed dams, barrages, reservoirs, lifts, tunnels, and canals are actually built, IRLP may not be viable from a financial viewpoint. As noted earlier, the government is yet to make public a detailed financial cost-benefit analysis of the project. However, available information indicates that its financial rate of return is likely to be low, if not negative. As noted earlier, the IRLP is expected to generate 34,000 MW of power and expand irrigation by 35 million hectares. However, much of the hydropower generated will have to be spent in lifting water across hills and mountains, so that the net gain in the power supply will be much less. So far as irrigation is concerned, we saw earlier that publicly financed surface water irrigation projects generally lead to misdirected crop choice and wasteful use of water, so that the net increase in crop production is not that large. As noted earlier, Duflo and Pande (2007) found that the contribution of dams to increased agricultural output was only 8 percent. To the extent that IRLP will involve a lot of cross-engineering, its net benefit is likely to be less while its financial costs will be much higher.

Eighth, the economic rates of return of IRLP are almost certain to be negative. In computing these, as noted in chapter 3, both marketed and non-marketed inputs and outputs need to be taken into consideration and economic (or shadow) prices are to be used to evaluate even the marketed inputs and outputs, particularly in cases they involve externalities, both positive and negative.[26] Needless to say, the negative externalities of IRLP will be serious, both for Indian states from which water will be withdrawn and for Bangladesh, in particular, whose ecology and economy will suffer irreparable damage. Furthermore, as noted earlier, IRLP is likely to harm the political relationship between Bangladesh and India, entailing negative consequences in a much broader arena, including trade and security. In view of this, it is difficult to see how IRLP can have positive rates of economic return.[27]

Many scholars involved in the critique of IRLP realize that it is necessary to go beyond it and take a deeper look into the philosophy that gives rise to this type of project. For example, Ramaswamy Iyer, the noted Indian river scholar, urged to look into what lay behind IRLP and behind similar projects in other countries and identified the spirit of gigantism, an adversarial approach to nature and a certain definition of development to be the main underlying factors (Iyer 2002, 2004a). He offered an elaborate critique of the thinking that underpins IRLP-type projects and outlined a "transformation" of thinking that was necessary for better utilization of water resources. He therefore presented a 25 point declaration, which reflected much of what has been described in this book as the Ecological approach to rivers (see Iyer 2004b).

8.5.3 Growing demand for demolition of the Farakka Barrage

The spread of the Ecological approach to rivers in India can be seen in the growing demand in the country for demolition of the Farakka Barrage. The Indian establishment has always maintained and continues to maintain that the Farakka Barrage was very useful for national interests. For a long time, it was therefore considered a taboo subject for Indian nationals, and speaking out against Farakka was seen to be unpatriotic.

The situation in this regard seems to have changed in recent years, mainly because of the increasing severity of the negative consequences of the Farakka Barrage for India itself. As noted in chapter 5, by causing upstream sedimentation, the Farakka Barrage has now become a cause of upstream flooding and riverbank erosion, particularly in the Indian state of Bihar. As a result, politicians representing the people of the affected areas, as well as experts and policymakers, are now calling for demolition of the barrage.

The most prominent among these politicians is Nitish Kumar, the former chief minister of Bihar, who has been calling for demolition of the Farakka Barrage for quite some time now. However, his calls have become louder and are getting more resonance in recent years. On July 17, 2016, Kumar raised the Farakka issue at the 11th Inter State Council meeting in New Delhi—chaired by Prime Minister Narendra Modi and attended by Union (Central) Ministers and Chief Ministers of the States—and demanded removal of the barrage. He declared that the disadvantages of the Farakka Barrage were greater than its benefits.[28] He also sought the intervention of the Indian central government to ensure uninterrupted flow of water from the upper co-basin states, so that the entire stretch of the Ganges had continuous flow of water even during the lean season. On August 21, 2016, during heightened flooding in Bihar, Kumar, at a press conference in Patna, the capital of Bihar, reiterated his demand for removal of the Farakka Barrage, noting, "current flood like situation in 12 districts including Patna has been caused by huge siltation in the river Ganga. This situation is the result of silt being deposited in Ganga due to construction of Farakka dam in West Bengal. The only way to remove silt from the river is to remove the Farakka dam (Kumar 2016)."

Taking things further, the Bihar state government organized a two-day conference, February 25–26, 2017, in Patna, on "Incessant Ganga," to discuss the state of the Ganges River and the impact on it of the Farakka and other dams and barrages. Attended by a large number of prominent river experts and activists, the conference identified siltation of the Ganges, caused by Farakka and other dams and barrages, as the most important problem.

Rajendra Singh, called the "Waterman of India," informed the conference that he was against the Farakka Barrage from the very beginning and now felt vindicated by the fact that it caused major problems for both upstream and downstream regions without yielding much benefit. He observed that on the basis of the evidence presented at the conference, it was clear that Farakka was inauspicious (*ashubha*) for Bihar. He characterized it as a curse (*abhishap*) which needed to be removed, because unless and until it was removed, it would not be possible to move forward.[29]

Another prominent figure, Himanshu Thakker, Coordinator of the South Asian Network on Dams, River and People (SANDRP), speaking at the same conference, expressed the urgent need for a review of the Farakka Barrage which, he claimed, had failed to fulfil any of the purposes—irrigation, hydroelectric power, water supply—for which it was built. Referring to the practice of advanced countries, where dams and barrages are reviewed every 20 years, he noted that no such review has been conducted for Farakka, even though it is now 42 years old (Kumar 2016).

At this conference, Nitish Kumar reiterated his demand for removal of the Farakka and other barrages and for restoration of the natural flow of the Ganges. He noted that the Indian central government's goal of a "Nirmal (clean) Ganga" cannot be achieved unless the natural flow of the Ganges is restored. He therefore opposed the planned new barrages in Uttar Pradesh.[30] The conference adopted an 11 point Patna Declaration embodying these demands. The Patna conference was followed by a two-day conference in Delhi in May 2017, at which similar demands were made.

Of course, the demand by Nitish Kumar and others for demolition of the Farakka Barrage is not sufficient for the Indian central government to adopted this decision. However, these calls have broken the taboo and have shown that the demand for removal of the Farakka Barrage is not an unpatriotic demand for Indians. Voicing of this demand by the chief minister of a state of India, and others, shows that many within the Indian political establishment hold contrary views on the Farakka Barrage. These developments may encourage others in India to join the opposition to this barrage and other similar water diversionary structures.

The opposition to the Farakka Barrage by Indian politicians may not be rooted in a profound change of philosophy regarding rivers. Instead, they may be just responding to the grievances of the people (i.e., their electorate) affected by the flooding and erosion caused by the Farakka Barrage. However, this does not undercut the genuineness and potency of their position. Instead, it shows that people can arrive at positions conforming to the Ecological approach via different routes.

8.5.4 Ecological approach in northeastern states of India

The Ecological approach to rivers is also finding favorable ground in India's northeastern states—Assam, Meghalaya, Tripura, Mizoram, Manipur, Nagaland, and Arunachal—the "Seven Sisters." These states are connected with the rest of the country through the 22 mile wide "goose neck" between Nepal and Bangladesh. They lie in the basins of the Brahmaputra and Meghna Rivers, both of which then pass through Bangladesh. Because of its generally mountainous and hilly terrain, this region has considerable hydropower potential, and India has already built dams and barrages to make use of this potential and has plans to build several hundred new dams and other river intervening structures in this region in the coming decades, as noted in chapter 2. The region is also rich in other natural resources, such as timber and various minerals, including coal and oil. India plans to exploit these mineral resources too.

However, India's massive plan to exploit the hydropower potential and other natural resources of the northeastern states has also generated resistance on the part of the peoples of this region (Kumar 2008). As elsewhere, there are two main interrelated sources of grievance (Gill 2015; Hassan 2008; Hebber 2014). One is the feeling of deprivation, arising from the perception that the central government (or the rest) of India is benefitting more from the natural resources of the region than the peoples of the region themselves. The other is the destruction and despoliation of the environment that the drive to exploit the natural resources inevitably causes. As a result, there is strong opposition in this region to the Commercial approach and a favorable view of Ecological approach (Sharma 2012).

Many people of the Indian northeastern states are also more sympathetic to Bangladesh's demand for restoration of the natural flows of shared rivers. This is partly because these states require transit and transshipment facilities through Bangladesh for their economic development, and they understand that it will be politically easier for Bangladesh to allow these facilities if the Farakka Barrage and other diversionary barrages are removed (Abdin 2020).

The opposition to the Commercial approach in the Indian northeastern states may be motivated more by the sense of deprivation and injustice than by a new philosophy regarding rivers. However, as noted earlier, people may arrive at the Ecological approach conforming positions through diverse routes.

8.6 Ecological approach in other Asian countries

8.6.1 Nepal and Bhutan

The influence of the Ecological approach to rivers can be seen in both Nepal and Bhutan, where several dam projects have been either cancelled or put on hold in recent years. One example of a large dam project cancelled due to protests is the West Seti Dam on the Seti River in western Nepal. This was a 750 MW project to be constructed at an estimated cost of $1.6 billion. The license was given in 1994 to the Australian multinational company Snowy Mountains Engineering Corporation (SMEC). The project's terms of reference allowed 90 percent of electricity to be exported to India and about 600,000 additional hectares of land in Uttar Pradesh of India to be irrigated based on water stored in Nepal. Nepal would receive the remaining 10 percent of the electricity and a nominal royalty from SMEC's profit, even though about 30,000 local Tharu people would be displaced by the project (Bhandari 2011, p. 1). However, the proposed division of costs and benefits was considered unfair by many in Nepal. In addition, it was also suggested that the project was approved without necessary ratification by Nepal's parliament. A campaign based on these economic, legal, human, and environmental arguments succeeded in getting the dam project cancelled.[31] Similarly, the Arun 3—a 900 MW run-of-the-river dam on the Arun River—was cancelled due to protests responding to similar concerns as caused the cancellation of the West Seti Dam. The World Bank withdrew from financing the Arun 3 in 1995 in face of the protests, and, as a result, the project had to be cancelled.

Bhutan finds itself in a situation similar to that of Nepal and the Indian northeastern states with regard to dams and barrages, built under joint venture with India. As noted in chapter 5, there is also a general feeling in Bhutan that India benefits more from these river intervening structures, while Bhutan bears most of the environmental and human costs (Walker 2015). This sense of national deprivation is prompting many in Bhutan toward a favorable view of the Ecological approach to rivers. However, greater appreciation for unspoiled natural surroundings also plays an important role in causing the people of Nepal and Bhutan to be in favor of the Ecological approach. This is particularly the case for Bhutan, which, as noted earlier, has put forward the idea of the Gross National Happiness (GNH) as an alternative to the conventional Gross National Product (GNP) as a measure of welfare. Harmony with the nature is an important element of GNH.[32] Needless to say, the Ecological approach is more conducive to achieving harmony with nature than the Commercial approach.

However, as is the case with India's northeastern states, the greater social support for the Ecological approach in Nepal and Bhutan has not yet stopped the governments of these countries from construction of dams and barrages. Thus,

Bhutan is going ahead with plans for construction of another ten dams under joint venture with India.[33] Nepal also has not given up the idea of reaping its hydroelectric potential through construction of more dams and barrages. However, the greater social support for the Ecological approach in these countries is certainly acting as a restraining factor in plans for future interventions in the rivers of these countries.

8.6.2 Pakistan

Reflections of the Ecological approach can also be seen in the growing opposition to many of the new proposed dams in Pakistan. As elsewhere, other concerns often get intertwined with ecological ones in determining the approach taken to rivers or particular river-related projects. An example is provided by the opposition to the proposed Kalabagh Dam, which has been at the center of inter-provincial disputes in Pakistan for a long time. It has been proposed three times since the 1980s but postponed because of vigorous opposition.[34] It remains a controversial proposal, though the central government of Pakistan, dominated by the Punjab province, is still pushing for it. As noted earlier, the former government of Pakistan, with Nawaz Sharif as the prime minister, included Kalabagh Dam in the list of new dams that it planned to construct. One of the arguments for this new dam is that it will help to compensate for the loss in water storage capacity suffered by the Tarbela Dam reservoir due to sedimentation.[35] However, there is considerable opposition to this new dam, both because abstraction of water has already exhausted the flow of the Indus River (as noted in chapter 5), and because it will enhance the control of the Punjab province over the river and the benefits that province derives from it, at the expense of other provinces, particularly the downstream province of Sindh.

The debate regarding the Kalabagh Dam flared up again following the unusual flood of 2014, which killed about 500 people, displaced hundreds of thousands, and destroyed considerable amounts of crops and property. Many proponents of the Commercial approach used this event to argue again for Kalabagh and other new dams.[36] However, others have used the experience of the 2014 flood as a warning of what is to come in the future due to climate change and as a proof that dams will prove ineffective. They have drawn attention to the adverse effects of dams, such as damage to river morphology; loss of land and resources due to submergence; wastage of reservoir water; etc. With respect to the Kalabagh dam, the additional arguments concerning regional grievance have also been reiterated. In particular, it has been argued that instead of directing more water toward wasteful irrigation, it is important to save the Indus delta, save Karachi city, and prevent salinity intrusion in the Thar Desert.[37]

8.6.3 Bangladesh

The Ecological approach to rivers has found a strong foothold in the civil society of Bangladesh. In fact, there is now a strong civic movement advocating
this approach, led by Bangladesh Poribesh Andolon (BAPA), a nation-wide environmental movement, and Bangladesh Environment Network (BEN), a global
network of non-resident Bangladeshis dedicated to protection of the environment. As will be noted in chapter 10, the thrust of this movement is against lateral interventions, which are the common form of river intervention in deltaic
Bangladesh. However, it is also waging an active campaign against dams and
barrages that India has constructed and plans to construct on the shared rivers
along the border of Bangladesh. At the 2006 National Conference on Rivers,
BAPA and BEN adopted a Resolution on Rivers (RoR), formalizing their position in favor of the Ecological approach. These organizations are also playing an
active role in rallying river activists of the South Asian region in support of the
Ecological approach, as noted below.

8.6.4 Region-wide river movements in South Asia

In addition to the country-based movements described above, there are efforts
to forge a region-wide movement in South Asia, advocating the Ecological approach. A regional movement makes particular sense for South Asia, where—as
mentioned before—river basins became politically fragmented when country
boundaries were drawn on the basis of political considerations. Regional cooperation is therefore particularly necessary in South Asia in order to ensure
basin-wide care for the rivers. Unfortunately, no such regional cooperation has
materialized in South Asia so far. Instead, as noted in chapter 5, negotiation and
cooperation regarding rivers has remained limited to a bilateral framework,
and almost all such cooperation efforts have been based on the Commercial
approach, which generally created and aggravated conflicts. Protection of the
South Asian transboundary rivers therefore requires regional cooperation based
on the Ecological approach.

Though the governments of South Asian countries, by and large, continue
to adhere to the Commercial approach, there are civil society initiatives toward promoting the Ecological approach and cooperating on that basis.[38]
An important initiative in this regard is the South Asian Network on Dams,
Rivers, and People (SANDRP), established in 1998 and headquartered in
Delhi, capital of India. SANDRP has been playing an important role in disseminating information regarding the adverse effects of the Commercial approach to rivers and in popularizing the ideas of the Ecological approach. An

important tool that it uses for this purpose is its journal, *Dams, Rivers and People*, which is available in its website.[39]

There have been efforts toward regional organization building too. As part of this effort, BAPA and BEN—together with other pro-environment organizations—held in 2004 the International Conference on Regional Cooperation on Transboundary Rivers (ICRCTR) to discuss the impact of the IRLP. Attended by a large contingent of river experts and activists from India (including Medha Patker), Nepal, and other countries, this conference made possible one of the most detailed discussions of the IRLP and its possible consequences. In 2013, BAPA and BEN, together with other pro-environment organizations, held a follow up conference on Water Resources of South Asia—Conflict to Cooperation (WRSA-CC), in which a large number of river experts and activists from India, Nepal, and other South Asian countries participated and discussed ways to enhance cooperation following the Ecological approach.

These conferences provided a stimulus to building region-wide organizations devoted to the protection of rivers and promotion of the Ecological approach. For example, initiatives were taken to form the South Asian Alliance for Peoples' Movements, advocating policies reflecting the Ecological approach, and the South Asian Solidarity for Rivers and Peoples. A more recent organization initiated with similar goals is HIMSAR (South Asian Network for Himalayan Rivers). It remains to be seen whether these and other such organizations can survive and mature in the future. However, given its long tradition of revering rivers, South Asia can be at the forefront of the world-wide movement toward the Ecological approach. Provided it garners enough support among the people, such a movement can then persuade the governments of the countries of the region to switch to the Ecological approach to rivers.

8.6.5 Myanmar

The Ecological approach is gaining ground in other Asian countries too. As seen earlier, a variety of factors often converge to produce outcomes that conform to the Ecological approach. For example, in Myanmar, President U Thein Sein suspended in 2011 the 6,000 MW Myitsone Dam Project on the Ayeyawady (Irawaddy) River for the duration of the tenure of his government.[40] This is one of the eight dams proposed on the headwaters of the Ayeyawady River under a deal signed between the prevailing Myanmar military government and China's State Power Investment Corporation (SPIC) in December 2006. Under the terms of the agreement, SPIC would operate the dams for 50 years, generating revenue from the sale of electricity. Construction of the Myitsone Dam began in December 2009, despite significant local and national opposition.

The Myitsone Dam's location is at the Mali and N'Mai Rivers' confluence, considered a sacred site, with a special place in the national imagination. Further, the area is inhabited by the Kachin people, who have been waging a war against the Myanmar central government for the right to self-determination for quite some time. The 1994 ceasefire agreement between the Kachin Independence Army (KIA) and the Myanmar government collapsed in part due to the latter's plan to build the dams.[41] For a while, the Myanmar government tried to confront the opposition to dams by force, sending large number of troops to the area.

However, it is not only the Kachin people who are opposed to the Myitsone Dam. According to opinion polls, 90 percent of Myanmar's people oppose the dam, reflecting the special place that the dam site holds in the Myanmar psyche. A national "Save the Ayeyawady" campaign was launched in 2010. In a government workshop on the environmental impacts of the Ayeyawady hydropower dams, held on September 17, 2011, it was revealed that there were key information gaps regarding the Myitsone Dam's social and environmental impacts, including information about the risks owing to the seismicity of the region. The Environmental Impact Assessment (EIA) report, that SPIC had to release, confirmed these gaps. This chain of events led the then President U. Thein Sein to suspend the Myitsone Dam project, even though approximately 10 percent of the work on the dam had been completed. It is the first high-profile Chinese overseas dam project to be suspended during construction.

Among the reasons President U. Thein Sein cited for his decision, first and foremost were the "natural beauties" of Myitsone, which were a gift of nature and a landmark not only for Kachin state but also for Myanmar and which might disappear. He noted that the dam could have a "devastating effect" on the Ayeyawady River. Other well-known arguments were also mentioned, such as "loss of livelihood" of local people due to upstream submergence and risks of dam collapse due to earthquake and excessive precipitation caused by climate change.[42]

With the installation of the new government of the National League for Democracy (NLD), led by Aung San Suu Kyi, many hoped that the Myitsone Dam proposal would be finally cancelled. However, the NLD government instead commissioned a new Strategic Environment Assessment (SEA). The necessary research was conducted by ICEM, an Australian environmental assessment group, and the findings were approved by the Myanmar government's Ministry of Natural Resources and Environmental Conservation as well as the Ministry of Electricity and Energy (MOEE). The SEA also revealed problems with the Myitsone dam, warning that its construction will "break river connectivity, trap sediment, and alter river flow on a wide scale." It also recommended against other proposed dams in the Ayeyawady watershed, finding that "the plan of five large dams on the main stems of Ayeyawady and Thanlwin (Salween) rivers would completely alter the river system's hydrology, sediment transport, and geomorphic functioning (Fawthrop 2019)."[43]

Despite these strong recommendations, the NLD government is yet to cancel the Myitsone Dam project. According to some observers, outside pressure is playing an important role, and SPIC appears to have embarked on a vigorous public relations campaign to build support within Myanmar for the dam (Fawthrop 2019).[44] For example, diplomats sympathetic to SPIC visited the Kachin capital and held a meeting with the leaders of the Kachin people in December 2018.[45] Aung San Suu Kyi continues to be silent on the issue and has not made public the report on all dam projects on the Irrawaddy River, submitted to the president's office in November 2018 by the 20 member commission led by the speaker of the parliament U. T. Khun Myat. It is reported that the Myanmar energy ministry is now preparing a white paper, spelling out a policy that will leave enough wiggle room for the Myitsone, or a revised version of it, to slip through.

Despite the above, the mere fact that about 90 percent of the Myanmar people opposed the Myitsone Dam and forced the government to stop and postpone the project is remarkable and shows that some of the arguments for the Ecological approach to rivers found considerable resonance in the country.[46]

8.6.6 Laos

The situation in Laos regarding the Ecological approach is somewhat similar to that of Myanmar. On the one hand, the country is proceeding with a plan to construct numerous dams and barrages, mostly with technical expertise and finance from China, as noted in chapter 2. On the other hand, it is also facing some opposition to dam construction, most prominently to the Xayaburi Dam on the mainstream Mekong River. Planning for this dam began in 2007, and it was to be constructed in cooperation with Thailand, whose construction giant Ch. Karnchang was the dam builder and whose banks, including the state-owned Krung Thai bank, were financing the project. Furthermore, the Electricity Generating Authority of Thailand (EGAT) had already signed a contract regarding purchase of power generated by the dam, specifying its construction deadlines.

Construction began in March 2012, but shortly after, on July 15, 2012, the Laos government announced postponement of the Xayaburi Dam pending further environmental studies.[47] The postponement was the result of considerable opposition from within Laos and from the downstream countries of Vietnam and Kampuchea. The opposition from within Laos was mainly on environmental grounds, while downstream countries were worried about the decrease and interruption in water and sediment flow that Xayaburi would cause.

The dam was also opposed by the Mekong River Commission (MRC), which, in September 2010, had published an SEA on Mekong mainstream dams, recommending a 10 year moratorium on all dam building on the Lower Mekong River to allow time for more studies and evidence-based decision making. Also, under the 1995 Mekong Agreement, prior consultation was required before any dam construction by the countries of the river basin. The Xayaburi Dam, as the first test of this provision, signified a failure of the agreement.[48] All these factors and concerns played a role in the postponement of the Xayaburi Dam.

However, the postponement did not last too long. Under the pressure of various concerned parties, particularly the Thai construction company, financing bank, and electricity buying authority, the Laos government resumed construction later in 2012, and the dam is to be completed and become operational in late 2019.

Thus, as was the case with the suspension of Myitsone Dam in Myanmar, the postponement of the Xayaburi Dam in Laos did not signify a general shift from the Commercial to Ecological approach at the government level. In the end, the Laos government not only proceeded with the Xayaburi Dam but is moving ahead with many other dam projects on the Mekong River and its tributaries.[49] The Xayaburi experience also shows that basin-wide initiatives, such as the setting up of the MRC, cannot succeed in minimizing interventions in a river unless the initiative is based on the Ecological approach and upheld by all the participating countries.

8.6.7 South Korea

South Korea, as noted earlier, is a major builder of dams. However, opposition to the Commercial approach has developed there too. For example, a long national and international campaign led by local residents and the Korean Federation for Environmental Movement led President Kim Dae-jung to scrap in June 2000 the proposed multipurpose dam on the Tong River and to designate the area as a nature reserve.[50]

8.6.8 China

As is clear from the discussion so far, China is currently the main promoter of the Commercial approach to rivers. It alone accounts for about 40 percent of all large dams in the world, and it is fast constructing more dams inside China. Furthermore, China is the largest dam builder overseas, having already constructed several hundred such dams and planning to construct several hundred more. China's Belt and Road Initiative (BRI), though focused on communication infrastructure, has been expanded to include river-intervention projects. Thus, many dam projects in Myanmar and Laos, constructed with Chinese assistance, are considered to be part of the BRI.

However, the Ecological approach is gaining support in China too. For example, there was some domestic opposition to the Three Gorges Dam, which involved resettlement of a large number of people. In some cases, the opposition to the Commercial approach is even bearing fruit. For example, in November 2007, local authorities in China cancelled the planned Megoe Tso Dam in Eastern Tibet (also known as the Mugecuo Dam), which would have caused significant damage to a lake considered sacred by the local population and recognized as important for its biodiversity.[51]

There is now significant awareness in China about the harm that has been done to rivers and other waterbodies through industrial pollution. Until recently, China was almost solely focused on GDP growth, resulting in considerable environmental damage, both globally and locally. Globally, China became the largest emitter of GHG, surpassing the United States, in terms of total volume. In per capita terms, it surpassed the European Union. Sulphur dioxide emissions from China's coal-fired power plants also cause global damage, while locally, smog in Chinese cities is a visible example of pollution. The damage to rivers from pollution is also significant.

However, there is now a growing civic awareness about pollution of rivers and other waterbodies. The government too, now that economic success has been achieved, is paying more attention to environmental issues and taking measures to curb industrial pollution of waterbodies and to clean up the pollution that has already been caused. China is also now trying to reduce GHG and sulphur emissions and solve its smog problem, and it has embarked on the largest afforestation program in the world.

However, the greater awareness about the environmental consequences of economic growth and various steps to mitigate them has not yet translated into a re-examination by the Chinese government of the country's general approach to rivers. Recognition of the negative consequences of the Commercial approach and appreciation for the Ecological approach is so far limited mostly to civic groups. Still, given China's importance in promoting the Commercial approach both inside and outside China, such a policy change—as has occurred with respect to climate change—has to occur soon, if rivers are to be protected from further large-scale interventions across the world.

8.7 Ecological approach in Latin America

The Ecological approach is spreading in Latin America, home of the Amazon, the largest river system in the world. It is also organically connected with the Amazon forests, the world's largest rainforest and considered to be the "lungs of the earth." Protection of the Amazon river system is therefore necessary for protection of the Amazon forests too. As Brazil contains the major part of both the Amazon forests and the river system, it has the main role in this regard and indeed has been a leader in promoting sustainable development with a focus on environmental protection. As noted earlier, it hosted both the 1992 Rio Conference, where the *Agenda 21* was adopted, and the 2012 Rio + 20 conference, where the proposal for formulation of SDGs was adopted. In view of this commendable record, it can be expected that Brazil will play an active role in promoting the Ecological approach.

The Ecological approach should find a favorable ground in Latin America in view of the general move of many countries of the region toward the *Buene Vivir* philosophy and recognition of the rights of nature, as noted in chapter 7; according to this worldview, rivers cannot be regarded as mere objects and properties. Instead, they need to be treated as subjects, having rights of preservation.

The *Buene Vivir* philosophy appears to be more popular in such northern Andean countries as Ecuador, Bolivia, and Venezuela, among both the public and governments. As a result, there is more sensitivity in these countries toward river-intervention projects. For example, in December 2006, the Bolivian government protested against construction of the Santo Antonio and Jirau dams on the Madeira River in Brazil, which were intended to generate 6,450 MW electricity. A tributary of the Amazon River, the Madeira River is a hub of biodiversity, including 33 endangered mammal species. Apart from general concerns regarding global biodiversity, Bolivia had direct reasons to be concerned, because these dams would have submerged a large part of its forest area.[52]

The influence of the Ecological approach can be seen in other parts of Latin America too, including its southern cone. For example, in September 1997, the northeastern province of Entre Rios ("Between Rivers") of Argentina passed the "Anti Dam Law" that declared the province free of any new dams on the Parana and Uruguay Rivers. The law was prompted by the proposed massive Parana Medio Dam that would have created the world's second largest reservoir, flooding about 760 sq km of wetlands, islands, and forests.[53]

As noted, Brazil has a crucial role to play in promoting the Ecological approach in Latin America. There is indeed a strong civic movement in Brazil for the Ecological approach. It will be necessary for this movement to prevail and gain further strength in the coming years.[54]

8.8 Ecological approach in Africa

The Ecological approach is also spreading in Africa, the home of the world's longest river, the Nile and other powerful rivers, such as the Congo River, the second largest in the world in terms of volume of flow. It is therefore important that the Ecological approach makes progress in Africa, and it is encouraging to see some evidence of this happening.

As noted in chapter 3, dams and barrages have caused serious damage to the continent's floodplain wetlands by eliminating seasonal flooding. Several African countries are now trying to reverse this damage through artificial flood releases from reservoirs as a new method of wetland management.

For example in Nigeria, flood releases from the Tiga and Challawa Gorge Dams have been used in an effort to restore the Komadugu-Yobe basin's ecosystems, with positive results, indicating that long-term recovery may be achieved if the process is continued.[55] Similarly, it has been shown that damages done to the floodplains of the Senegal River below Manantali Dam can be reversed through additional releases from the reservoir that would be possible if the hydropower targets of the dam can be lowered by a small amount. This could happen if more social forces rallied behind the Ecological approach to rivers and overcame the political power of dam supervisors who serve primarily the interests of the elite who benefit from the additional electricity at the expense of the farmers and fisherfolk downstream.[56]

Similarly, researchers have shown that the damage done to the Zambezi river floodplains and its delta can be mitigated by increased releases from the Cahora Bassa Dam with little loss of its electricity generation capacity. Such releases could improve the quality of wildlife grazing lands and restore floodplain fisheries. Modest flow releases, sufficient to spread water across the river's floodplains and timed to coincide with the local flood-based cropping pattern, can go a long way toward reviving the Zambezi Delta. Furthermore, releases early in the rainy season can create greater space in the reservoir to hold water and thus help avoid bigger floods like those witnessed in 2001.[57] However, these changes require greater influence of the Ecological approach on dam operation decisions.

The above steps toward releasing more reservoir water for downstream stretches of rivers to repair damage caused by dams are certainly welcome, and they reflect the Ecological approach. However, as noted in chapter 2, most African countries remain enamored with the Commercial approach and are busy building dams and barrages, often with foreign assistance. However, it may be hoped that with time African countries too will see the merits of the Ecological approach and switch to it in a more comprehensive way.

8.9 Conclusions

After several centuries of domination by the Commercial approach, the Ecological approach is now gradually spreading across the world. Progress is more visible in developed countries, where new construction of dams and barrages is now rare. Instead, many existing dams are being decommissioned and demolished, freeing up rivers. This, in turn, has allowed the restoration of rivers and river basin ecology, in some cases leading to return of fish previously prevented by the dams. Many deltas that were desiccated due to over-abstraction of river flows are now witnessing efforts to revive them through restoration of baseline and pulse flows. In Europe, the progress toward the Ecological approach has been more comprehensive, promoted by, for example, the EU Directive on Waters, issued in 2000, declaring water as "not a commercial good."

The Ecological approach is progressing in developing countries too. In many of them, major dam projects have been cancelled or postponed under pressure from peoples' movements grounded on ecological considerations. In other cases, people are still protesting plans for further interventions in rivers. In many cases, groups have come together to ask for demolition of dams and barrages that have already been built.

Some developing countries have also taken steps that are more advanced than those taken by developed countries. A prominent example in this regard is the inclusion of the rights of nature in Ecuador's constitution of 2008. Several other developing countries, such as Bangladesh and Colombia, have declared rivers as "living entities" with the legal standing of a person. These steps restored to rivers the status of subject—a status they used to have during pre-industrial times. In some countries, the return of rivers to the subject status (from being objects) has been facilitated by cultures and traditions that have endured from pre-industrial times.

Despite the progress above, the Ecological approach is yet to become the dominant approach and accepted as official policy in most developing countries, and in many developed countries. However, the adverse consequences of the Commercial approach are becoming more clear with time, and climate change is aggravating them further. It is hoped that growing evidence of the incompatibility of the Commercial approach with sustainable development will persuade more people and governments to switch to the Ecological approach in the coming years.

Notes

1. See McCully (2001, p. xxvi). These data exclude most large dams in China. As WCD (2000, p. 9) put it, "the decline in the pace of dam building over the past two decades has been equally dramatic, especially in North America and Europe where most technically attractive sites are already developed."
2. The Cuyahoga River fire incident in Ohio in 1969 made it amply clear that rivers in the United States had been polluted enough. The Clean Water Act (1972, 1977) was

in a sense a result of this realization and signified the beginning of a new phase of US treatment of its rivers and other surface waterbodies.

3. As a result of these activities, rivers and other surface waterbodies in the United States are now cleaner than they were before.

4. As WCD notes (2000, p. 10), "Momentum for river restoration is accelerating in many countries, especially in the United States, where nearly 500 dams, mainly relatively old, small dams have been decommissioned. Since 1998, the decommissioning rate for large dams has overtaken the rate of construction in the US. Experience in North America and in Europe shows that decommissioning dams has enabled the restoration of fisheries and riverine ecological processes."

5. See McPhee (1989, 1998) for details.

6. It may be noted that UNESCO considers areas for designation as Biosphere Reserves only after the nation in which the site is located submits a nomination, and nations retain their sovereign jurisdiction over these areas after the designation.

7. This arrangement is, apparently, also aimed at increasing the efficiency with which the water can be used. In addition to facilitating renovation of irrigation canals in the Mexicali Valley to reduce leakage, it will allow release of 56 million cubic m of water to the delta on average per year.

8. For details, see Bosshard (2007, p. 23).

9. The set of criteria for classification of rivers into high status concerns (i) biological elements (such as composition and abundance of aquatic flora; composition and abundance of benthic invertebrate fauna; composition, abundance and age structure of fish fauna); (ii) hydromorphological elements supporting the biological elements (such as hydrological regime, meaning quantity and dynamics of water flow and connection to groundwater bodies); river continuity; morphological conditions (river depth and width variation, structure and substrate of the riverbed, as well as structure of the riparian zone); chemical and geophysical elements supporting the biological elements (such as thermal conditions, oxygenation conditions, salinity, acidification status, nutrient conditions); specific pollutants (EU 2000, pp. 34–35).

10. The WFD asks all member states to identify all waterbodies used for the abstraction of water intended for human consumption providing more than 10 cubic m a day on average or serving more than 50 persons and those waterbodies intended for such future use (EU 2000, p. 12). It calls for characterization of surface waterbodies, including rivers, noting altitude; latitude; size; catchment area; geology; energy of flow; mean water width, depth, and slope; form and shape; valley shape; river discharge category; transport of solids; acid neutralizing capacity; mean substratum composition; chloride; air temperature mean and range; and precipitation (pp. 23–24). It calls for establishment of type-specific reference conditions for surface waterbody types and identification of pressures, including

   a) Estimation and identification of significant water abstraction for urban, industrial, agricultural, and other uses;

   b) Estimation and identification of the impact of significant water flow regulation, including water transfer and diversion, on overall flow characteristics and water balances;

    c) Identification of significant morphological alterations to water bodies;

    d) Estimation and identification of other significant anthropogenic impact on the status of surface waters;

    e) Estimation of land-use patterns, including identification of the main urban, industrial, and agricultural areas, and, where relevant, fisheries and forests. (p. 28)

11. See EU (2000, p. 4 and p. 31).

12. As per the WFD, the information to be included in RBMPs for international river districts should cover (i) a general description of the characteristics of the river basin district; (ii) a summary of significant pressures and impacts of human activity on the status of surface water and groundwater; (iii) identification and mapping of protected areas; (iv) mapping of the monitoring networks; (v) list of environmental objectives; (vi) a summary of economic analysis of water use; (vii) summary of programs adopted under Article 11; (viii) a register of any more detailed programs and management plans for the river basin district dealing with particular sub-basins, sectors, issues or water types, together with a summary of their contents; (ix) a summary of public information and consultation measures taken, their results, and the changes to the plan made as a consequence; (x) a list of competent authorities; (xi) the contact points and procedures for obtaining background information . . . , in particular details of the control measures, and of the actual monitoring data (EU 2000, pp. 66–67).

13. Earlier, the European Environment Agency's report, *Environment in the European Union—1995*, presented an updated state of the environment, confirming the need for action to protect Community waters in qualitative as well as quantitative terms (EU 2000, p. 1).

14. The Murray-Darling Basin Agreement was adopted in 1982, with its full legal status enacted in 1993. A number of new organizations were formed under this, commonly known as the Murray-Darling Initiative, including the Murray-Darling Basin Ministerial Council and Murray-Darling Basin Commission (MDBC). The basic realization that prompted all this was that the river system has been overexploited and diversion of water needs to be cut in order to restore the river system's ecological health.

15. This guide is considered to be the first part of a three-stage process to address the problems of the Murray–Darling Basin.

16. More than 90 percent of those who took part in the plebiscite in January 2000 said no to a flood control dam across the Yoshino estuary in the city of Tokushima. Though the result was not legally binding, a policy panel of the ruling Liberal Democratic Party recommended, following the outcome of the plebiscite, that the dam be scrapped and alternative flood management measures be considered (McCully 2001, p. lix).

17. See, for more details, McCully (2001, p. 21).

18. It is claimed that the project can feed about 20 million people, provide domestic and industrial water for about 30 million, create jobs for about 1 million, and generate peak electric power to satisfy high unmet power demand.

19. The civic groups that combined to form NBA included Gujarat-based Arch-Vahini (Action Research in Community Health and Development) and Narmada

Asargrastha Samiti (Committee for people affected by the Narmada Dam), Madhya Pradesh-based Narmada Ghati Nav Nirman Samiti (Committee for a New Life in the Narmada Valley) and Maharashtra-Based Narmada Dharangrastha Samiti (Committee for Narmada Dam-Affected People) who either believed in the need for fair rehabilitation plans for affected people or who vehemently opposed dam construction despite a resettlement policy. Several non-governmental organizations (NGOs), along with local people, professionals, and activists with a non-violent approach also joined NBA, which solidified in 1989 under the leadership of Medha Patkar, who gave up her PhD studies to devote herself fully to the cause of saving the Narmada River. NBA pressed for the Ecological approach to rivers and an alternative model of development. Internationally, it aimed at building pressure on the World Bank. NBA's slogans included "Vikas Chahiye, Vinash Nahin!" (Development wanted, not destruction!) and "Koi nahi hatega, bandh nahi banega!" (We won't move, the dam won't be built!). See McCully (2001) for further details on NBA.

20. Medha Patker and her colleagues had visited the project site in 1985 and found that the work was suspended due to objections from the Ministry of Environment and Forests. Upon investigation, she found out that the project was not approved by this ministry and therefore wondered how the World Bank could have financed it. Her conclusion was that the World Bank officials ignored the environmental and human consequences of the dam project (see McCully 2001 for further details).

21. Among the important personalities who joined or expressed support for NBA, was Baba Amte, well known for his work against leprosy, who published a booklet titled *O Beloved Narmada* in 1989, protesting against construction of dams. Another was Amir Khan, the famous film actor. Film maker Ali Azimi made the film *Narmada—A Valley Rises* in 1994, documenting the *Jan Vikas Sangharsh Yatra* (Struggle March for People's Development) that was held over a five-week long period (December 25, 1990 to January 31 of 1991). Anand Parwardhan made the award-winning documentary film *A Narmada Diary*. Among the politicians supporting NBA were Alok Agarwal, current member of the Aam Admi Party. More prominent was Nitish Kumar, then chief minister of Bihar, who participated in an NBA rally at Rajghat on the Narmada bank on September 16, 2016, and appealed to Prime Minister Narendra Modi not to close the gates of SSD and drown a quarter of a million people. He urged rehabilitation of people not by giving cash, but by giving them alternative land/employment. This shows that NBA included both groups that were opposed to the dam and those who wanted to mitigate and redress unfair consequences for the people to be displaced, by making sure that they received proper compensation and resettlement. The film *Drowned Out* (2002) depicted the process of drowning one of the tribes that refused to move by the rising Narmada reservoir water. Another early documentary film on the dam was *A Narmada Diary* (1995) that portrayed the efforts by the NBA to seek social and environmental justice for those most directly affected by the dam. NBA received the Right Livelihood Award in 1991. Arundhuti Roy, the Booker Prize winning novelist from India, reflecting on Narmada and dams in general, wrote, "Big Dams are to a Nation's 'Development' what Nuclear Bombs are to its Military Arsenal. They are both weapons of mass destruction. They're both weapons

Governments use to control their own people. Both are Twentieth Century emblems that mark a point in time when human intelligence has outstripped its own instinct for survival. They are both malignant indications of civilization turning upon itself. They represent the severing of the link, not just the link—the understanding—between human beings and the planet they live on. They scramble the intelligence that connects eggs to hens, milk to cows, food to forests, water to rivers, air to life and the earth to human existence" (Roy 1999).

22. In the final line of the verdict, the court enjoined the government to make every endeavor to ensure that the project was completed as expeditiously as possible.

23. As McCully (2001, pp. 301-306) notes, NBA put forward a different notion of development and argued that no matter how large are the benefits, they cannot outweigh the social and environmental costs.

24. For example, mountains have higher elevation while valleys have low. That does not mean that mountains need to be cut down and valleys have to be filled up. Both mountains and valleys have their unique ecologies and they both need to be appreciated and cherished.

25. The narrow passage between Bangladesh and Nepal that connects the rest of India with her seven northeastern states.

26. These prices are supposed to reflect the true value to society of the inputs and outputs of the project. For example, the true value to society of the land and forests consumed by the canals may be much higher than what would appear based on their current market prices. Similarly, the market price of the additional crop produced through expanded irrigation may be less than its economic price because of the loss of water during transportation through canals or because of the deterioration of soil quality caused by irrigation in arid regions. Economic cost-benefit analysis requires consideration of opportunity costs of the resources to be used up in the project.

27. Both financial and economic cost-benefit analyses are based on the assumption that the costs and benefits can be enumerated and quantified. However, the discussion here makes it clear that for such a project as IRLP, it is difficult to enumerate all the cost and benefit items, let alone quantify them. This is particularly the case on the cost side. It is difficult to quantify the human costs suffered by the displaced people, the ecological damages that river basins (particularly the ones from which water will be transferred) will suffer, and the losses that, for example, both India and Bangladesh will suffer from further strain on their relationship due to IRLP.

28. See Kumar (2016). The remaining quotes and points of information in this paragraph are also from this source.

29. See, *The Hindu*, February 27, 2017, http://www.thehindu.com/news/national/other-states/farakka-barrage-a-curse-for-bihar-say-experts/article17373696.ece. He further added that discussion of Farakka often focused on its engineering and technological aspects. However, more attention needed to be given to the environmental, cultural, natural, spiritual aspects.

30. Kumar also opposed the Allahabad-Haldia "Waterway 1" project, demanding further study of its rationale and viability.

31. According to Bhandari (2011, p.1), "The project bypassed the constitutional pro-
    vision requiring parliamentary ratification of any project or agreement in which
    resources or its by-products were to be divided between Nepal and India." Im this
    context, he noted that more than 60 percent of the Nepali population did not have
    electricity, and those who did had to suffer up to 18 hours of blackout daily in the dry
    months. Finally, he informed, "The combination of local and international pressure
    and strong arguments against the project from legal, human rights, environmental
    and economic perspectives were key to the success of the West Seti campaign (to
    cancel the project)."
32. Bhutan has drawn international attention by proposing the concept of GNH, as noted
    earlier; the Ecological approach to rivers therefore finds a more favorable philosoph-
    ical ground there. Partly as a result of this, in addition to the physical terrain-related
    constraints, many of the dams in Bhutan are of the run-of-river type, minimizing the
    impact on river channels and basins.
33. http://www.vasudha-foundation.org/wp-content/uploads/Final-Bhutan-Report_
    30th-Mar-2016.pdf.
34. See McCully (2001, p. lx).
35. See McCully (2001, p. lx).
36. For example, Mirza Asif Baig, chairman of the Indus Water Commission, argued for
    more dam construction, lamenting that the "precious resource" (monsoon water)
    is getting wasted. In his view, "if we can control water by investing in big dams we
    can overcome the electricity crisis and improve our agriculture (Ebrahim 2014)." He
    drew attention to the fact that the Indus basin irrigates about 14 million hectares of
    land in Pakistan –the largest irrigated area in the world—for which a huge amount
    of water is needed. With only two existing major reservoirs in the Indus basin—the
    Mangla and Tarbela—Pakistan's storage capacity is only about 30 days, while most
    developed countries have 1–2 years' water storage capability. Akhtar Ali of Asian
    Development Bank echoed this view arguing that "the increased storage capacity
    from building new reservoirs could store floodwater for productive use and lessen
    flood peaks downstream (Ebrahim 2014)." Ghulam Rasul, deputy director of the
    Pakistan Meteorological Department, noted that "Pakistan's per capita water availa-
    bility is presently estimated at 950 cubic meters, quite a dip from 5,500 cubic meters
    in 1951. . . . If we continue business as usual, how will we meet the water needs of a
    growing population?" he asked (Ebrahim 2014).
37. For example, Mushtaq Mirani, director of the Centre of Engineering and
    Development at the Mehran University of Engineering and Technology in Sindh
    province, questions the rationale of the Kalabagh dam proposal. He thinks that dam
    proponents are ignoring the hydrologic changes that dams may cause—dams can
    heavily modify the volume of water flowing downstream. He notes that most dam
    proponents come from Punjab province, while the people of Sindh and Pakhtunkhwa
    oppose the proposed 3,600 MW Kalabagh Dam, because it will give Punjab control
    over their water. He also notes that big reservoirs require huge investment, displace-
    ment of local people, and loss of rich biodiversity. He adds that big dams can be-
    come easy targets for militants, as the Mosul dam recently became in Iraq. Mirani

also draws attention to the wasteful use of irrigation water provided by dams, noting that about 60 percent of water is wasted because of the outdated canal and irrigation system. He notes that the practice of flood irrigation itself is a sure recipe for wastage (Ebrahim 2014). For an early critique of the Kalabahg Dam proposal, see Khan (1999).

38. It is important to note that the Ecological approach cannot be fully successful unless all the co-riparian countries adopt it. From this perspective, regional initiatives for promotion of the Ecological approach in all countries of South Asia are particularly important. The encouraging factor in this regard is that the common people of South Asia are more receptive to the message of the Ecological approach. By contrast, it is genrally the elite of these countries, who, imbibing the ideas of the Commercial approach from the industrial west, imposes this approach from above.

39. See https://sandrp.in/. Being a large country with many states and many languages, India has many local and state level river movements. Not all of them are known at the national and international levels, NBA being an exception in this regard. There are however efforts to build a network of such movements so that these could find strength from each other and have a larger combined effect on national and international policy making.

40. See Fawthrop (2019).

41. The Kachin Independence Army have issued several warnings since 2006 against the plan to build the dams.

42. President U. Thein Sein wrote to the parliament: "Since the first phase of the Ayeyawady Myitsone hydropower project was implemented with investment from the People's Republic of China, we noted that there arose the following public concerns about the Myitsone project:

    a) Natural beauties of Myitsone, the gift of nature and a landmark not only for Kachin state but also for Myanmar may disappear;

    b) Possible loss of livelihood of national races and villages due to inundation at the upstream of the river;

    c) Commercially grown rubber and teak plantations which are heavily invested by private entrepreneurs may be destroyed;

    d) Melting ice from snow-capped mountains at the far north triggered by climate change, torrential rains or severe earthquake may destroy Myitsone Dam, claiming lives and property of the people in towns and villages downstream of the dam; and

    e) There may be devastating effect on the Ayeyawady River.

    As our government is elected by the people, it is to respect the people's will. We have the responsibility to address public concerns in all seriousness. So, construction of Myitsone Dam will be suspended in the time of our government." (Thein Sein 2011)

43. See https://thediplomat.com/2019/03/myanmars-myitsone-dam-dilemma/.

44. Efforts have included commissioning glossy flyers and short films, staging high profile visits to the resettlement villages, and offering training for Burmese dam engineers abroad.

45. According to press reports, following the visit, the website, favorable to the construction company SPIC, claimed that Kachin leaders had a positive attitude to the dam, even though they all had opposed the dam at the meeting. Accordingly to this website, Kachin political leaders and social organizations had a positive attitude toward the Myitsone Dam, and that the local people of the Kachin State do not oppose the Myitsone hydropower project. Others however have disputed this account. See Fawthrop (2019).

46. In June 2011, Tarpein Dam, developed by Datang Corporation, became the focus of fighting by rebel Kachin forces. In addition to local ethnic opposition, civil society and non-governmental organization activists based in Rangoon (Yangon) were able to build almost unanimous national opposition to the dam. At the same time, it should be noted that, while the construction of the Myitsone Dam could be suspended, this was not the case with the other seven dams of the proposed eight-dam project. This shows that the opposition to the Myitsone Dam was motivated more by the emotion attached to the proposed site of the dam than by a general realization of the adverse effects of dams on river morphology and river basins.

47. There was however some lack of clarity about the full implications of the decision, because the government also allowed the dam building company to continue "scheduled activities at the dam site, including the resettlement of affected villages (International Rivers 2012b)." Also, implementation of decisions regarding Xayaburi needed cooperation of neighboring Thailand, which—as noted—was financing and constructing the dam, and was supposed to be the main buyer of the electricity produced by the dam, and hence would also be impacted by these decisions.

48. See International Rivers (2012b).

49. One of these is the 260 MW Don Sahong Dam on the lower Mekong. Laos is going ahead with this dam project despite strong protests from the co-riparian countries of Kampuchea and Vietnam. Environmentalists have pointed out that the dam would block migratory fish routes and negatively affect the livelihoods and nutrition of large number of people across several countries. The controversial dam is being built by Malaysia's Mega First Corporation Berhad (Mega First) on Southeast Asia's key artery the Mekong River, just two km north of Kampuchea. Apparently, the 1995 Mekong Agreement, which led to the MRC's formation, was not of much help in influencing the decision regarding the dam. According to observers, the consultation process under the Mekong Agreement is "neither a right to veto the use nor unilateral right to use water by any riparian without taking into account's other riparian's rights." See: https://www.rfa.org/english/news/laos/dam-12242014141437.html.

50. See McCully (2001, p. lx).

51. See Bosshard (2007, p. 19).

52. See Bosshard (2007, p. 19).

53. See McCully (2001, p. lix).

54. Brazil's pro-environment position seems to have weakened in recent years, with the new government relaxing restrictions on development in the Amazon forests and planning to approve dam projects proposed earlier. There was also some talk about Brazil's withdrawal from the Paris Agreement on climate change. It remains to be

seen whether pro-environment forces of Brazil can prevent this retreat from the pursuit of environmental goals.

55. See IRN (2007, p. 19).

56. See IRN (2007, p. 19).

57. See for details Beilfuss (1999, 2012), who informs that prescribed flood releases from the Cahora Bassa Dam could provide millions of dollars in benefits from increased shrimp production, improved quality of wildlife grazing lands, restored floodplain fisheries, and other social and ecological benefits. He notes further that changing the management of Cahora Bassa Dam would reduce impacts for bigger floods as well. Accoding to him, with little or no reduction in total hydropower generation, the dam could release a prescribed flood early in the wet season that would, on the one hand, have great benefits for people and wildlife downstream, and, on the other hand, would increase the available reservoir storage for incoming flows and thereby reduce the risk of large floods later in the flood season. He thinks that such flow releases will be of modest size, enough to spread floodwaters into the floodplains of the Zambezi Delta, and timed to enable flood-recession cropping systems, but not huge to cause large, damaging floods. Beilfuss cautions that extreme flooding events, like that of 2001, are inevitable, albeit infrequent, because even large dams such as the Cahora Bassa and Kariba cannot capture Zambezi flows fully in years when rainfall in the region is very high. For the Zambezi floodplain restoration experience, see also Zvomuya (2017).

9

Cordon approach to rivers

The lateral version of the Commercial approach

9.1 Introduction

The lateral version of the Commercial approach is the Cordon approach, as explained in chapter 1. It is based on the premise that rivers should always remain confined to their channels and not be allowed to overflow onto their floodplains. This approach assumes more importance in the middle and lower reaches of rivers, where they flow through floodplains and deltas.

Most regions experience seasonal fluctuations in precipitation. Consequently, rivers flowing through these regions have seasonal variation in flows, with high flows in peak seasons and low flows in lean seasons. In many such regions, rivers overflow their banks during peak seasons and spread over their floodplains. The Cordon approach strives to prevent these overflows, which it essentially considers as a nuisance, so to speak. To stop river overflows, it promotes construction of continuous embankments (or levees or dykes) along river channels. It is thought that by cordoning off floodplains from rivers, it will be possible to grow more crops, expand settlement, undertake more economic activities on them, as a result of which the commercial value of floodplains will increase. Thus, cordons also represent the pursuit of commercial goals through the imposition of human will on rivers, forcing them to remain confined to their channels.

However, as with frontal interventions, lateral interventions entail many undesirable and unintended consequences. River channels and their floodplains constitute one organic whole. Rivers give birth to floodplains by depositing sediment carried from upstream reaches. Inundation of floodplains during peak seasons is therefore a normal, expected, and desirable phenomenon. The frequency and extent of inundation differ across rivers, depending, in particular, on the seasonality of precipitation. However, inundation is how rivers nurture their floodplains, a process similar to mothers nurturing their babies. By trying to stop regular river inundation, the Cordon approach ends up harming both river channels and floodplains.

Construction of low height, earthen, embankments was practiced by pre-industrial societies. However, the technological might, acquired through the Industrial Revolution, allowed human societies to raise embankment

Rivers and Sustainable Development. S. Nazrul Islam, Oxford University Press (2020). © Oxford University Press.
DOI: 10.1093/oso/9780190079024.001.0001

construction to a new level, aimed at containing rivers within their channels. Industrial era embankments gradually spread across the world, beginning with developed countries and moving on to developing countries.

The Cordon approach had its initial successes. In many places, it led to increased crop production, expansion of settlement, and the building of commercial establishments, infrastructure, and even whole towns and cities. However, with time it became clear that the short-term commercial benefits of the Cordon approach do not necessarily match up with its broader and long-term adverse consequences. Among the latter are harmful morphological effects on river channels, including riverbed aggradation, often raising the elevation of riverbeds above adjoining floodplains, and thus creating the danger of floods, which become catastrophic because of the below-flood-level settlement that the Cordon approach encourages on floodplains. Cordons also involve switching from natural gravity flow irrigation to pump-based irrigation, which often proves to be costly, inadequate, and ill-suited. Waterlogging is another important consequence: while preventing river water from getting in, the Cordon approach prevents rainwater that falls on floodplains from getting out. The Cordon approach also pitches people inside cordons against those remaining outside, and thus becomes a source of conflict. It leads to deterioration and decay of inland waterbodies, harming capture fisheries, shrinking waterways, and reducing water supply during dry periods. More broadly, the Cordon approach causes damage to the flora, fauna, and ecology of floodplains.

This chapter presents the Cordon approach, describes its methods, reviews its spread across the world, and analyzes its consequences. It is organized as follows. Section 9.2 discusses the general relationship between river channels and their floodplains and explains the nurturing functions of regular river inundations. Section 9.3 discusses the instruments of the Cordon approach. Section 9.4 explains the relationship between the Cordon and Polder approaches and also presents a classification of cordons into different types. Section 9.5 discusses the consequences of the Cordon approach, distinguishing between consequences for river channels and for floodplains. Section 9.6 provides an overview of the experience of the Cordon approach in different parts of the world, focusing on the United States, Europe, and India. Sections 9.7 and 9.8 present two case studies of the Cordon approach, examining the experience of the Mississippi levee system and the Huang He River embankments. Section 9.5 concludes.

9.2 River channels and floodplains

Rivers have lives of their own. Each river is distinct from others. Even a particular river differs in its various stretches and in various seasons of the year. Rivers also change over time. In short, rivers are different and dynamic.

In general, rivers originate in mountains and hills and flow down toward seas and oceans, either by themselves or by merging with other rivers or branching out into other rivers. In hills and mountains, river channels are usually narrow, and they are fast paced because of the steeper slope. Once out of hills and mountains, rivers usually slow down in speed and expand in width.

In flowing through relatively flat and low-lying lands, rivers often create alluvial plains, by depositing sediment they bring from the hills and mountains. Most rivers have seasonal variation of flow, depending on local and regional precipitation and glacier melting patterns. During the peak season, rivers may overflow their banks and spread over adjoining areas. In the process, they create ridges along their banks where a considerable amount of sediment settles because of sudden slowdown in the velocity of flow and decrease in water depth. The remaining sediment is deposited further inside the lowlands, away from the river. During the lean season, river flows recede to their channels. This is the process through which rivers create floodplains. Rivers may create floodplains in their middle reaches. For example, the Ganges, Indus, Mississippi, Nile, and Huang He Rivers have all created large floodplains along their banks long before they enter their lower reaches and meet the seas.

In their lower reaches, when rivers are close to seas and oceans, they generally expand in width and slow down further in speed. Also, tidal flows now interact with river flows, redistributing sediment across tidal plains. Rivers that carry a large amount of sediment often form deltas at their mouth. Well-known deltas of the world include the Amazon (Brazil), Huang He (China), Mekong (Viet Nam), Mississippi (United States), Nile (Egypt), and Rhine (the Netherlands). The largest and most active is the Bengal Delta, spreading across Bangladesh and India and created through the combined flow of three major rivers, the Ganges, Brahmaputra, and Meghna. Deltas comprise both floodplains and tidal plains and are generally crisscrossed by rivers and their distributaries. We already noted some of these deltas in chapter 5.

9.2.1 Floodplain nurturing functions of rivers

Rivers give birth to floodplains and continue to nurture them through regular inundations. This nurturing function has many dimensions, some of which are noted briefly here.[1]

- *Maintaining floodplain elevation:* During inundation, rivers deposit sediment, including silt (the part of the sediment comprising finer particles), on floodplains, raising their elevation gradually. Without regular sediment deposition, floodplain elevation would generally decrease due to compaction, resulting in subsidence. The extent of sediment deposition depends on the sediment load and other physical characteristics of rivers and their floodplains.

- *Fertilization:* The silt brought to floodplains by river inundation also works as a natural fertilizer and generally helps to sustain the quality and productivity of the soil. It is largely due to this natural fertilization process that pre-industrial societies could maintain agricultural productivity on floodplains and deltas for millennia without requiring chemical fertilizers.
- *Irrigation:* River inundation serves as the natural form of irrigation. The overflow provides water to crops grown during its duration. The moisture retained by the soil, following inundation, helps to grow crops even in the lean season, when rivers recede to their channels.
- *Recharging of surface waterbodies and underground aquifers:* River inundation recharges surface waterbodies located in floodplains and thus helps them remain healthy. The water retained in these waterbodies can be used to meet varied requirements, including irrigation, during the dry season. Inundation also facilitates recharge of underground aquifers through seepage and percolation.[2]
- *Fisheries:* River inundation helps to sustain the open-or capture fish stock, maintaining its diversity.
- *Preservation of flora and fauna*: River inundation helps to preserve the flora and fauna of river basins. Aquatic diversity depends directly on inundation, while terrestrial flora and fauna also depend on water and moisture it provides. Many unique plant and animal species depend on regular inundation.
- *Waterways:* By recharging floodplain waterbodies, river inundation helps to preserve waterways and facilitate their use.
- *Moderation of temperature*: River inundation exerts a temperature moderation effect: recharged waterbodies help to moderate extreme heat in summer and extreme cold in winter.
- *Cleansing:* River inundation has a great cleansing effect on the overall physical environment of floodplains.
- *Aesthetics:* River inundation serves an aesthetic function through its role in preserving flora and fauna, recharging inland water bodies, and through its general cleansing effect and the watery landscape it creates during the peak season.
- *Socially fair distribution:* River inundation also facilitates socially fair distribution of resources. Many of the bio-physical benefits of inundation listed above can be shared by all people living on the floodplain. Those not owning cultivable land can benefit from open fisheries, which are generally treated as a common property resource. Similarly, waterways can provide a livelihood for many who lack other assets.

This list makes it clear that the health of floodplains depends crucially on the nurturing functions of river inundations. But river channels also depend on floodplains in several important ways. First, floodplains provide the space for rivers to spread their peak season flows, for which river channels do not have adequate holding capacity. Second, floodplains provide an additional passage for river water to proceed downstream and ultimately to the sea and so can be regarded as reserve channels of rivers. Third, floodplains provide the space for rivers to deposit the sediment they carry. Fourth, part of the water stored in floodplain surface waterbodies and in underground aquifers flows back to river channels during the lean season, providing greater balance of river flows across seasons and helping to ensure environmental flow throughout the year.

Thus, a river and its floodplains constitute one organic whole. Each is necessary for the other. Each is incomplete without the other. It is therefore wrong to view a river as consisting of its channel only. Floodplains are as much a part of the river as its channel. Floodplains are a creation of rivers and belong to them. Rivers can claim them when needed. Similarly, floodplains depend on river overflows. Given their mutual dependence, it is not surprising that the Cordon approach, by trying to separate floodplains from river channels forcibly, causes harm to both.

9.3 Instruments of the Cordon approach

The Cordon approach employs different instruments to insulate floodplains from river inundations. The main instruments are briefly noted in the following.

9.3.1 Embankments

The most important instrument of the Cordon approach is embankments, also called dykes (in Europe) and levees (in the United States). These are generally constructed using earth excavated from nearby land and reinforced through additional measures, such as revetments, made possible by industrial-era technologies. Embankments differ regarding their height and length, depending on the characteristics of the river that is to be contained. Embankments also differ regarding "set-back distance," which is the distance between the riverbank and the embankment. Looked at from the perspective of the river channel, the greater this distance, the greater the room for the river to expand during its high stage. On the other hand, looked at from the perspective of floodplains, as the set-back distance is reduced, more of the floodplain is cordoned off. Usually embankments with a lower set-back distance require greater height. Conversely, embankment height can be lower when the set-back distance is greater. Embankments therefore have two faces, the river-side face and the floodplain-side face (sometimes also called the country-side face). Depending on the characteristics of rivers and floodplains, embankments may need to be riveted, particularly on the river-side face. The achievements of the Industrial Revolution play an important role in the type of revetments that can now be deployed as compared to what was possible during pre-industrial era. The technicalities of embankment construction are the subject matter of a subdiscipline of civil engineering.

9.3.2 Floodwalls

Embankments are often accompanied by floodwalls, which are of two types. Along some stretches of rivers, particularly in urban areas, the banks are paved to become walls of concrete or brick: the paved banks of the River Seine in Paris or the Moscow River in Moscow, for example, are well known. Brick walls constructed on top of earthen embankments, in certain stretches of some rivers where the flood stage exceeds the height of the embankments, constitute the other type of floodwall.

9.3.3 Channelization

The cordoning off of floodplains through construction of embankments often involves channelization of rivers. The basic purpose of channelization is to straighten river channels, as much as possible. The idea is that straightening will improve the pace of the river flow, help navigation, and increase the area of cordoned off floodplains. River channels come in different types and shapes. In floodplains, rivers often have a meandering and braided pattern. Under channelization, the meander loops are often completely cut off to straighten up river channels. For braided rivers, the widths of river channels are often reduced by the cutting of secondary channels.

9.3.4 Canalization

Canalization is similar to construction of floodwalls along riverbanks. But with canalization, even the river bottom is paved, so that the river channel becomes a paved canal. A prominent example of this is the Los Angeles River, most of which (70 km out of its total length of 80 km) is now a canal, cemented on all sides, including the riverbed.[3]

As noted earlier, the main purpose of these instruments is to cordon off as much of the floodplain as possible by confining the river flow to as narrow and straightened a channel as possible. However, the Cordon approach creates many adverse consequences, as we shall see.

9.4 Cordon approach versus Polder approach and different types of cordons

9.4.1 Cordon approach versus Polder approach

In discussions regarding embankments, one often encounters the term polders. It is therefore useful at this stage to clarify the relationship between cordons and polders, on the one hand, and between the Cordon approach and the Polder approach, on the other.

Cordoning floodplains from river channels may or may not require construction of embankments on all sides. For example, Mississippi levees represent a quintessential application of the Cordon approach. However, these levees were not constructed all around the floodplains. Instead, these were constructed along one side of the floodplains (along the riverbank) only. The same is true of the Huang He River embankments of China, another important application of the Cordon approach. In fact, construction of embankments along only one side of floodplains is common with middle stretches of rivers. However, in lower stretches, particularly in deltas, embankments may need to be constructed on all sides to cordon off particular tracts of floodplains or tidal plains. In such cases, cordoned off areas are sometimes called polders. Thus the Polder approach implies construction of embankments on all sides to cordon off parts of the delta. This approach found its most prominent application in the Netherlands, where polders were constructed even to reclaim land from the sea. In fact, about one-third of the country's area currently comprises such reclaimed land.

However, it is clear from this discussion that the Cordon approach is not the same as the Polder approach. The latter is a particular variant of the former and generally applies to deltas, as we shall see in more detail in chapters 11 and 12. In other words, the Cordon approach subsumes the Polder approach. What applies for the former also applies generally for the latter, though not the converse. Also, in the following, I use the word cordon to refer to both the embankments used to insulate floodplains from river channels as well as the areas that are insulated. The expression cordon, when used in the latter sense, is a broader category than polder. In other words, polders are a particular type of cordons; what applies to the former also applies to the latter, though not the converse.

9.4.2 Different types of cordons

Cordons vary widely and may be classified from different viewpoints (Table 9.1). For example, from the viewpoint of the land-use pattern, cordons may be classified into rural and urban. From the viewpoint of extent of closure, they may be classified into full and partial. From the viewpoint of whether they face the sea/ocean, cordons may also be classified into inland and coastal or sea/ocean facing.

Rural cordons encircle rural areas and are mainly aimed at preventing flood and providing pumped irrigation. Their main purpose is to help raise agricultural output. Urban cordons, on the other hand, encircle urban areas, and their main aim is to protect houses and other structures from inundation. However, what began as a rural cordon can over time become an urban cordon, if sufficient settlement takes place inside the embankments. At intermediate points of time during this conversion, the cordon may acquire a mixed character, with part of it remaining rural, while the remaining becoming urbanized.

Coastal cordons encircle areas located in coastal regions. Often these areas are rural, so that coastal cordons can be a type of rural cordon. However, the distinctive mark of these cordons is that protection of encircled areas from tidal saline water is one their important goals.

The distinction between full and partial cordons depends on the degree of insulation. This distinction is not the same as that between cordons resulting from embankments on one side and polders, which result from embankments on all sides. The full/partial distinction refers to the purpose of the embankments, namely whether they aim at sealing off cordoned areas completely or partially. As we shall see, some cordons do not aim at sealing off completely and instead leave openings for water to flow either in one direction or in both directions. On the other hand, sometimes cordons aim at complete sealing off, with no openings at all. We call the former partial cordons and the latter, full. Needless to say, full cordons are the classical cordons and embody the spirit of the Cordon approach to its fullest extent, whereas partial cordons represent a compromise. It may be noted in this context that openings of partial cordons are generally supposed to be regulated. However, as we shall see, these regulators often malfunction and

Table 9.1 Classification of cordons

Classification criterion	Types	
Land-use pattern inside cordon	Rural	Urban
Extent of closure of cordons	Full	Partial
Whether it faces the sea/ocean	Inland	Coastal

Source: Author

sometimes even cease to function, so that intended partial cordons end up being actual full cordons.

9.5 Consequences of the Cordon approach

Rural cordons, as noted earlier, are generally expected to raise agricultural output by increasing the cropping intensity and by facilitating the switch to high-yielding varieties of crops that require controlled irrigation, together with the use of chemical fertilizers and pesticides. The urban cordons, on the other hand, are expected to facilitate construction of houses, infrastructure, and commercial establishments inside the cordons. Thus, both rural and urban cordons are expected to raise the commercial value of the cordoned floodplains. Many cordons do meet some of these expectations, at least initially. However, with time, many undesirable and harmful effects surface, undercutting the benefits and putting into question the very rationale of cordons. The following provides a brief review of some of these adverse effects.

9.5.1 Effects on river channels

9.5.1.1 Effects of embankments
Just as frontal interventions do, lateral interventions inspired by the Cordon approach also have adverse effects on the morphology of rivers. Cordon structures are generally built for particular stretches of rivers. The morphological effects can therefore be distinguished into three parts: (i) what happens upstream; (ii) what happens to the stretches that are confined; and (iii) what happens downstream. These effects are difficult to generalize, because they depend on the characteristics of the riverbed, floodplains, volume and type of sediment carried, seasonality, set back distance, etc., and on whether rivers are embanked on one-side or on both sides. However, some observations, as in the following discussion, can be made.

The upstream effects are easier to predict. There, cordons generally have, what is often called, the backwater effect. As cordons restrict river passage width, river water backs up and increases the lateral pressure on riverbanks upstream, leading to bank erosion and widening of the channel. Thus, the backwater effect may also lead to the "funnel effect." Also, as cordons force the pace of upstream flow to slow down, more sediment settles on the riverbed, causing upstream riverbed aggradation and adding to the funneling effect.

The effect on river morphology in the embanked stretches is more difficult to pin down, because it is the outcome of a more complicated process of fluid dynamics and depends on the specific conditions of a river, including the texture of the riverbed and riverbanks, sediment volume and type, etc. In general, Cordons force a river to flow through a more confined channel. On the one hand, this may lead to faster speed and more scouring and erosion, as the river tries to keep the volume of the channel intact by increasing its depth. This is possible particularly where riverbeds consist of soft materials that can be easily scoured. On the other hand, cordons may create a congestion effect, slowing down the river and increasing the amount of sediment collected on the riverbed, raising its elevation. This process of riverbed aggradation, accompanied by its counterpart process of floodplain elevation loss (to be discussed later), leads to dangerous situations where the riverbed is higher than the adjoining floodplain, creating so-called elevated, suspended, or flying rivers. Aggradation of riverbeds in embanked stretches is quite common.

What happens downstream, as the river emerges from the confines of the embankments, depends to some extent on what happens within the embanked stretch. The river tries to widen and regain its natural width and configuration (regarding meandering and braiding). It also slows down (assuming that its pace increased in the embanked stretch). If the cordons led to more scouring and increased the sediment volume, the river will deposit some of that sediment as it emerges from their confines.[4]

9.5.1.2 Effects of channelization

Channelization also has harmful effects on river morphology and river basins. The most obvious is the loss of wetlands. The slow-moving water in the meanders and secondary channels, and the marshes and wetlands that often accompany them, serve several important purposes.[5] One is to serve as the habitat of many aquatic and terrestrial species. Second, they serve as filter for surface water sources. Third, they allow adequate percolation and hence help regeneration of groundwater tables. Once rivers are channelized, meanders, secondary channels, marshes, and wetlands are disconnected from the rivers and fail to perform their natural role as parts of the river systems. Instead, they become pools of stagnant water and ultimately often dry up. Second, the straightening resulting from channelization often increases the pace of water flow and leads to more erosion. Third, by allowing larger volume of water to flow at a more rapid pace, channelization often causes flooding at points where the channelization ends. Fourth, by reducing the river width, cutting off meanders and subsidiary channels, etc., channelization also harms capture fish stock.[6]

9.5.1.3 Effects of canalization

Canalization, as noted earlier, robs a river of its natural properties. With the river channel converted into a canal that is cemented all around, there is nothing left of its natural morphology. Whether or not it will increase the speed of the water flow depends on the width and depth of the canal relative to that of the original river. Canalization may prove self-defeating too, especially when canalization slows down the river, so that more sediment is deposited on the bottom, nullifying some of the goals that canalization was supposed to achieve.[7]

9.5.2 Effects on floodplains

9.5.2.1 Below-flood-level settlement

One of the important consequences of cordons is encouragement of below-flood-level settlement. In the absence of cordons, people generally remain aware of the possible extent and height of annual inundation and hence take these into consideration in deciding where and how to construct houses and structures. They therefore either limit such construction to lands that are above the flood level or follow the dig-elevate-dwell principle of settlement on floodplains. Under this principle, people dig ponds and other waterbodies and use the earth therefrom to raise the elevation of the lands before constructing houses and structures on them. Another option for settlements on floodplains is to design houses and structures in such a way as would allow these to stay above the flood level. For example, houses may be built on pillars and roads may be constructed as causeways. In short, without cordons, dwellings and structures would generally have been either above the flood level or flood proof. By contrast, cordons create an artificial and false sense of dryness. People then tend to forget about the flood level and start constructing houses and structures below flood level. This happens in both rural and urban cordons.

9.5.2.2 Subsidence of floodplains

Cordons contribute to subsidence of floodplains by preventing sediment from being deposited on them. Inundation, as noted earlier, brings not only water but sediment too. By preventing inundation, cordons prevent the annual sedimentation process that is necessary to sustain and raise the elevation of floodplains. In addition to loss of sedimentation, floodplains also experience compaction. As a result, cordons often cause subsidence of floodplains.

9.5.2.3 Danger of catastrophic flooding

Aggradation of the riverbed and subsidence of the floodplains, on the one hand, and expansion of below-flood-level settlement, on the other, create the conditions for catastrophic flooding. In the event of any failure of embankments, floodplains are inundated with water to a higher level than would have been the case if the embankments had never been built.[8] The below-flood-level houses and structures that were built get completely submerged, causing huge damage to life and property. Thus, while cordons can prevent regular flooding, they create the danger of catastrophic flooding, of deluge, and greater damage.

Overall, evidence shows that building cordons is an ineffective strategy for dealing with floods. First, the average annual number of major floods occurring around the word has increased from between seven and nine during the period from 1950s to 1970s to 20 in the 1980s and 34 in the 1990s.[9] Second, the cost of damages from floods has also increased over time.

9.5.2.4 "From lower to higher embankment!" vicious cycle

The Cordon approach leads to the "Lower to higher embankment!" vicious cycle. This happens due to the path dependence of the consequences of the Cordon approach and responses to them. As noted earlier, cordons promote below-flood-level-settlement. People invest in constructing houses, structures, and other establishments and create a life around them. They adopt agricultural practices and initiate other economic activities, assuming continued dryness. Consequently, when the danger of catastrophic flooding arises, and deluges actually occur or threaten to occur, their natural response is to protect these investments. They therefore demand embankments be made taller and stronger. Thus, a vicious cycle develops, whereby initial embankments lead to the demand for higher, stronger, and longer embankments (Figure 9.1). One consequence of this process is that the cost of embankment maintenance rises with time.[10] Often, the necessary money proves to be long in coming and perhaps short when it arrives, causing embankments to deteriorate, thus increasing the danger of catastrophic flooding further and also the demands on the maintenance budget. Thus, a subsidiary vicious cycle regarding maintenance expenditure also develops (Figure 9.2).

9.5.2.5 Increased flooding outside cordons – Transfer flooding

An important consequence of cordons is increased flooding in parts of floodplains that are not cordoned off. (For brevity I call these parts non-cordon areas). This follows from simple arithmetic. Cordons cannot reduce the total volume of river flows. All they can do is alter the spatial distribution of these flows, as they eventually move to the sea. Failing to spread out onto cordoned off areas, the river flow has two options. One is to remain constricted to river channels, thus raising the river stage. The other is to spread out to non-cordon areas. The first increases the risk of catastrophic flooding, and the second increases flooding in non-cordon areas. A particular form of this phenomenon is the backwater effect (mentioned earlier), whereby flooding spreads to upstream reaches, which are outside cordons.[11] However, aggravation of flooding in non-cordon areas is more general and applies to all stretches of rivers. Enhanced flooding in non-cordon areas, caused by embankments, is sometimes called

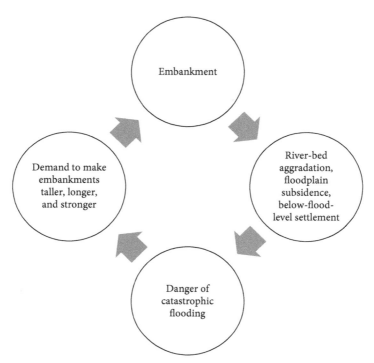

Figure 9.1 "From lower to higher embankment!" vicious cycle
Source: Author

"transferred flooding," and is generally more acute with rivers embanked on one-side and particular stretches only.

9.5.2.6 Conflicts among co-riparian communities

A particularly unwarranted consequence of cordons is generation and aggravation of conflicts among co-riparian communities. As noted earlier, while people inside cordons "enjoy" artificial dryness, people outside cordons face increased flooding. The latter therefore feel themselves as victims of cordons. Driven by these grievances, non-cordon people often take it upon themselves to cut cordons so that river water can spread out more evenly. However, resistance by cordon people to such acts often leads to bloody encounters. Thus, cordons create divergence of interests among co-riparian communities of the same river basin and pitch them against each other. In situations where cordon and non-cordon people belong to different ethnicities, religions, etc., these conflicts acquire more ominous characteristics. Cordon vs. non-cordon conflicts often add to urban-rural conflicts because urban areas are generally considered to be more deserving

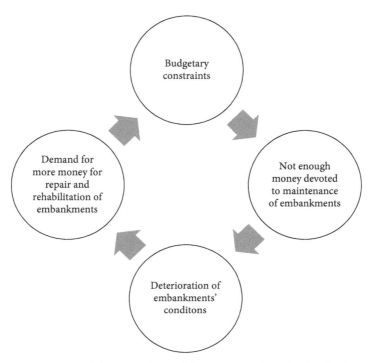

Figure 9.2 Vicious cycle between demand for money and repair of embankments
Source: Author

of being cordoned, leaving surrounding rural areas as non-cordons and thereby unprotected from flooding. In such situations, river policies add to the discrimination that rural people generally face vis-à-vis the urban population.[12]

9.5.2.7 Substitution of gravity-based drainage and irrigation by pump-based drainage and irrigation

Cordons require introduction of artificial, pump-based drainage and irrigation, replacing the previous natural, gravity-based drainage and irrigation. As noted earlier, embankments prevent river water from coming in and providing natural irrigation. As a result, river water now has to be pumped across the embankments and then channeled to fields through systems of irrigation canals that often need to be newly created. Needless to say, pumped irrigation proves costlier than gravity irrigation.

At the same time, embankments also prevent rainwater falling inside cordons from moving out to rivers. Thus, cordons disrupt the natural, gravity-based drainage system too and require its replacement by the artificial, pump-based drainage system. For this purpose, an additional system of canals, commonly known as drainage canals, needs to be constructed inside cordons in order to collect the rainwater and bring it to the pump heads. Needless to say, pumped drainage also proves more costly than natural, gravity-based drainage. Clearly, the irrigation issue is more important for rural cordons, while the drainage issue applies to all cordons.

9.5.2.8 Waterlogging inside cordons

Cordons often end up creating new problems for people living inside them. One of these is waterlogging. The substitution of gravity drainage by pumped drainage often proves inadequate due to a variety of technical and financial reasons. So far as pumps are concerned, it often proves difficult to ensure adequate pumping capacity. This is particularly true where precipitation is highly seasonal. In such areas, it becomes necessary to install a large pumping capacity, which then remains idle for most of the year. Even when a large pumping capacity is installed, precipitation may exceed the projected level. Also, it is common for some of the pumps to malfunction and become inoperative, because of inadequate maintenance and other reasons, giving rise to a vicious cycle similar to that depicted in Figure 9.2. Finally, the power supply needed to run the pumps often proves either inadequate or costly. So far as canals are concerned, it generally proves difficult to maintain two systems of canals in proper conditions, for many technical and financial reasons. This is particularly true for the system of drainage canals, the demand for which is less sharp than that for irrigation canals. The drainage canals therefore often fall into disuse, are encroached upon and filled up, leaving no route for rainwater to reach pump heads. A combination of these problems often results in waterlogging inside cordons,[13] becoming perennial in many places. Waterlogging is a particularly serious problem for urban cordons.

9.5.2.9 Toxification of soil and water

One of the arguments for replacing gravity-irrigation by pump-irrigation is the greater certainty that the latter provides regarding volume and timing of irrigation, both of which are somewhat uncertain in case of the former. It is this certainty that allows farmers inside cordons to switch from traditional varieties to high-yielding varieties (HYV) of crops that require controlled irrigation. HYVs however generally also require chemical fertilizers and pesticides. Thus, construction of cordons and replacement of gravity irrigation by pump irrigation have gone hand-in-hand with a rise in the use of chemical fertilizers and pesticides. This in turn leads to toxicity of soil and groundwater. The problem is aggravated by the fact that cordons also imply the end of the natural fertilization of floodplains by silt brought by river water. As a result, soil quality deteriorates, various mineral deficiencies become acute, dependence on chemical inputs increases over time, and the run-off becomes more toxic.

9.5.2.10 Adverse effects on surface waterbodies
By preventing river water from entering cordons, embankments harm waterbodies inside the cordons. As noted earlier, one of the beneficial effects of regular river inundation is recharge of the floodplain waterbodies, such as ponds, lakes, canals, marshes, and other wetlands. These recharged waterbodies usually serve as an important source of water during the dry season, for both household as well as agricultural purposes. Without the annual recharge, these waterbodies decay over time and often dry up and cease to be a source of water during the dry season.

9.5.2.11 Adverse effects on fisheries
Adverse effects on waterbodies imply adverse effects on capture fisheries. Many indigenous fish species disappear over time. Though farm fisheries can compensate for the volume of fish output, they usually focus on a few particular fish species, often alien ones, and hence cannot prevent the loss of fish and aquatic biodiversity.

9.5.2.12 Adverse effects on waterways
The decay of surface waterbodies also implies loss of length, navigability, and importance of waterways over time. Water-based transportation, though slower in speed, generally costs less and has many important advantages, particularly in shipping bulky agricultural produce. Water transportation is also generally less polluting.

9.5.2.13 Adverse effects on underground waterbodies
By preventing river water from entering cordons, embankments also reduce the rate of recharge of groundwater bodies. This rate depends on geological conditions and varies across floodplains. However, there is no doubt that it decreases when river water fails to spread across floodplain. Less recharge of the groundwater table reduces availability of groundwater during the lean season. It also reduces subsurface water flow back to rivers, which is important for maintaining environmental flow during the lean season.

9.5.2.14 Adverse effects on flora and fauna
By putting an end to regular river inundation, the Cordon approach adversely affects floodplain flora and fauna and overall ecology. Many plant and aquatic species depend on the impulse they receive from river overflows during the peak seasons. The general loss of moisture in the floodplains lead to important and unwarranted changes in vegetation.

9.5.2.15 Adverse effects on cleansing, hygiene, and aesthetics

An end to regular inundation also implies an end to the annual cleansing effect, along with its positive impact on health and the environment. Damage to flora and fauna, noted above, and loss of the cleansing effect harm the beauty of the landscape, particularly of rural areas. The resulting damaging impact on aesthetics is also an important negative accompaniment of the Cordon approach.

Overall, cordons not only hamper the beneficial functions of river inundation, they create many new problems.[14] Cordons therefore often prove to be a lose-lose proposition and a waste of valuable resources, as is illustrated by the experience of cordon projects around the world. Before examining these effects in the context of several renowned cordon projects, I review briefly the general expansion of the Cordon approach across the world.

9.6 Cordon approach across the world

The Cordon approach, as with the Commercial approach in general, found its vigorous application initially in developed countries and then spread to developing countries. In some cases, the Cordon approach was practiced by developing countries in earlier times, as we shall see in the case of the Huang He River of China.

9.6.1 United States

The Cordon approach has been implemented in the United States with much vigor, and as a result, an extensive levee system has been constructed in its major river valleys. In many cases, US levee systems have been part of greater schemes that include construction of dams and barrages, so that the frontal and lateral versions of the Commercial approach were combined to tame and utilize particular rivers as economic resources.[15]

According to some estimates, about 160,000 km of levees—including coastal levees—have been constructed in the United States, so that 43 percent of its people live in a county with at least one levee.[16] Levees along major rivers in the United States are estimated to have a total length of about 40,000 km.[17] The most important of the US levee systems is that of the Mississippi River, which will be discussed later. Other major levee systems include those constructed for the Sacramento and San Joaquin Rivers in California and the Kissimmee River in Florida. Apart from river-specific organizations, such as the TVA, the US Army

Corps of Engineers (USACE) and the BuRec have played an important role in implementing the Cordon approach in the United States.[18]

In many cases, the application of the Cordon approach in the United States involved channelization and canalization of rivers. Thus, meander bends have been cut, secondary channels have been severed, and similar other measures have been taken to "improve" the shape of the rivers, make them run faster, and shorten the route to the sea. Much of the channelization in the United States was done by the USACE. In west Tennessee, almost every river has been channelized, either partially or completely. Considerable channelization has been done on the Mississippi River, which has lost more than 300 km of its original length due to the cutting-off of its meander bends.[19] As noted earlier, there has been canalization of rivers in the United States too, the most prominent example being the canalization of the Los Angeles River.

The experience of levees, channelization, and canalization in the United States illustrate the consequences of the Cordon approach described earlier. Almost everywhere, levees have promoted below-flood-level settlement, led to aggradation of riverbeds, created the danger of catastrophic flooding, and caused enormous damages when the levees actually failed. Thus, despite huge progress in levee construction, the value of flood damages in the United States has increased over time. It is estimated that the US government spent more than $123 billion on flood control schemes since the 1920s, and yet the average annual inflation-adjusted cost of flood damage has tripled to $6 billion. (This is without taking into account damage caused by Hurricane Katrina, estimated to be at least of $100 billion.[20]) Using another benchmark, the inflation-adjusted cost of flood damages in the United States has doubled since the Flood Control Act of 1937.[21] The number of people killed by floods has also not decreased.

The main reason for the rising cost of flood damage has been the rise in below-flood-level-settlement that levees have encouraged, as noted earlier. Many high-value dwellings and structures have been built, and the depth of the flooding has been higher, as river water had a smaller area to spread over. In face of the disappointing figures on flood damages, the USACE try to justify their structures by arguing that the flood damage would have been greater ($19 billion a year) if these had not been built. These claims however do not take into account the fact that had the flood control structures not been built, people would not have constructed high-value structures in flood-prone areas and then suffered huge damages when levees failed.[22]

As a result of channelization and other river-constricting measures, it is estimated that about 87 million hectares of wetlands, mostly freshwater marshes, have been destroyed in the United States since colonial times. As a result, less than 100 million hectares of wetlands remain.[23] The harmful effects of channelization can be seen prominently in the case of Florida's Kissimmee River, where it has led to considerable loss of wetlands, soil erosion, and downstream flooding, as larger volumes of water now reach more rapidly choke points, which were not designed to hold so much water arriving almost all at once. The channelization of the Kissimmee River has also led to declines in river fish populations. Similarly, studies have shown that channelization of Chariton River in northern Missouri has resulted in about a 30 percent loss in the number of fish species and about an 80 percent loss in the biomass of the fish caught. This has been the result of reduction of habitat, elimination of riffles and pools, greater fluctuations in water level and temperature, and shifting substrates. The experience of the United States also points to the long-lasting nature of the damages done by channelization. The rate of recovery for a stream once it has been dredged is extremely slow, with many streams showing no significant recovery 30 to 40 years after channelization.[24]

Canalization of rivers in the United States has destroyed their natural properties. The Los Angeles River, for example, has lost its organic connection with the soil, surface and groundwater bodies, and the vegetation of its basin.

9.6.2 Europe

The Cordon approach has found extensive application in Europe too. Elaborate dyke systems have been built along the Danube, Loire, Maas/Meuse, Po, Rhine, Rhone, Scheldt, and Vistula, rivers. As a result, floodplains have been severed from rivers, encouraging below-flood-level settlement; marshes have been drained and forests have been cleared to give way to agriculture and settlement. It is estimated that about 90 percent of the former floodplain of the Rhine River in Germany has been drained and developed. In general, European floodplains have undergone radical changes through the application of the Cordon approach.

Apart from construction of embankments, Europe has also carried out considerable channelization of rivers. In fact, the Rhine River has lost 80 km of its length due to channelization, and has suffered a drastic reduction in its width. Previously, it was at places about 12 km wide, with braided watercourses, islands, backwaters, and meanders which served as rich wildlife habitat. In some places, embankments and channelization have reduced these stretches to only 250 m.[25]

Unfortunately, as in the United States, the adoption of the Cordon approach has not resulted in reduction of flood and flood damages. The shortened, straightened, and narrowed river channels, with less vegetation along their banks, have less capacity to hold water and allow flood water to move downstream quickly, without giving adequate time for downstream communities to prepare for it.

For example, floods from Basel (Switzerland) on the upper Rhine can travel to Karlsruhe (Germany) in half the time as before. Also, the probability of simultaneous cresting of the Rhine and its important tributaries has increased.[26] This has resulted in rising flood damages in Europe. For example, flood losses in Switzerland have increased four-fold over the past 35 years, though there have been large increases in flood control investments during this period.[27] Climate change is aggravating the harmful effects of the Cordon approach in Europe, as in other parts of the world.[28] The experience of the Po River in Italy, meanwhile, testifies to the "From embankment to higher embankment! syndrome noted earlier (Figure 9.1). Studies show that embankments, in many cases, have led to significant elevation of the riverbed, so that they need to be raised higher to prevent future inundations.[29]

9.6.3 India

Among developing countries, India has implemented the Cordon approach with considerable vigor. During pre-colonial times, embankments of local significance were sometimes constructed by local feudal lords. However, during the late nineteenth century, following disputes with local lords, the colonial government assumed the right to construct embankments along rivers.[30] Following independence in 1947, India engaged in a massive embankment building effort, as a result of which about 35,200 km of embankments have been built along its rivers, protecting, as per the claim of the authorities, 2,802 towns and 18 million hectares of land from floods.[31] Not surprisingly, 39,710 km of drainage canals also had to be dug. Despite this massive undertaking, the area affected by floods has increased from 25 million hectares in 1952 to about 50 million hectares in 2010, showing that investments in embankments is proving largely counter-productive.[32]

Particularly instructive in this regard is the experience of the Indian state of Bihar, through which many important tributaries flow from Nepal to meet the Ganges River, whose middle stretch lies in this state. The length of embankments in Bihar increased from a mere 160 km in 1952 to 3,430 km in 1998, while its flood-prone area increased over the same period from 2.5 to 6.9 million hectares.[33] The average annual damage (inflation-adjusted) caused by floods in Bihar almost quadrupled between the 1950s and 1970s.[34]

We noted earlier that embankments generate conflicts.[35] A particular variant of these conflicts is that between people on the country-side and those on the river-side of embankments. This type of conflict is quite acute in Bihar, where a large number of people find themselves trapped on the river-side, because of the embankments' wide set back distance. Mishra (2005) provides a poignant description of the plight of about 1.5 million people of 380 Indian and 34 Nepali villages located on the river-side of the Kosi River embankment.[36] Apparently, they were given dwelling land on the country-side but were supposed to continue to cultivate land that lay on the river-side of the embankment. However, the distance between their proposed dwellings and their cultivable lands was so great (about 5 km on average) that this was an absurd proposition. Moreover, their new dwellings were soon affected by waterlogging created by the embankments. These people therefore moved back to their original villages and are flooded every year, with the depth of flooding now being much higher, because the embankment prevents the Kosi water from spreading out over wider areas of the floodplains. In such a situation, breaches in the embankments—often created deliberately by the river-side people—provide some relief to them, though they create havoc for the country-side people, who are often their friends and cousins. The Kosi embankment therefore provides a tragic example of how the Cordon approach divides people and creates conflicts among them.[37]

The Kosi embankment also illustrates the danger of catastrophic flooding that the Cordon approach creates. This became clear during the Kosi flood of 2008 when the embankment breached at Kusaha, resulting in a deluge that cost 540 lives and affected about 3.4 million people in five districts and 1,067 villages. The floodwater spread over 415,000 hectares of land, of which 176,000 hectares were cropped. The deluge destroyed 244,128 houses and killed 31,995 animals. The estimated total loss was about $2.3 billion. The Kosi embankment was supposed to protect 214,000 hectares of land from flood. The area submerged by the 2008 deluge was therefore about double what the embankment intended to protect.[38] It may be noted that breaches of Kosi embankments are quite common, and the 2008 breach was the eighth such breach since they were was constructed. In some cases, the breaches were the result of sheer negligence.[39]

The fact that more embankments end up causing more catastrophic floods and flood damages is borne out by the experience of other parts of India too. Table 9.2 presents data on the area flooded, number of people affected, and estimated damage suffered in the Indian part of the Brahmaputra Basin, another area that has witnessed considerable expansion of embankments in the past decades. The data show that the average annual number of flood-affected people and value of flood-induced damages increased between 1953–1959 and 1999–2005 by 5.3 and 122.3 times, respectively. The data also show that the percentage of cropland affected by flooding has also increased sharply over time.

India's experience also validates the "From embankment to higher embankment!" vicious cycle; even government agencies are forced to acknowledge this reality. This can be seen in the candid view conveyed by the government of the Indian state of West Bengal in a report to the central government's National Flood Commission: "where river water carries heavy silt discharge, [a] vicious

Table 9.2 Flood damage trend in the Brahmaputra Basin in India

Period	Average annual area flooded (m ha)		Flooded crop area as % of total inundated area	Average annual number of people affected (million)	Average annual damage (million rupee)
	Total	Cropped			
1953–1959	1.01	0.10	9.9	0.86	58.6
1960–1969	0.75	0.16	21.3	1.52	75.7
1970–1979	0.87	0.18	20.7	2.00	151.8
1980–1988	1.43	0.40	28.0	4.55	1445.2
1999–2005	1.07	0.38	35.5	4.59	7171.7

Source: Thakkar (2007).

race starts . . . between the rise of the river bed and raising of the embankments, in which the latter has not even a remote chance to win" (Thakkar 2006).

9.7 Mississippi levee system: A case study of the Cordon approach

The brief review above shows how the Cordon approach has spread across both developed and developing worlds and how measures inspired by this approach have often failed to achieve the stipulated goals, and in fact have created more problems. To further illustrate the pitfalls of the Cordon approach, I present two case studies, namely the Mississippi levee system of the United States and the Huang He River embankments of China—two of the largest and most prominent applications of the Cordon approach in the world.

The United States, a leader in industrialization and in implementing the frontal version of the Commercial approach, is also a leader in implementing the Cordon approach, as already noted. Among the many US rivers that have been subjected to this approach, the most prominent is the Mississippi River, the largest in the country. Over time, the United States built a massive levee system to protect Mississippi floodplains from the river overflow.

9.7.1 Physical characteristics of the Mississippi River

The Mississippi and Missouri rivers together comprise one of the longest river systems in the world. The Mississippi alone ranks as the fourth longest and fifteenth largest river in the world. Rising in northern Minnesota and flowing through or bordering the states of Minnesota, Wisconsin, Iowa, Illinois, Missouri, Kentucky, Tennessee, Arkansas, Mississippi, and Louisiana, it travels 3,730 km to reach the Gulf of Mexico, 167 km downstream from the city of New Orleans. It represents the largest drainage system of North America, draining all or parts of 31 US states and two Canadian provinces between the Rocky and Appalachian Mountains. The catchment basin extends over 3.22 million sq km and covers about 40 percent of the continental US landmass. It takes about 90 days for Mississippi water to travel from Lake Itasca, on its headwaters in Minnesota, to the Gulf.

The Mississippi river is generally divided into three parts: upper, middle, and lower. The upper part extends from its origin up to St Louis in Missouri state, where it merges with the Missouri River. This part of the river has many lakes. The middle part extends from St. Louis to Cairo, Illinois, where it meets with the Ohio River. This part is 310 km long, and the river is free flowing here, with its elevation falling by 67 m over 290 km, suggesting an average rate of 0.23 m per km. At its confluence with the Ohio River in Cairo, the river is still 96 m above sea level. The lower Mississippi extends from Cairo to the gulf.

At its mouth, the river has created the unique Mississippi delta with its bayous, which are basically abandoned distributaries of the river.[40] In the process, the coastline of south Louisiana has advanced toward the gulf from 24 to 80 km over the past 5,000 years. The currently active delta lobe is often called the Birdfoot Delta, because of its resemblance to a bird's foot.[41] The annual average discharge of the Mississippi River to the gulf ranges from 7,000 to 20,000 cumec. This makes it the fifth largest river by volume of discharge, though it is only about 8 percent of the Amazon River's wet season flow, which is about 200,000 cumec.

9.7.2 Evolution of the Mississippi River levee system

Structural interventions in the Mississippi river began in the upper part of the river, primarily for the purpose of navigation, and so levees in this part of the river were often constructed in combination with dams and locks. Between Lake Itasca and St. Louis, there are 43 dams, 14 of which are in the headwaters region above Minneapolis and are geared to ward power generation, recreation, and other purposes. The other 29, starting from the one in downtown Minneapolis, have locks and were meant primarily to improve commercial navigation, by ensuring the minimum draft of 2.7 m necessary for ships and barges. The USACE, established in 1802, was in charge of the construction of these structures, which were undertaken mostly during the 1930s, following the Flood Control Act of 1928, authorizing the Mississippi River and Tributaries Project. Needless to say, the upper part of the Mississippi River and the ecology of the basin there is dominated by these 43 dams.

The Mississippi levee system (including embankments and winged dams) begins from around St Louis (from Cape Girardeau, Missouri, to be precise) and continues to the city of New Orleans and the Mississippi Delta.[42] The levee construction actually began from the south, around New Orleans, and gradually moved up river. The process began with French settlers in Louisiana, building levees in the eighteenth century to protect the city of New Orleans. The first Louisiana levees were about 1 m high and covered a distance of about 80 km along the river.

The USACE, in conjunction with the Mississippi River Commission, extended the levee system beginning in 1882 to cover the riverbanks from Cairo, Illinois, to the mouth of the Mississippi delta in Louisiana. By the mid-1980s, this system had reached its present extent, comprising over 5,600 km of levees and including flood-bypasses, regulators, and other control structures (Figure 9.3). The height of the Mississippi levees averages 7.3 m, though in some places they are as high as 15 m. The system includes some of the longest continuous individual levees in the world. One such levee extends southward from Pine Bluff, Arkansas, for a distance of about 610 km. While building the levees, the USACE also carried out considerable channelization, cutting necks of horseshoe bends, etc., thus reducing the river length, as noted earlier, by about 300 km. The idea was that straightening the river channel would allow water flow faster and help reduce flood heights.

9.7.3 Consequences of the Mississippi levee system

In view of the earlier discussion, the consequences of the Mississippi levee system should not be surprising. First, it promoted below-flood-level settlement in the floodplains. Entire new cities and towns grew up, relying on the protection of the levees. Roads, highways, power plants, etc. were constructed to serve these settlements and their economic activities. As expected, this gave rise to the danger of catastrophic flooding.

Second, the levees have also led to "transfer flooding," whereby construction of levees in one place increase flooding in other places.[43]

Third, the levee system promoted chemical input-based agriculture, leading to an increase in agricultural output but also toxic runoff. The flood-free conditions reduced the scope for mixed farming based on a combination of aquatic and land-based agricultural activities. It has stopped natural irrigation through river overflow and made drainage dependent on pumps. It also promoted monocrop agriculture that is more demanding of chemical inputs and damaging to soil quality.

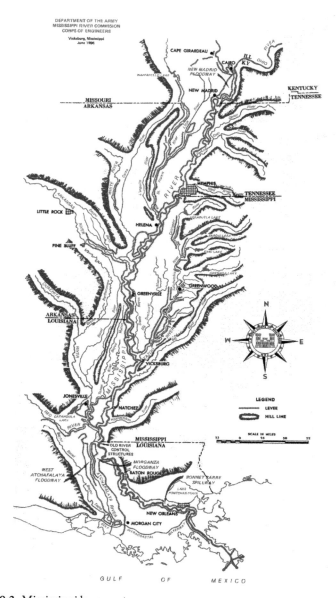

Figure 9.3 Mississippi levee system

Source: Mississippi River Commission, and Mississippi Valley Division Corps of Engineers.

Fourth, the levees have led to riverbed aggradation. As the river cannot overflow on to floodplains, more sediment gets deposited on the riverbed. Aggradation of the riverbed is particularly prominent in the lower part, near the city of New Orleans, where the riverbed has a higher elevation than of much of the city itself, creating a risky situation, as we shall see later.

Fifth, the volume of sediment carried by the Mississippi River to its delta has declined from about 400 million tons in 1900 to about 145 million tons in recent decades. This has been the combined result of several factors. One is the numerous dams and control structures built on the Mississippi and its tributaries, such as the Missouri and Ohio rivers; these structures also include river training and bank revetments. Another factor is the channelization of the Mississippi and shortening of its length so that the river now has less access to sediment materials. Thus, the reduction in the volume of sediment carried to the sea by the Mississippi River is a net result of many processes working in different directions and triggered by both the frontal and lateral interventions in the river and its tributaries and their watersheds and basins.

Sixth, the diminution of water and sediment flows of the Mississippi to the Gulf had serious consequences for the Mississippi River Delta and the marine environment near the estuary.[44] The Mississippi coast is now facing erosion and salinity intrusion. According to some estimates, the Mississippi River Delta and coastal Louisiana are losing a football field of wetlands every 100 minutes. The cumulative loss, since the 1930s, amounts to more than 5,180 sq km, which is roughly the area of the state of Delaware.[45] There are many reasons for this rapid loss, including sea level rise, oil and gas related activities, etc. However, there is no doubt that diminution of water and sediment flows, as noted earlier, has been the major cause. Thus, the Cordon approach, applied to the middle and lower stretches of the Mississippi River, in combination with the frontal version of the Commercial approach (embodied in dams and reservoirs), applied to its upper part, has created a serious crisis for Mississippi River Delta.[46]

The most consequential outcome of the Mississippi levees has been catastrophic flooding. Despite the engineering prowess and resources spent on construction and maintenance of the Mississippi levee system, it failed and is failing more frequently in recent years. I next consider a few recent instances of such flooding.

9.7.4 Catastrophic flood: Great Mississippi and Missouri rivers Flood of 1993

The Great Mississippi and Missouri Flood of 1993 is a vivid example of the catastrophic flooding that a levee system can create. The flood covered about 1 million sq km, extending over nine states and lasted in some locations for about

200 days.[47] Fifty people died, and thousands had to be evacuated, in some cases for many months. The total damage is estimated to be about $16 billion (about $26 billion in 2018 dollars).

The flood resulted from several factors. The first was the unusually wet 1992 fall that meant the soil was already wet and less capable of absorbing and retaining water from rainfall in 1993. Second, for the same reason, upstream dam reservoirs were full and near full, with less capacity to hold additional water from 1993 rainfall. The third was of course the excessive rainfall in 1993. Between June and August, rainfall exceeded 12 inches in the eastern Dakotas, southern Minnesota, eastern Nebraska, Wisconsin, Kansas, Iowa, Missouri, Illinois, and Indiana, with some areas receiving more than 4 feet of rainfall. In many areas, rainfall was 4 to 7.5 times the average. Unusually high amounts of rain also fell in the catchment areas of many of the Mississippi River's tributaries, such as the Skunk, Iowa, and Des Moines rivers, and their excessive flows contributed to the flooding.

Under the impact of the higher levels of river flow, the levee system gave way. According to the National Weather Service, hundreds of levees failed along the Mississippi and Missouri Rivers.[48] Because the levees had promoted below-flood-level settlement in Mississippi floodplains, the flood proved to be catastrophic, going through the roofs in many places (Figure 9.4). The USACE reportedly argued that the levees helped to prevent a loss of $19 billion. However, as noted earlier, such claims ignore the fact that levees themselves had promoted below-flood-level settlement, which was the main cause of the actual and potential losses.[49]

9.7.5 Catastrophic flooding in subsequent years

Following the 1993 flood, there was some awareness about the danger of below-flood-level settlement, and some efforts were made to resettle people to areas that were above the flood level. However, validating the "From low to higher embankment!" syndrome (Figure 9.1), these resettlement ideas soon gave way to a focus on extending levees to new areas and strengthening of existing levees, by making them taller and stronger, to withstand higher stages of the river. As a result, there was more development behind the levees in the Mississippi basin. According to Pinter (2005), during the decade since the 1993 flood, about $2.2 billion was spent on new development in the St. Louis area alone, allowing for the building of 28,000 new homes as well as 26.8 sq km of commercial and industrial development. Similar new development has occurred in many other parts of the Mississippi basin that were inundated in 1993. As a result, Mississippi floods have assumed a recurring catastrophic character.

Figure 9.4 Flooding up to the roof: Great Mississippi Flood of 1993
Source: USGS Central Midwest Water Science Center.

Another recent example of catastrophic flooding occurred in 2011, and, in many ways, exceeded even the 1927 flood. Affected states included Illinois, Missouri, Kentucky, Tennessee, Arkansas, Mississippi, and Louisiana; cities such as Cairo, Baton Rouge, and New Orleans were all seriously threatened. Parts of Louisiana faced inundation with 20–30 feet of water. Many emergency measures were taken to avoid such calamities. The Birds Point-New Madrid Floodway of the Mississippi levee system was opened on May 3 using explosives to create a 3 km crevasse in the frontline levee, inundating 530 sq km of farmland in Mississippi County, Missouri, to save the city of Cairo, Illinois, and the rest of the levee system. Also, on May 14, for the first time in 37 years, the Morganza spillway was opened, flooding 12,000 sq km of rural Louisiana to protect the cities of Baton Rouge and New Orleans.

Apart from the flood of 2011, there have been several other major floods in the Missouri-Mississippi basin since 1993, in 2008, 2016 and 2019, with estimated damages of $12.0, $10.7, and $12.5 billion, respectively. The number of people killed was 24 and 13 in 2008 and 2016, respectively.[50]

Catastrophic floods with multi-billion-dollar damages, therefore, are no longer a rare event and unfortunately, are likely to become more frequent with climate change. As is known, one consequence of climate change is that extreme weather events—such as unusual amounts of precipitation in unusual periods—will become more frequent, intense, and larger in scope. The fact that floods caused by unusual precipitation patterns are proving to be catastrophic, entailing huge amount of damages, is a direct result of the Cordon approach to rivers.[51]

9.7.6 Catastrophic flood: Hurricane Katrina and New Orleans

The danger that extreme weather events is creating for the Mississippi levee system was brought to the fore by Hurricane Katrina in 2005. The City of New Orleans developed originally along the high ridges of the Mississippi River. However, embankments and floodwalls constructed along the Mississippi River to the south and Lake Pontchartrain to the north created a bowl-shaped, below-flood-level terrain where the city expanded. The elevation of both the Mississippi River and that of Lake Pontchartrain was higher than of most of the city, creating a risky situation. The USACE built an elaborate system of canals for drainage and navigation purposes, with levees and floodwalls preventing water from coming in and pump stations to evacuate the rainwater falling inside. Prominent among these are the 17th Street Canal and the London Avenue Canal, which are mainly for drainage; the Inner Harbor Navigation Canal, popularly called the Industrial Canal, connecting Lake Pontchartrain with the Mississippi River; and the Mississippi River-Gulf Outlet (MR-GO), which connects the Gulf with the Industrial Canal via the Intracoastal Waterway.

However, this elaborate system of levees, canals, floodwalls, spillways, outlets, etc. couldn't prevent New Orleans city from inundation during Hurricane Katrina. Breaches in levees and floodwalls occurred at about 53 points, particularly along the 17th Street Canal, London Avenue Canal, the Industrial Canal, and the MR-GO (which alone had about 20 such breaches). As a result, about 80 percent of the city went under water, the depth of which exceeded 6 m in some areas. The number of fatalities rose to 1,833, and the damage is estimated to be $125 billion (in 2005 prices). The hurricane also exposed the concentration of poverty in parts of New Orleans, where many people simply did not have the resources to move out even though they were asked to do so. According to Lawrence Roth, deputy director of the American Society of Civil Engineers, the flooding of New Orleans was "the worst engineering catastrophe in US History (Roth 2007)."[52]

There were many reasons for the catastrophe. Some have pointed to the inadequate design capacity of levees, floodwalls, and pumping facilities and poor construction and management of the canals, among other problems, for the debacle.[53] However, these cannot obviate the underlying problem of the Cordon approach, namely promotion of below-flood-level settlement behind levees and floodwalls. In this regard, it is instructive to note that even during the Katrina debacle, areas settled before 1900—that is, before the levee system was constructed—experienced either no flood at all or no serious flooding, partly because of their higher natural elevation and partly because of the architecture, which included a raised design to cope with floods. These areas comprise higher ground along the river front (such as Old Carrollton, Uptown, the Old Warehouse District, the French Quarter, Old Marigny, and Bywater) and the natural ridges (such as Esplanade Ridge, Bayou St. John, and Gentilly Ridge).

As has been the case with other parts of the Mississippi levee system, its failure at New Orleans prompted diverse responses. Some of these conformed to the "From low to high embankment!" syndrome. Thus, it was demanded that the New Orleans protection embankments and floodwalls be upgraded to be capable of withstanding hurricanes of category 5—instead of category 3, as was the case during Katrina. The USACE has spent about $14.45 billion since Katrina on repairing and improving hundreds of levees and pump stations; constructing the Gulf Intracoastal Waterway West Closure Complex, claimed to be the world's largest water pump station, with a capacity of pumping 3,800 cubic m per minute and costing about $1 billion; building the Inner Harbor Navigation Canal Lake Borgne Surge Barrier, the longest storm surge barrier in the United States, for protection against surges coming from the direction of Lake Borgne; and constructing the Seabrook Floodgate, a floodgate at the connection of Lake Pontchartrain with the Industrial Canal, for protection from surges coming from the direction of the lake; and other such projects. These structures by and large represent continuation of the Cordon approach.

The fact that even the Mississippi floods of 1993 and 2011 and Hurricane Katrina of 2005 could not move the US water establishment from the Cordon approach is illustrated by the renewed levee construction and development behind them near the Missouri-Mississippi confluence. Consequently, many fear that East St. Louis will soon become another New Orleans, in terms of perennial vulnerability to catastrophic flooding.[54]

9.7.7 Special problems of the lower Mississippi River

Levees have a created a special situation not only with the city of New Orleans but with the entire lower Mississippi River. As it flows south, the Mississippi River has two shorter routes to reach the Gulf of Mexico than the current "bird's foot." One is via the Atchafalaya River from below the point where the tributary Red River meets the Mississippi. The other is through Lake Pontchartrain and Lake Borgne. The USACE is waging a huge battle to prevent such a change of course. US government scientists determined in the 1950s that the Mississippi River was switching to the Atchafalaya River as its main channel because this provides a shorter and steeper path to the Gulf of Mexico. This would however leave the cities of Baton Rouge and New Orleans high and dry. To prevent this outcome, control structures were constructed at the junction in order to regulate the amount of the Mississippi River that can flow down to the Atchafalaya River. Named the Old River Control Structure, it was completed in 1963 and is used to allow only 30 percent of Mississippi water flow into the Atchafalaya River. Similarly, taking the Pontchartrain route to the Gulf would mean that the Mississippi River would bypass the main (downtown) parts of New Orleans. The Bonnet Carre Spillway has been added to the Mississippi levees in part to prevent that outcome. It is used to regulate the amount of Mississippi water that can flow into Lake Pontchartrain. However, it is difficult to say how long it will be possible to fight against nature and prevent the Mississippi River from another avulsion.

9.7.8 Mississippi levee system—overall

The Mississippi levee system represents probably the greatest industrial might that human societies could direct against a river. The goal was to confine the river to its channel and prevent it from overflowing on to its floodplains. However, the very success in doing so contained the seed of its undoing. The apparent flood-free conditions led to huge expansion of below-flood-level settlement on the Mississippi floodplains. The levees also led to riverbed aggradation. These two processes set the stage for catastrophic floods. Climate change has now added its force to this man-made vulnerability. Catastrophic floods can now happen in the Mississippi basin in any year. The Cordon approach has led to a special, almost intractable problem for the south Mississippi region, including the Mississippi Delta.

9.8 Huang He River: Another case study of the Cordon approach

The Huang He River of China provides another prominent example of application of the Cordon approach. It has a certain uniqueness in the sense that construction of embankments along the Huang He River predates the industrial era.[55] This is due to the fact that, unlike major dams, levees could be constructed with pre-industrial technology. Just as the Chinese constructed the Great Wall during the pre-industrial period, they also built embankments along the Huang He River. However, the pre-industrial embankments were upgraded to the industrial level during modern times.

9.8.1 Physical description of the Huang He River

With a length of 5,464 km, the Huang He River is the third-longest river in Asia, after the Yangtze River and Yenisei River, and the sixth-longest in the world (Figure 9.5). It originates in the Bayan Har Mountains in Qinghai province, western China, and flows through nine provinces before emptying into the Bohai Sea near the city of Dongying in Shandong province. The basin of the river extends about 1,900 km east-west and 1,100 km north-south, and has a total area of 752,443 sq km. The river is generally divided into three parts: the upper part, lying in the northeast of the Tibetan Plateau; the middle part, comprising the Ordos Loop; and the lower part, lying in the North China Plain.

The upper part extends from the source in the Bayan Har Mountains to Hekou Town in Inner Mongolia. It is steep, with the river dropping 3,496 m, with an average grade of 0.10 percent. In the first segment of this upper part, the river flows across the plateau, with its clear lakes, pastures, swamps, and knolls. The second segment—the valley section—extends from the Longyang Gorge in Qinghai to Qingtong Gorge in Gansu province. The river here is narrow and steep, with about twenty gorges, providing attractive sites for hydroelectric plants. In the third segment, the river flows through the Yinchuan and Hetao alluvial plains, comprising mostly deserts and grasslands having less steep gradient. This is also where Huang He River water historically was used for irrigation.

The middle part is also known as the Ordos Loop and extends for 1,206 km from Hekou Town of Inner Mongolia to Zhengzhou of Henan province. The elevation drops by 890 m, with an average gradient of 0.074 percent. Thirty large

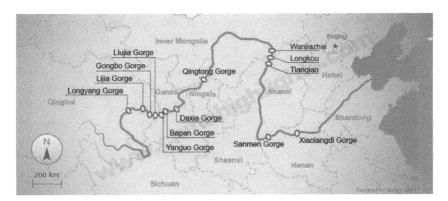

Figure 9.5 Map of the Huang He River
Source: China Highlights Travel.

tributaries join the river along this stretch and the flow of the river increases by 43.5 percent. Here the river passes through the Loess plateau, where it picks up the yellow colored mud and sand, comprising 92 percent of the river's silts, which once made the Huang He River the most sediment-laden in the world, with a record 3.91 billion tons in 1933.[56] In this middle part, the river also passes through continuous valleys, collectively known as the Jinshan valley.

The lower part stretches for about 786 km, flowing northeast across the North China Plain, from Zhengzhou, Henan to its mouth at the Bohai Sea. With no addition to water flow and sharp decrease in gradient (averaging only 0.012 percent), the river deposits here the silt that it picked up from the Loess plateau. This raised the elevation of the riverbed, sometimes above the elevation of the neighboring plains, so that river became unstable. The overflowing river had no reason to revert back to the channel and would often cut new channels and thus create havoc, devouring villages and towns and causing deaths of sometimes millions of people, earning the river such names as the "Sorrow of China" or the "Scourge of the Sons of Han."

Traditionally, it is believed that the Chinese civilization originated in the Huang He River basin. The Chinese therefore refer to the river also as "the Mother River" and "the cradle of Chinese civilization." During China's long history, the Huang He River thus has been considered a blessing as well as a curse and has been nicknamed both "China's Pride" and "China's Sorrow."

9.8.2 Controlling the Huang He River: "From lower to higher embankment!" vicious cycle

The Huang He River is probably the most prominent example of path dependence of river related structures and both the "More water, more thirst!" and the "From lower to higher embankment!" syndromes. Many researchers argue that the Huang He River originally was a stable system. Though it passed through the Loess plateau, it did not pick up as much sediment as it does now because the Loess plateau was largely uninhabited and covered with vegetation. However, Chinese emperors encouraged more Han Chinese to settle on the Loess plateau, so that they could serve as a buffer between the Chinese empire and the northern Mongol invaders. Construction of the Great Wall led to a further influx of people into the area. Use of iron ploughs led to baring of more soil on the plateau, denuding the vegetation cover, and thus increasing the sediment load of the Huang He River. The sediment level rose to a record 3.91 billion ton in 1933, as already noted. Once the sediment load increased to an extreme level, it set off a destabilization process. More recently, Huang He River carried about 1.6 billion tons of sediment at the point where it descended from the Loess Plateau and, left unobstructed, would carry 1.4 billion to the sea annually. According to Tregear (1965), 1 cubic m of river water contained 34 kilograms of sediment as compared to 10 kg for the Colorado River and only 1 kg for the Nile River. Currently, Huang He's sediment load is exceeded by only the Ganges-Brahmaputra and Amazon Rivers (Qian and Dai 1980).

Apparently, ways to deal with the Huang He River problems were debated back in ancient China, with two opposing schools of thought in this regard. On the one hand, there was the Confucian "contractionist" school, arguing for constricting the river between high river-side embankments. Using the concepts and terminology developed in this book, this would be the Cordon approach. On the other hand, there was the Taoist "expansionist" school arguing for low embankments with wide set-back distances so that the river had more space to overflow.[57] This would conform more to the Open approach, discussed in the next chapter. According to this narrative, the contractionists had the ear of the emperor, and China embarked on the Cordon approach and started building levees along the riverbanks. Some of the original embankments were in part natural ridges formed by the river itself and bolstered by human efforts. Thus began the race between the elevation of the riverbed and height of the embankment. Not surprisingly, embankments encouraged below-flood-level settlement and as a result, when they breached, floods became catastrophic.[58] Particularly devastating floods occurred in 1887 and 1931, which together led to the death of about 6 million people.[59]

Modern China embarked on the enterprise of controlling the Huang He River with considerable gusto. To this end, it undertook structural interventions of both frontal and lateral types. First, it built numerous dams in the upper and middle parts of the river to produce hydroelectricity, expand irrigation, and control the river flow.[60] Second, to keep the river confined and prevent floods, it built 800 km of levees, some of which are huge. However, as the riverbed has risen due to sedimentation, the height of the embankments needed to be increased. In many places, the riverbed elevation now exceeds that of the surrounding floodplains by more than 20 feet.[61] As the eminent journalist Edgar Snow (1961) recounted from his visit, the Huang He riverbed was twenty to twenty-five feet above the surrounding countryside, and he had watched ships sailing overhead at that height. At Kaifeng, Henan, the river is 10 m (33 ft) above ground level.[62] Currently, the Huang He River is above the adjoining plains for much of its last 500 miles to the sea, and the riverbed continues to rise at an alarming rate of 4 inches a year.

9.8.3 Degradation of the Huang He Delta

Modern China constructed many dams and barrages on the Huang He River, as noted earlier. Water from many dam reservoirs is used for irrigation and other commercial purposes. The use for irrigation increased by five times since 1959, and water from the Huang He was used to irrigate 74,000 sq km and served 140 million people by 1999. However, as a result of this impounding and abstraction, there is now little water left in the river to reach the sea. Since 1971 the river has been dry for most part of the year, for example, for 226 days in 1997.

The decrease in water flow was accompanied by decrease in sediment flow. As noted earlier, previously about 1.6 billion tons of soil flowed into the Huang He River each year, and about 1.4 billion ton reached the Yellow Sea, building the river's delta, which was increasing by 22 sq km per year (Figure 9.6). As a result of the interventions described earlier, the sediment flow reaching the delta has now declined to 0.018 billion tons, less than 1 percent of the previous volume (Yang, Milliman, Galler, Liu, and Sun 1998). As a result, the Huang He River Delta is shrinking.[63] According to Shandong Provincial Bureau of Geology and Mineral Services, its size has been decreasing by an average of 7.6 sq km per year since 1996. Thus, Huang He River is another example of a major river that has been thoroughly exhausted by the application of the Commercial approach. Needless to say, this exhaustion is affecting the coastal and marine environment near the estuary.

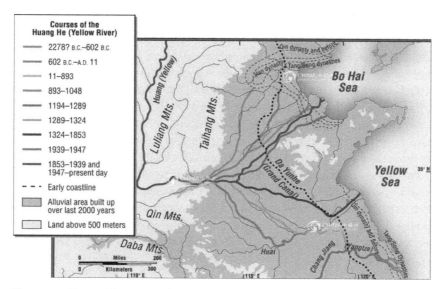

Figure 9.6 Huang He River Delta
Source: China Connection Tours (CCT).

9.8.4 Huang He River—overall

The Huang He River is a prominent example of continuation of the pre-industrial Cordon approach to its industrial version. The misdirected introduction of agriculture on the Loess Plains led to the river's unusually high sediment content, which also became a source of destabilization of the river. Since then there was no way back from the "From lower to higher embankment!" vicious cycle. As more embankments were built to contain the river, more sediment settled on the riverbed, raising its elevation, requiring in turn the raising of the height of the embankments. Impounding and abstraction of the river flow through the construction of numerous dams and reservoirs have in a sense mitigated the risky race between elevation of the riverbed and height of the embankment. Unfortunately, this mitigation is achieved by exhausting the river, which now basically dies before reaching the sea. Together with the river, its delta is dying, now shrinking each year. In recent years, the Chinese government has imposed some restrictions on the amount of abstraction so as to let some of Huang He flow reach the sea in order to halt the process of decay and erosion of the delta. It remains to be seen to what extent these measures prove successful.[64]

9.9 Conclusions

The Commercial approach takes the concrete form of the Cordon approach in areas where rivers pass through floodplains and deltas. The basic idea behind this approach is to cordon off floodplains from river overflows by constructing continuous embankments along riverbanks. Apart from construction of embankments and floodwalls, the approach also involves channelization and canalization of rivers. The goal of this approach is to create flood-free conditions so that more productive agriculture and development can take place in floodplains and thus increase their commercial value. Some embankments were also constructed in pre-industrial times. However, large-scale embankment construction, using industrial era technology, began in the nineteenth century.

While the implementation of the Cordon approach indeed helped to raise agricultural output and promote settlement, it also led to many negative consequences. It deprived floodplains of the nurturing functions of regular river inundations, triggering a long-term process of degradation of their soil and water bodies. Deprived of the sediment that river water brought to them, elevation of floodplains ceased to increase and in fact in places started to decrease because of soil compaction, withdrawal from underground aquifers, etc. Meanwhile, more sediment settled on riverbeds, raising their elevation. As a result, in many places riverbeds are now higher than their adjoining floodplains, creating a risky situation. More importantly, due to the below-flood-level settlement the Cordon approach promotes, floods—caused by breaches or overtopping—now prove to be catastrophic, entailing huge damages. The Cordon approach therefore generally ends up increasing flood-related damages rather than decreasing them. Also, by severing the links with the river, the Cordon approach disrupts the natural processes of irrigation and drainage, requiring both to be substituted by costly methods using pumps. However, such substitution often fails and leads to drainage congestion and waterlogging.

The industrial era of the Cordon approach began in developed countries and then spread to developing countries. The experience in both parts of the world provides ample evidence of the negative consequences of this approach. In most cases, the application of the Cordon approach took place in combination with the application of the frontal version of the Commercial approach. Thus, dams and barrages have been built in upper stretches of rivers, while embankments have been constructed along their middle and lower stretches. The combined effect has been particularly detrimental for deltas and estuaries. Reduction of water and sediment flow has led to shrinkage of deltas and deterioration of the ecology of the estuaries. The Mississippi River of the United States and the Huang He River of China offer prominent examples of the application of the Cordon approach and its consequences.

Notes

1. See Islam (1999, pp. 9–10) for details. See Smith and Ward (1998) for a related discussion of floods.
2. Water in underground aquifers can also be used during the dry season; some can return to the river.
3. See McCully (2007, p. 8).
4. "As a rising river flows between levees, it becomes deeper and faster and gains more erosive power. The water then exits the leveed stretch of river with more destructive potential than had the levees not been built" McCully (2007, p. 8). For effects of embankments on the consequences of dam failure, see Brown, Chapman, Gosden, and Smith (2008).
5. Marshes and other types of wetlands help keep rivers healthy in other ways, such as by naturally filtering pollutants and excess nutrients from sewage and agricultural runoff, trapping sediment, and providing habitat for fish and other riverine organisms (McCully 2001, p. 191).
6. See McCully (2001) for further details. See also Brooker (1985).
7. As McCully (2001) notes, "In urban areas, rivers are often dredged to increase their capacity, and may be cleared of vegetation and lined with concrete or rocks to stabilize their banks, and to increase the speed of flood flows. These alterations can prove

self-defeating (as well as destroying the ecological and aesthetic value of the river). Normal flows will slow down in the expanded channel and deposit more of their sediment load, eventually returning the channel to its original capacity and flow regime."

8. Embankments may fail for various reasons. For example, the flood level may exceed the design level for which the embankments were built. Furthermore, levees may break even if the flood level did not cross the design level, because of shoddy construction, poor maintenance, or even by deliberate breaches caused by people for whom the embankments were increasing flood depth. In fact, a panel of the US National Academy of Sciences concluded in 1982 that "it is short sighted and foolish to regard even the most reliable levee system as fail-safe." See, for details, McCully (2007, pp. 5–6).

9. See McCully (2007, p. 3). Increase in floods is due in part to increase in the annual sediment load of the rivers from about 9 billion tons to about 45 billion tons (see McCully 2001, p. 191). See Parker (ed.) (2000) for further discussion on floods.

10. As McCully (2007, p. 8) notes, "Another consequence of proliferating levees is a huge increase in the costs of monitoring and maintaining them. Where levees cause the riverbed to rise or the floodplain to subside, they must be constantly raised. As they grow (and as they age) their maintenance costs soar. Except for soon after major floods, allocating adequate funds for levee up-keep is often not a political priority, and so the levees steadily deteriorate."

11. "While a well built and maintained levee can protect the areas behind it from floods smaller than its design flood, it may do so only at the cost of magnifying flood damage in the unprotected areas. As rising river flows into an embanked stretch . . . with more destructive potential than had the levees not been built" (McCully 2007, p. 6).

12. "Because communities living downstream of a newly built set of levees will start to experience worse floods, they will frequently demand protection themselves. Thus, a vicious cycle of levees requiring more levees is set in motion. A similar pattern happens on tributaries. The rise of the bed of the mainstream causes flooding around the mouth of the tributaries, and creates political pressure for these stretches of river to be embanked. And so, at great cost, the levees spread like a cancer throughout the basin, offering a sense of false security to those living behind them, and increasing the flood damage suffered by those living in 'unprotected' areas" (McCully 2007, p. 8).

13. This was vividly revealed during Hurricane Katrina—to be discussed later— when pumps in New Orleans failed to evacuate the accumulated water into Lake Pontchartrain.

14. Mishra (2001) provides, in the following, a good exposition of the flaws of the Cordon approach: "Floodwaters contain sediments and floods spread them over a vast area. This is how the rivers perform their duty of land building. Any change in this behaviour leads to problems that are not easy to handle. Spreading of floodwater ensures fertility of the soil because of the nutrient silt and also maintains the level of groundwater. It also lowers the flood levels, spreading it thin over vast areas. Preventing floodwaters from spreading results in anomalies. Arrangements should be made to ensure that floodwaters spread. As a corollary to this, the drainage should be improved so that floodwater does not stay for long at any place."

Touching upon the reasons why Cordon approach is adopted despite its flaws, Mishra notes, "Unfortunately, the engineering that is taught in colleges is confined within a framework and prevents engineers from thinking beyond or to even extend the frame. They also have a contempt for traditional wisdom and very conveniently forget that people have been living in flood prone areas for centuries without any help from the outside. It is impossible to think that the people have not developed the means to face the situation that they live in. The people, in turn, ignore engineers as people with ulterior motives. This stalemate must be resolved; sooner the better."

Mishra also provides, in the following, a nice exposition of the vested interests that propel the Cordon approach and don't allow the country to move away from it: "Governments depend on their engineers and go by their advice and are averse to any suggestions from outside, even from their retired engineers. Engineers don't talk to people as such. The dialogue, if any, exists between the politicians and the public. It becomes very easy for the politicians to say that they have taken the advice of the best available technical brains, and the engineers often suggest that they have to work under politicians and are not the final authority. The public never gets an opportunity to talk to both of them together. The works proceed without any dialogue with the public. Then it can endlessly wait for the proposed dams in Nepal or Interlinking of Rivers or any such proposal. The debate whether these works will be worth the cost of investment is relegated to the background."

15. Prominent among such integrated schemes are the Tennessee Valley Authority (TVA) and the Mississippi Levee System overseen by the Mississippi Commission. The TVA was created by Congressional charter on May 18, 1933 as a federally owned corporation, aimed at providing navigation, flood control, and most of all, electricity generation. It includes 29 hydropower facilities. Under the TVA a huge network of dams and levees has been built. It has often been offered as a model of basin-wide development and has been imitated by other developing countries, such as India (in creating the Damodar Valley Corporation in West Bengal, discussed later, for the utilization of the Damodar River). The Sacramento is another river illustrating levee construction in the United States.

16. See http://www.air-worldwide.com/Publications/AIR-Currents/2014/The-United-States%E2%80%99-Aging-River-Levees/. For more detailed information on US levees, see the National Levee Database (NLD) of the US Army Corps of Engineers (USACE) website https://www.usace.army.mil/Missions/Civil-Works/Levee-Safety-Program/National-Levee-Database/. This site contains information on levees constructed by USACE.

17. See McCully (2007, p. 5).

18. By the end of the last century, the USACE had spent over $25 billion on 500 dams and 16,000 km of embankments (McCully 2001, p. 146).

19. See McCully (2007, p. 8).

20. See McCully (2007, p. 4).

21. During the first half of the 1990s, the average annual flood damage has been estimated to be $3 billion (McCully 2001, p. 146).

22. See Pinter (2005) and McCully (2007, p. 4). Tobin (1995) describes the history of flood control in the US as an "undying affair with levees."

23. See McCully (2001, p. 191), who also states that nearly four-fifths of the floodplain hardwood forests along the lower Mississippi have been lost to agriculture.

24. A study by Illinois State Water Survey found that every 1 percent increase in watershed area covered by wetlands decreased flood peaks in streams by nearly 4 percent. See McCully (2001, p. 191).

25. As a result, many wildlife-rich, interweaving, braided watercourses, islands, backwaters, and meanders have been lost. See McCully (2007, p. 8).

26. See McCully (2007, p. 8).

27. See Willi (2006) and McCully (2007, p. 3)

28. In the United Kingdom, many have realized that development permitted on floodplains has exposed more properties to flooding, and that substitution of concrete for natural strata has sped up the run-off of water, increasing the danger of flooding downstream (McCully 2001).

29. See McCully (2001) for details.

30. The colonial government of India passed the Bengal Embankment Act in 1866 and 1873, assuming the power to build and maintain embankments.

31. See Mishra (2001) who also states that 7,713 villages were raised above the highest flood level and 65 raised platforms were constructed.

32. As Mishra (2001) observes, "The investment in the flood control sector is doing more harm than good but the plans are afoot to pursue the same policy that has resulted in this adverse result."

33. This information was provided by Chandrashekhar, the secretary of a Supaul-based NGO, while speaking at the conference on Agony of Floods: Flood Induced Water Conflicts in India: A Compendium of Case Studies, organized by India's Forum for Policy Dialogue on Water Conflicts in India and Megh Pyne Abhiyan, on October 22 in 2013 in Patna, the state capital of Bihar. The continued building of river embankments along rivers in Bihar was heavily criticized at this conference. See Anand (2013) for a report on this conference. See Somanathan (2013) for insightful discussion of effectiveness of embankments as a strategy of flood control.

34. See Comments by H. Thakkar in Green, Parker and Tunstall (2000, p. 149).

35. Embankment-induced conflicts was the main theme of the Agony of Floods conference (see n.33). As the forum's statement put it, ". . . capturing, understanding, and disseminating knowledge, and initiating a dialogue around flood induced conflicts in different locations in India . . . understanding the correlation between floods and the conflict perspective will not only broaden the existing discourse around floods but also strengthen the effort to showcase the extended impact of human induced disaster in the country leading to better strategies in dealing with them." Speaking at the conference, Mishra noted that embankments not only worsened floods but also heightened the possibilities of conflict between groups of residents. He noted that on some occasions, people themselves breached the embankments to let the flood waters flow from their areas, leading to violence between communities on

either side of the embankments. He also drew attention to the drainage problems (waterlogging) that embankments create. As he noted, "Proper drainage of the flood waters is needed but these embankments actually prohibit the flow of water." See Anand (2013).

36. The Kosi River of India is one of those rivers that have formed inland deltas or fans. The Kosi fan is below Siwaliks, extending about 15,000 sq km, across which the river has shifted its course over the years. See Gole and Chitale (1966). This tendency to shift courses makes the application of the Cordon approach particularly challenging for the Kosi River.

37. See Mishra (2005). See also McCully (2007, p. 8) and Thakkar (2006).

38. See Mishra (2001).

39. In fact, Mishra (2015) shows that even the 2008 breach was due to negligence. He notes that the 2008 breach occurred when the discharge level was 4,700 cumec, while the designed capacity for the embankment at that point was 26,900 cumec. According to Mishra, negligence was an important contributing factor to the Kosi breach. He recounts that the nose of the spur at 12 km upstream of the Kosi barrage was attacked by the river on August 5, 2008, so that there was plenty of time to take preventive measures, but nobody cared to do so. The negligence is highlighted further by the fact that the project authorities had been assuring residents as late as August 17 that the embankment was safe, when in fact it breached the next day (August 18)! There has been some talk about applying a basin-wide approach to the Kosi River. However, not much progress has been made, and a basin-wide approach based on the Cordon approach is not likely to provide the desired results, as the floods of 2019 have demonstrated again. For more discussion of Kosi embankments, see Devkota, Crosato, and Giri (2012), Dixit (2009), Sinha, Gupta, Mishra, Tripathi, Nepal, Wahid, and Swarnkara (2019), and Wahid, Kilroy, Shrestha, and Bajracharya (2017).

40. "Through a natural process known as avulsion or delta switching, the lower Mississippi River has shifted its final course to the mouth of the Gulf of Mexico every thousand years or so. This occurs because the deposits of silt and sediment begin to clog its channel, raising the river's level and causing it to eventually find a steeper, more direct route to the Gulf of Mexico. The abandoned distributaries diminish in volume and form what are known as bayous. This process has, over the past 5,000 years, caused the coastline of south Louisiana to advance toward the Gulf from 15 to 50 miles (24 to 80 km). The currently active delta lobe is called the Birdfoot Delta, after its shape, or the Balize Delta, after La Balize, Louisiana, the first French settlement at the mouth of the Mississippi" (Changnon 1996).

41. It is also called Balize Delta, after La Balize, the first French settlement at the mouth of the Mississippi.

42. Below St. Louis, the Mississippi is relatively free flowing, in the sense that there are fewer frontal interventions in the form of dams and weirs. Changnon (1996, p. 264) provides the following further details regarding the history of the Mississippi levees:

"Construction of levees along the Mississippi by farmer groups began in the 1850s, and then levees along the river were enhanced by federal levee program in 1917 and again in the late 1920s."

"Before the Great Mississippi Flood of 1927, the Corps' primary strategy was to close off as many side channels as possible to increase the flow in the main river. It was thought that the river's velocity would scour off bottom sediments, deepening the river and decreasing the possibility of flooding. The 1927 flood proved this to be so wrong that communities threatened by the flood began to create their own levee breaks to relieve the force of the rising river."

"Environmental interests claimed the widespread development of levees by the Corps had been a "military campaign against nature," and they argued for different land management practices and return of some floodplains to natural wetlands. In the post-flood debates, others argued for a fiscal approach involving a mix of levees and wetlands, depending on the best economic solution." See also Shabman (1994) and Faber and Hunt (1994).

43. See Pinter (2005) and Pinter et al. (2008 and 2016) for discussion of both catastrophic and transfer flooding caused by Mississippi levees.

44. According to some claims, sediment flows to Lower Mississippi River have declined by 70 percent since 1850. See, for example, http://mississippiriverdelta.org/our-coastal-crisis/land-loss/ (accessed on May 23, 2019).

45. http://mississippiriverdelta.org/our-coastal-crisis/land-loss/ (accessed on May 23, 2019). See also Grist Staff (2008) .

46. As Sparks (2006) explains, in addition to the deterioration of floodplains, the Commercial approach applied to the Mississippi River has led to the deterioration of the Mississippi Delta. The reservoirs created in the upper reaches of the Missouri River now trap about half of the sediment that once flowed down the Mississippi. As a result, the Mississippi delta stopped growing around 1900 and in fact has been shrinking since then. More than 4,900 sq km of Louisiana's coastal wetlands—almost twice the area of Luxembourg—have been eroded by the waves of the Gulf of Mexico since the 1930s. According to the estimates by the US Geological Services, the current rate of loss of the delta equals to about two soccer pitch's worth of wetlands every hour. As McCully (2007, p. 3) notes, areas along the Louisiana coast look on a map more like watery lacework than landscape. The loss of wetlands and offshore barriers islands has increased significantly the vulnerability of coastal communities to hurricanes. It has also caused damage to local ecosystems.

47. See for information: https://www.weather.gov/dvn/071993_greatflood.

48. According to other estimates, about 700 levees failed.

49. As Changnon (1996, p. 264) notes, "The Corps of Engineers calculated that the flood-protection works (reservoirs and levees) on the Upper Mississippi had actually prevented an additional $19.1 billion in damages. Environmentalists countered, arguing that had the floodplains largely been left in their natural state, the 1993 flood would have been of lesser magnitude and the damages due to unwise occupancy of the floodplains would have been negligible. In the last 200 years, 57 percent of all the wetlands in the nine states with flooding have been destroyed." See also IFMRC (1994).

50. Estimates for 2008, 2011, and 2016 are from NOAA Centers for Environmental Information, available at https://www.ncdc.noaa.gov/billions/events.pdf (accessed May 21, 2019). The figures are adjusted for inflation. The damage figure for 2019 pertains only to the spring floods of the year and are from Cusick (2019).

51. Evaluating the relative roles of climate change and the Cordon approach in recent catastrophic Mississippi floods, Andrew Newman, an NCAR expert in hydrology and hydrometeorology, concluded that land-use changes, stormwater flows, and channelization of the Mississippi are "probably overwhelming climate change" as drivers of flood risk. Climate change, Newman and his colleague Andreas Prein believe, acts like an accelerant on flood conditions (see Cusick 2019).

52. See https://web.archive.org/web/20071015234208/http://eng.auburn.edu/admin/marketing/seminars/2007/l-roth.html.

53. According to Pinter (2006, p. 3), the flooding of New Orleans was "a levee disaster: the result of a flood-protection system built too low and protecting low-lying areas considered uninhabitable through most of the city's history." See Hersher (2018) for continuation of Mississippi levee building activities despite flooding getting worse over time.

54. See https://grist.org/article/gertz2/.

55. Some of the embankments along the Huang He River go back to 3,000 years or more into the past. See Kidder and Liu (2014) for more on the geological history of the Huang He River.

56. Sometimes there is so much sediment in the river it looks like chocolate milk.

57. See McCully (2007, p. 4). See also Needham (1971, p. 235).

58. Researchers have now furnished convincing evidence that the catastrophic flooding was in part a result of the levee construction or the Cordon approach. It is this system that converted a stable river into an unstable one. See, for example, Iacurci (2014).

59. Since historians began keeping records in 602 BC, the river has changed course 26 times and produced 1,500 floods that have killed millions of people. When the river's dykes were breached in 132 BC, floods occurred in 16 districts and a new channel was opened in the middle of the plain. Ten of millions of peasants were affected. The break remained for 23 years until Emperor Wu-ti visited the scene and supervised its repair. In AD 11, the Huang He River breached its dykes near the same place, and the river changed course and forged a new path to sea, 100 miles away from its former mouth. Repair work took several decades. In a tactic intended to halt the southward movement of Japanese soldiers from Manchuria before World War II, Chiang Kai-shek ordered his soldiers to breach the levees of the Huang He River and purposely divert its flow. At least 200,000, maybe millions, died, millions more were made homeless and the Japanese advanced anyway. Sometimes when the Huang He River floods it becomes like a flowing mudslide. The river normally carries an enormous amount of silt and the amount increases when it floods. During a 1958 flood, sediment levels were measured at 35 pounds per square foot, causing the river surface to become "wrinkled" (See Forbes 2015 and McCully 2007, p. 4).

60. Below is a list of hydroelectric power stations built on the Huang He River, arranged according to the first year of operation (in brackets):

Sanmenxia Dam (1960; Sanmenxia, Henan)

Sanshenggong Dam (1966)

Qingtong Gorge hydroelectric power station (1968; Qingtongxia, Ningxia)

Liujiaxia Dam (Liujia Gorge) (1974; Yongjing County, Gansu)

Lijiaxia Dam (1997) (Jainca County, Qinghai)

Yanguoxia Dam (Yanguo Gorge) hydroelectric power station (1975; Yongjing County, Gansu)

Tianqiao Dam (1977)

Bapanxia Dam (Bapan Gorge) (1980; Xigu District, Lanzhou, Gansu)

Longyangxia Dam (1992; Gonghe County, Qinghai)

Da Gorge hydroelectric power station (1998)

Li Gorge hydroelectric power station (1999)

Wanjiazhai Dam (1999; Pianguan County, Shaanxi and Inner Mongolia)

Xiaolangdi Dam (2001) (Jiyuan, Henan)

Laxiwa Dam (2010) (Guide County, Qinghai)

Yangqu Dam (2015) (Xinghai County, Qinghai)

Maerdang Dam (2016) (Maqên County, Qinghai)

As reported in Deng (2000), the seven largest hydropower plants (Longyangxia, Lijiaxia, Liujiaxia, Yanguoxia, Bapanxia, Daxia and Qinglongxia) had a total installed capacity of 5,618 MW.

However, due to the extremely high sediment content of the Huang He River, reservoirs behind the dams on this river generally are quickly silted up. For example, the reservoir behind the Sanmenxia Dam, built in 1960 with Soviet design and help, silted up in just two years. Various commercial diversions and uses reduced the flow of both water and sediment going downstream.

61. "China's Huang He River is the best-known example of suspended river. It is in places as high as 20 meters above the surrounding land—the height of a six-story building" (McCully 2007, p. 6).

62. See Leung (1996).

63. The delta appears to have shrunk from about 36,272 sq km used to about 8,000 sq km. See Min, Yao, Wang, Zhang, and Stive (2016) for discussion on Hunag He Delta.

64. The Huang He River has also suffered atrocious pollution. According to the Huang He River Conservancy Commission, waste and sewage discharged into the system in 2006 totaled 4.29 billion tons, with industry and manufacturing accounting for 70 percent; households for 23 percent; and other sources for about 6 percent. From its survey of more than 13,493 km of the river in 2007, the commission found that more than one-third of the river system was worse than UN-EPA's category Level 5, which implies that the river is unfit for drinking, aquaculture, industrial use, or even agriculture. See Branigan (2008) for details.

10

Open approach to rivers

The lateral version of the Ecological approach

10.1 Introduction

The adverse consequences of the Cordon approach gave rise to the Open approach, which is the lateral version of the Ecological approach. While the Cordon approach proceeds from the premise that river flows should always be confined to their channels, the Open approach argues that floodplains should remain open to river channels. The Open approach considers the river channel and its floodplains an organic whole and the artificial separation of the two—as under the Cordon approach—detrimental for both. Based on this realization, the Open approach discourages lateral interventions, such as embankments, channelization, canalization, etc. Instead, it advocates preservation and expansion of the natural links that connect rivers with their floodplains.

The Open approach was the norm for pre-industrial societies. This was in part because they lacked the technological capacity to carry out large-scale lateral interventions in major rivers. It was also because pre-industrial societies had a different attitude toward and philosophy regarding rivers. Instead of trying to prevent river overflows, they tried to benefit from inundations, making use of their various nurturing functions.

The post-industrial Open approach differs from the pre-industrial Open approach in that it does not result from a lack of technological capacity. Instead, it is, in a sense, an outcome of high level of technology acquired through the Industrial Revolution. It is the result of the realization that the irrational use of powerful technologies to disconnect floodplains from river channels does not serve the best, long-run interests of rivers and the people living in their basins.

Apart from increasing awareness about the general necessity for environmental protection, the accumulating evidence on the adverse consequences of the Cordon approach, as noted in chapter 9, played an important role in the emergence of the Open approach. In particular, increasing incidence of embankment failures, with devastating consequences, and rising costs of flood damages have convinced many that constructing embankments in an effort to prevent floods may not be the best strategy. Climate change, with increasing and erratic precipitation in many places, has accelerated this change in thinking. The European

Rivers and Sustainable Development. S. Nazrul Islam, Oxford University Press (2020). © Oxford University Press.
DOI: 10.1093/oso/9780190079024.001.0001

Union's 2007 Directive on Floods, the initiatives of the USACE to "loosen up" the Mississippi levee system, the "Room for River" project of the Netherlands, and other such initiatives reflect this paradigm shift.

The idea behind the Open approach is sometimes expressed using other terminologies, such as "flood management" (as opposed to flood prevention or elimination), "living with floods," "adoption of flood-proofing measures," "adoption of soft (as opposed to hard) measures," etc. As noted in chapter 9, the Open approach has also been referred to as the "expansionist" approach, in the sense that it expands the area for rivers to spread their peak season flows. This is in contrast to the "contractionist" approach, which tries to contract this area and conforms to the Cordon approach. The basic idea behind these different terms and expressions is similar, namely that regular river inundations are a normal, natural, and desirable phenomenon, and it is better to accept them and make use of them, instead of trying to prevent them.

This chapter provides an overview of the Open approach, focusing on its merits, progress, and prospects. The discussion is organized as follows. Section 10.2 discusses how the Open approach can be more conducive to sustainable development. Section 10.3 shows that the Open approach is not a passive approach but requires sustained activities along many dimensions. The rest of the chapter follows the progression of the Open approach. Section 10.4 reviews the progress of the Open approach in Europe, focusing on the European Union's Directive on Floods and its implementation. Section 10.5 does the same for the United States, including the recent policy changes with regard to the Mississippi levee system. Section 10.6 reviews the progress of the Open approach in other parts of the world. This part of the review is short, in part, because the contrast between Cordon and Open approaches is discussed in more detail in the context of the Bengal Delta in chapters 11 and 12. Section 10.7 concludes.

10.2 Open approach and sustainable development

Sustainable development, as discussed in chapter 1, has three dimensions, namely economic growth, social development, and environmental protection. The fact that the Open approach is more conducive to protection of the environment is quite obvious in the light of the discussion in earlier chapters. However, it is important to recognize that the Open approach is helpful for economic growth and social development too.

The general points of the discussion that follows have been made in the discussion in chapter 7 showing how the Ecological approach can be conducive to sustainable development. However, here I strive to make that discussion more specific to the Open approach and the different ways it can be helpful for sustainable development.

10.2.1 Open approach and sustainable agriculture

The Open approach helps agriculture to be sustainable in several ways.

First, it allows sedimentation of floodplains to continue. River sediment includes silt, which, as a natural fertilizer, helps to revitalize the soil and reduce dependence on chemical fertilizers and the toxic runoff that they create.

Second, the Open approach can help sustain groundwater by allowing natural irrigation through river overflow to continue and reduce the necessity for extraction of groundwater for irrigation. As noted earlier, groundwater is often a limited geographical resource and not fully rechargeable. Reliance on river overflow also helps to avoid the salinity and soil degradation that results from excessive dependence on irrigation using groundwater. The Open approach can thereby help crop cultivation to be sustainable.

Third, the Open approach can help to promote sustainable, open fisheries by preserving waterbodies inside floodplains. The reduction of chemical runoff that the Open approach can bring about can be of further help in this regard. It can also help to preserve the indigenous fish stock and allow even closed fisheries to focus on a wide variety of fish species, rather than on just a few, often alien, ones.

Fourth, by allowing regular river overflows, the Open approach can promote mixed farming, combining crop production with fisheries and combining cultivation of a variety of crops rather than fostering monocrop agriculture. Mixed farming, in which waste from one type of production serves as input for another, can help to achieve a circular and sustainable economy.

Fifth, the Open approach is also conducive to organic farming by relying more on silt as fertilizer and by promoting mixed farming, which generates organic fertilizer in the form of crop waste, and by offering natural pest control.

Finally, the Open approach promotes sustainability of agriculture not only from the environmental viewpoint but also from the social viewpoint. By decreasing the necessity of purchased inputs and by preserving common property resources, the Open approach helps smallholders and farmers with less capital to succeed in agriculture. Similarly, by promoting capture fisheries and reducing the capital required for farm fisheries, the Open approach is more favorable to artisan fisherfolk and fishing communities with less capital.

10.2.2 Open approach and sustainable urbanization and industrialization

The Open approach can be conducive to sustainable urbanization and industrialization by promoting settlement and expansion of industries only on grounds having an elevation higher than the flood level. This can help them to avoid the problem of catastrophic flooding. Second, it can also help urbanization and industrialization to be geographically more compact, saving resources in many ways. Third, the Open approach can help urbanization and settlement to depend on the natural availability of water rather than having to depend on pumped water, which may not be a sustainable option, as noted earlier.

10.2.3 Open approach and climate change

An important merit of the Open approach is its ability to help the world cope better with the effects of the climate change. As mentioned earlier, climate change is increasing, in most places, both the total volume and the seasonal variation of precipitation. It is also making precipitation patterns erratic, in terms of both timing and volume. The result is more untimely and unexpected floods, which embankments often fail to contain. The Open approach can cope with this new situation more readily, because it creates more space for river overflows and hence greater capability to accommodate wider amplitudes of precipitation variation.

Another important way in which the Open approach can help cope with climate change is by protecting deltas against sea level rise. As noted earlier, the Open approach allows rivers to deposit more sediment on coastal plains and raise their elevation, thus countervailing sea level rise. The extent of this countervailing effect depends on the volume of sediment carried to the delta, the size of the delta over which the sediment spreads, and the rate at which sea level is rising near the delta. Through the same process, the Open approach also helps to resist salinity intrusion caused by sea level rise.

Apart from these direct ways, there are many indirect ways in which the Open approach can help the world cope with climate change. For example, by reducing dependence on chemical inputs, the Open approach can help to reduce GHG emissions that result from production of these inputs. Second, it also helps to conserve electricity required for pump-based irrigation and drainage. Third, it can promote water-based transportation, which is less GHG intensive than land-based and airborne transportation. Fourth, it can help conserve resources spent on construction and maintenance of embankments and make the freed up resources available for other climate change mitigation and adaptation activities.

10.3 Open approach as an active strategy

As is true for the Ecological approach in general, the Open approach is an active strategy and should not be regarded as passive. Though it discourages construction of embankments, it requires activities of other types, which are often more encompassing, require more persistence, and involve more participation of the people living in river basins. These activities may be of both the "hard/structural" and the "soft/non-structural" variety. As noted earlier, it is not accurate to equate Cordon and Open approaches with hard/structural and soft/non-structural approaches, respectively. While the Cordon approach focuses mostly on hard/structural measures, the Open approach is more flexible, allowing scope for both hard/structural and soft/non-structural measures. In presenting the activities recommended by the Open approach, we follow here a more analytical distinction and classify them into two groups. The first are flood-proofing measures, which are activities undertaken to better cope with floods and can be both hard/structural and soft/non-structural. The second group can be considered flood-regulating measures, comprising activities that are aimed at regulating the extent and timing of flooding and entail mostly hard/structural measures, such as construction of discontinuous, temporary and low height embankments and construction of canals.

10.3.1 Flood proofing measures

Flood proofing measures, as just explained, are aimed at coping with and reaping the benefits of flooding. Activities that can help achieve this goal are described in the following.

10.3.1.1 Dig-elevate-dwell settlement principle

One of the most effective ways to live with floods is to adjust the settlement pattern. First, it is better to limit settlement to areas that are above the flood level. As noted earlier, this may lead to more compact settlement, which has many other benefits. In particular, it may help in conserving agricultural land, which is a scarce resource in many countries. Compact settlement may also make it easier to provide residents with modern utilities and other services and amenities. However, if for some reason it is absolutely necessary to settle on places that are below flood level, it is useful to follow the "dig-elevate-dwell" rule. Under this rule, people dig ditches to get the soil, which is then used to raise the elevation of the land on which to construct dwellings and other necessary structures. The ditches become "ponds" which serve as reservoirs of water necessary for various household and production purposes. Dig-elevate-dwell is the rule that pre-industrial societies used in settling on floodplains. The same rule can be followed in the post-industrial era, whose technologies can be used to help follow it more effectively and on a wider scale. In places where it is either difficult or impossible to follow the dig-elevate-dwell rule, dwellings may be constructed on pillars to ensure that settlement is not below-flood-level.

10.3.1.2 Minimization of obstruction to waterflows on floodplains

The Open approach requires not only keeping floodplains open to river channels but also minimizing obstructions on them to both cross-sectional and longitudinal waterflows. Unobstructed cross-sectional flows are required for river overflows to spread across floodplains and then to recede to river channels when the river stages fall. Unobstructed longitudinal flows are required so that floodplains can serve not only as storage space for river overflow during the high season but also as an additional passageway for river water to move down to the sea. Floodplains can play this role because their gradients in general are also toward the sea. Thus, floodplains can help the river overflow to move down to the sea much more quickly than if river channels alone had to perform this carrying function.

Of course, it may be difficult to keep floodplains completely unobstructed. However, there are many ways in which the obstructions can be minimized. First of all, the policy of compact settlement, noted earlier, can help to keep more of the floodplain free from obstructions. Second, highways and railways can be aligned with the direction of rivers so as to reduce the obstruction of longitudinal flows. Also, enough bridges and culverts may be allowed in constructing highways and railways to reduce obstruction to cross-sectional flows. Third, highways can be constructed as causeways to allow free flow of water underneath. This will also help to conserve land, which is a scarce resource, as noted above, particularly in densely populated countries. Thus, the Open approach does not rule out settlement and development on floodplains. However, it urges that these be carried out in a way that minimizes obstructions to the free flow of flood water.

10.3.1.3 Re-excavation and dredging

Dredging and re-excavation of waterbodies inside floodplains is an important line of activity under the Open approach. Dredging of canals is necessary to let river water reach deep inside floodplains. The same canals can also serve for drainage, taking the rainwater falling inside floodplains to rivers. Dredging of lakes, ponds, and other closed (during the lean season) waterbodies is necessary to create more space for water in high season, keep the flood level low, and store water for use during the lean season. As noted earlier, healthy, annually recharged waterbodies of ample size and depth can be of significant help for sustainable crop cultivation and fisheries; maintaining and using waterways; and attaining greater temperature balance across seasons. They can also be aesthetically pleasing.

10.3.1.4 Land leveling and terracing

Land leveling and terracing is another important line of activity under the Open approach. These are helpful for quick passage of water across floodplains and also for optimal management of regulated flooding and for reaping its benefits.

10.3.1.5 Re-direction of crop research

An important way to reap the benefits of the Open approach is to make use of modern genetic science to invent high-yielding crop varieties that can be cultivated under flood conditions. Until recently, more research was directed to crops grown under controlled irrigation and requiring chemical fertilizers and pesticides. Relatively less attention was given to crops to be grown under natural irrigation and siltation. The Open approach requires redirection of crop research toward the latter crops. Many pre-industrial societies, through centuries of crop selection, devised crop varieties that can withstand deep flooding. However, their yields are generally low. Modern science can be applied to improve their yields and to strengthen their flood-resistant properties. Instead of discarding indigenous, pre-industrial crop varieties, the Open approach advocates protecting indigenous crop diversity and making those crops more productive and more suitable to the changes brought about by climate change.

10.3.1.6 Enhancing groundwater recharge rate

Modern technology can be used to increase the rate of recharge of groundwater tables. This rate varies depending on geological and hydrological conditions and can be limited in many areas. Increasing the recharge rate can improve the use of underground aquifers as a vast storage space for peak season overflows that can then be used during the lean season. Similarly, modern ground sensing technology may be used to determine the location and level of groundwater tables and plan how river overflows can be used to replenish them. Modern technologies can also be used to inject more of the river overflow into underground aquifers.[1]

10.3.1.7 Better use of weather forecasts and other scientific information

The achievements of modern science can help the Open approach in many other ways. One example is the advances in weather forecasts due to satellite observation facilities. These forecasts may be used to better inform the population living in floodplains about oncoming floods and help them prepare for the floods.

The discussion above shows that flood-proofing measures under the Open approach can be both soft/non-structural and hard/structural. For example, re-direction of crop research or better use of weather forecasts to predict flood levels are non-structural or soft measures. On the other hand, consolidation and elevating the ground level of settlements, construction of causeways, etc. are examples of hard/structural measures. Land leveling and terracing may also be considered as structural measures. In all cases, the goal is to keep floodplains as open as possible to river overflows and to make the best use of the benefits of overflows. Thus, flood-proofing measures are not directed at influencing the river overflow. However, the Open approach also allows measures directed at regulating the extent and timing of river overflows, in particular through construction of discontinuous or low-height embankments, as discussed next.

10.3.2 Flood-regulation measures: embankments under Open approach

The Open approach is cognizant of the fact that untimely and unusual inundation can do considerable damage to the floodplains, and it is necessary to do as much as possible to prevent such events and the damage they cause. It allows use of embankments for this purpose. However, embankments under the Open approach are meant for regulating instead of severing connections between river channels and floodplains. Their purpose is to control the timing and extent of inundation and not to prevent inundation. The goal of these embankments is flood management and not flood prevention or flood elimination. As is clear from earlier discussions, river overflows, under the Open approach, are, in general, a welcome phenomenon and regarded with a positive attitude.

In view of the above, embankments under the Open approach differ from those under the Cordon approach in several ways. First, these embankments are generally discontinuous, with adequate openings for river water to flow in to floodplains and for rainwater accumulating in floodplains to flow out to river channels. In other words, openings in embankments under the Open approach are meant for river inundation to take place on a regular basis, as would have occurred had there been no embankments. The openings could be equipped with appropriate regulators to control the timing and extent of inundation. Design, construction, operation, and maintenance of these embankments obviously require considerable effort. By contrast, embankments under the Cordon approach are generally continuous, that is, without breaks. In cases where breaks are allowed, they are generally for drainage and not for inundation. Also, because of the general philosophy of viewing rivers as outsiders, even these drainage openings soon become dysfunctional, making these embankments totally closed.[2]

Second, embankments under the Open approach can be of low height, to hold off the inundation only for a certain period, until, for example, the crop is harvested. Low height embankments may also be used to retain high season river flow in the floodplains to be used during the lean season. Third, embankments under the Open approach may also be seasonal, removable, and temporary. Rubber dams are an example of a river intervention that can be of temporary nature.[3] Similar technology can be used to regulate openings of discontinuous embankments under the Open approach.

This discussion makes it clear that the Open approach is not a passive approach. It entails both flood-proofing and flood-regulating activities, both of which include hard/structural measures as well as soft/non-structural measures. The Open approach therefore requires considerable initiative, imagination, and effort.

It is encouraging that after many of decades of efforts to contain rivers to their channels only, policymakers in different parts of the world are realizing the importance of restoring and preserving the connections between rivers and floodplains and allowing river overflows to spread over floodplains. In other words, they are moving away from the Cordon approach and moving toward the Open approach. In the following, I provide a brief review of this process.

10.4 Open approach in Europe

Europe is currently at the forefront of the switch toward the Open approach. As noted in chapter 9, about 90 percent of European floodplains have disappeared since the first Industrial Revolution due to the spread of the Cordon approach. As a result, flooding has intensified in the remaining 10 percent of floodplains that are open to rivers.[4] Furthermore, the value of flood damages has increased over time, because, in the event of breaches in embankments, more built-up areas now go under water. Meanwhile, climate change is aggravating the situation. Both peak flows that cause floods and low flows that cause draught, are becoming more extreme.[5]

The major river floods of the 1990s were indications of what is to come, as were the floods of the first decade of the twenty-first century, including the 2002 catastrophic flooding along the Danube and Elbe Rivers, followed by the severe 2005 flood that provided the background for the EU Flood Directive of 2007. Though the knowledge and recognition of the beneficial relationship between floods and ecosystems existed before, these recent events forced this recognition up to the national policymaking level.[6]

10.4.1 European Union Flood Directive, 2007

The European Union's Flood Directive, issued on October 23, 2007, signifies an important advance toward the Open approach to rivers. Formally known as "Directive 2007/60/EC of the European Parliament and of the Council on the Assessment and Management of Flood Risks" (EU 2007), it begins by recognizing that "floods are natural phenomena which cannot be prevented" (p. 1). It recognizes floodplains as "natural retention areas" for river overflows and points out that human settlement and construction on floodplains, in combination with climate change, are increasing the likelihood and damages of floods (p. 1).

As noted in chapter 6, the EU Water Framework Directive (WFD) required each river basin district to develop river basin management plans (RBMPs), aimed at achieving "good ecological and chemical status" (EU 2000). The Flood Directive goes further and asks that member states, in developing policies related to water and land uses, take into account the likely effects of these policies on flood risks and the management of flood risks (EU 2007). It classifies floods into several categories, namely river floods, flash floods, urban floods, and floods from sea in coastal areas, and asks for the preparation of "flood hazard maps"[7] and "flood risk maps"[8] showing potential adverse consequences under different flood scenarios. It calls for the formulation of Flood Risk Management Plans (FRMPs) based on these maps. In particular, it requires RBMPs to take into account "flood conveyance routes" and to protect "areas which have the potential to retain flood water, such as natural floodplains" (p. 5). It asks Member States to give rivers "more space," and to that end, not only to preserve (maintain) the floodplains but also "restore" floodplains that have been lost, that is, "built up." It calls for "sustainable land use practices," improvement of water retention, and "controlled flooding" (p. 5). It enjoins RBMPs to focus on "non-structural initiatives" and to review the "effectiveness of existing man-made flood defense infrastructures" (p. 5).[9] It asks for RBMPs to be reviewed periodically in view of climate change.[10]

The EU Flood Directive also advocates basin-wide strategies and calls for the formulation one single plan for a river's entire basin. It notes three possibilities in this regard. First, the entire river is contained in only one Member State. Second, the river extends to more than one Member States. Third, the river extends to countries that are not in the European Union. In the case of the first, it puts the responsibility on the Member State to prepare the single plan. In case of the second, it asks relevant Member States to produce one single international FRMP or a set of FRMPs, coordinated at the level of the international river basin district. In the third case, it asks EU Member States to cooperate with non-EU countries to produce a single plan or a set of FRMPs, coordinated at the level of the international river basin district (pp. 5–6).

With respect to international rivers, the EU Flood Directive conforms with the spirit of the 1997 UN Convention regarding rivers and follows its "no harm" (to co-riparian countries) principle. In fact, it advocates "solidarity" and asks member states to refrain from including in their FRMPs any measures or actions which significantly increase the risk of flooding in other Member States, unless these measures have been coordinated and an agreed solution has been found among the Member States concerned (p. 5).[11] It is clear that the assessments and recommendations of the EU Flood Directive, by and large, agree with the Open approach to rivers.

10.4.2 Reflections of the Ecological approach in the EU flood risk study

Europe's move toward the Open approach becomes more evident from the more recent follow up on the Flood Directive. For example, the 2016 European Environmental Agency (EEA) Report on Flood Risks and Environmental Vulnerability (RFREV), titled *Exploring the Synergies between Floodplain Restoration, Water Policies, and Thematic Policies* (EEA 2016), reflects the propositions of the Ecological approach more clearly. It is based on the data and information contained in Preliminary Flood Risk Assessments (PFRA) that EU Member States produced in following up on the Flood Directive of 2007.[12] Aiming to support the implementation of the Flood Directive, the RFREV examines the synergies between floodplain restoration activities and other EU environmental policies, such as the Bird Directive (1979), Habitat Directive (1992), the *Natura* 2000 program (1992), Biodiversity 2020 Strategy, and the Water Blueprint (2012). (See Box 10.1 for a brief description of these various policy initiatives.)

Box 10.1 Pro-environment initiatives of the European Union

One of the important messages of the EU Flood Directive is that flood risk and river management activities need to be carried out in conjunction with the implementation of other EU pro-environment directives. To better understand the implications of this message, it is useful to be aware of these directives. To that end, some of the recent EU pro-environment directives and initiatives are listed here, along with annotations.

Bird Directive (1979): Concerned with the decline of bird species, Member States unanimously adopted Directive 79/409/EEC in April 1979. The oldest piece of EU legislation on the environment and one of its cornerstones, it requires establishment of Special Protection Areas (SPAs) for birds. On its amendment in 2009, it became Directive 2009/147/EC.

Bern Convention (1982): The Convention on the Conservation of European Wildlife and Natural Habitats, popularly known as the *Bern Convention*, is a binding international legal instrument that aims at conserving wild flora and fauna and their natural habitats. It is particularly concerned about protecting natural habitats and endangered species, including migratory species. A few African countries are also signatories of this convention.

Habitat Directive (1992): More formally, the Directive on the Conservation of Natural Habitats and of Wild Fauna and Flora, this requires the setting up of Special Areas of Conservation (SAC), meant to protect 220 habitats (particular types of forest, grasslands, and wetlands) and approximately 1,000 species (other than birds) listed in annex I and II of the directive. The SACs are to be selected from the Sites of Community Importance (SCIs) determined at the national level by the Member States.

Natura 2000 (1992): This is the network of protected sites across Europe that include SPAs (under the Bird Directive) and SACs (under the Habitat Directive). It is, in turn, a part of the Emerald Network of Areas of Special Conservation Interest (ASCIs) launched by the Council of Europe as part of its work under the Bern Convention.

Biodiversity 2020 Strategy: This follows from EU commitments under the UN Convention on Biological Diversity (CBD), complemented by additional commitments made by European leaders in 2010. The broad goal is to halt the loss of biodiversity and degradation of ecosystem services in the European Union by 2020 and restore them in so far as feasible, while stepping up the EU contribution to averting global biodiversity loss. It aims at maintaining and enhancing ecosystems and their services by 2020 by "establishing green infrastructure and restoring at least 15 percent of degraded ecosystems."

Water Blueprint 2012: It outlines actions focused on better implementation of current water legislation, integration of water policy objectives into other policies, and filling the gaps regarding water quantity and efficiency. The goal is to ensure sufficient quantity of good quality water to meet the needs of people, the economy, and the environment throughout the European Union.

The RFREV begins by offering useful conceptual discussions that conform with the propositions of the Ecological approach.[13] For example, it distinguishes three types of services that floodplains offer, namely (i) regulating services, such as protection against floods and water purification; (ii) provisioning services, such as fertilization of soil, supply of nutrients, and sustenance of open fisheries; and (iii) cultural, such as recreation, tourism, etc. (EEA 2016, p. 6). Most of these services are included in the list of floodplain nurturing functions performed by regular river overflows, presented in chapter 7.

The RFREV observes that floodplains have come under pressure from both existing and new activities, such as urban settlement, infrastructure building, or agricultural developments (p. 60). It distinguishes between "grey" and "green" infrastructures aimed at dealing with floods. The former refers, by and large, to the structures built under the Commercial and Cordon approaches, while the latter are basically those promoted by the Ecological and Open approaches.[14] The RFREV notes that the supply of ecosystem services can be optimized using green infrastructure, and relying on them is compatible with many other EU directives, strategies, and policies, such as the Habitats Directive

(EEC 1992, Art 10) and Biodiversity strategy 2020 (EU 2011) (see EEA 2016, p. 21). It emphasizes the necessity of reducing "hydraulic resistance"—a term that it uses to refer to the obstruction to water flows caused by structures built on floodplains (p. 49). These are also the propositions of the Open approach.

The RFREV emphasizes the important point—noticed earlier in this book— that seasonal variations in river flows are necessary for river ecology. As it puts it, "ecosystems usually need varied intra- and inter-annual flows to remain functionally intact" (p. 12). It notes that good ecological status requires going beyond the minimum discharges (during dry periods) and taking into account "the full range of discharges, from base flows (including low flows) to flood regimes with different magnitudes, frequency and duration" (p. 12). It also notes that regular floods serve as ecological "refueling" events, while extreme floods can serve even as "reset" buttons because of "the drastic way they change the landscape (p. 61)."[15] The RFREV thus undercuts an important rationale of dams, namely that they stabilize river flow by doing away with seasonal variations of water flows. It also questions the "degree to which the water level is managed within narrow boundaries" to support navigation, hydropower and other economic activities (p. 13).

The RFREV notes the important connection between the flood management approach of a country and the Sendai Framework for Disaster Risk Reduction adopted at the third UN World Conference on Disaster Risk Reduction 2015–2030 (United Nations 2015b), held in Sendai, Japan, in March 2015. It draws attention to the recognition in the Sendai Framework that reconstruction of ecosystems and nature-based solutions are crucial in the protection against disasters.[16] Similarly, it notes that the EU Flood Directive and other policy initiatives are conducive to achieving the objectives of the Ramsar Convention (UNESCO 1971), because many wetlands that this convention aims to protect require preservation of floodplains.[17] It also notes that "ecological-economic assessment," as recommended in the Flood Directive, is a basic requirement of sustainable development (EEA 2016, p. 33).[18] Reflecting the 1997 UN Convention on International Watercourses, the RFREV points to an important criterion that a "basin-wide" approach to rivers should satisfy, namely that it should "avoid passing on negative consequences further downstream" (p. 52). The RFREV also refers to the Flood Directive's emphasis on "preparedness measures, such as flood forecasting and warning" (p. 7).[19]

The RFREV notes that it is in implementation of measures where various directives and policies come together. In this context, it recognizes the useful role of the Natural Water Retention Measures (NWRM) program (EU 2014), included in the EU Flood Directive and representing a "nature-based approach to pursue the objectives of water management, providing a variety of co-benefits in terms of biodiversity enhancement, greenhouse gas mitigation, energy saving or rural development opportunities" (EEA 2016). RFREV notes further that NWRM calls for an integration between the Water Directive (EU 2000) and the Flood Directive (EU 2007) as well as between nature legislation and all policy fields requiring careful coordination of water and land planning needs, and that it calls for better environmental options (such re-meandering of rivers) for flood risk management and alternatives to hard (grey) infrastructure (EEA 2016, p. 34).

The RFREV further notes that similar implementation mechanisms were often called by other names, such as "building with nature," "Room for River," "green measures," etc. It observes that no matter what they are called, they basically advocate "working with natural processes" to "maximize the common goals and objectives of water management, economic development, nature conservation, and ecosystem services" (p. 64). The RFREV observes that while levees and dykes have the greatest impacts on a river's ecosystem, these are relatively more easily reversible than the morphological changes that rivers undergo (p. 50).

In view of the above, the RFREV calls for Strategic Flood Risk Management (SFRM), which includes maintaining, restoring, and strengthening the long-term health of all associated ecosystems, societies and economies. It notes that restoration of former floodplains and habitats is an important task under SFRM (p. 31). The RFREV calls for natural flood risk-reduction measures that would contribute to the restoration of the characteristic hydrological and geomorphological dynamics of rivers and floodplains and ecological restoration for biodiversity (p. 33).[20] The SFRM, advocated by the RFREV, suggests that working with NWRMs needs to be a part of the core toolkit of all flood risk managers. The RFREV calls for safeguarding ecosystem services and adoption of soft-path measures (e.g. land use changes or wetland restoration). Overall, it points out that flood risk management should work *with* nature, rather than *against* it.[21] Needless to say, the flood risk management strategy suggested by RFREV resonates well with the Open approach put forward in this book

10.4.3 River restoration activities in Europe

Flood risk reduction activities along the lines just described have often, not surprisingly, coincided with river restoration activities. The RFREV observes that the Flood Directive was implemented differently across the EU and even within the same country, so that progress was not uniform.[22] However, it provides a detailed list and description of river restoration activities undertaken in Europe in recent years (see EEA 2016, Boxes 2.5 and 2.6, p. 35).

Interest in river restoration had already increased in Europe in the 1990s. In fact, a European Centre for River Restoration (ECRR) was established in 1995. A RiverWiki-database for the RESTORE (Rivers: Engaging, Supporting and Transferring knOwledge for Restoration in Europe) project offers information on almost 1,000 river restoration case studies. It notes that initially restoration activities focused on river channels and on aquatic ecology. However, following the FLOBAR2 (Floodplain Biodiversity and Restoration) project (Hughes 2003) and the establishment of ECRR, more attention was given to floodplain restoration "as an essential part of sustainable water management." Restoration activities so far have focused on its floodplain reconnections, riparian tree planting, and removal of exotic plants as well as creating wetlands and backwaters.

Many river restoration projects in Europe were initiated with help from LIFE, the EU program for environment and climate action (EEA 2016, p. 35).[23] River restoration actions have also been taken under NWRM, noted earlier. Maximization of the use of NWRM is one part of the specific objectives of the Water Blueprint (EC 2012). The NWRM initiative catalogues about 125 case studies illustrating hydromorphology measures aimed at river wetland and floodplain restoration and management, restoration and reconnection of seasonal streams or oxbow lakes, elimination of riverbank protection, re-naturalization of polder areas, etc.

10.4.4 River Mur—a case study

One successful case of river restoration is that of the River Mur, which flows through Austria, Slovenia, Hungary, and Croatia before reaching the Tiver Drava, a tributary of the Danube River (Figure 10.1).[24] The organization managing the Mur basin was awarded the European River Prize during the sixth European River Restoration Conference (ERRC) in October 2014. The Upper Mur is considered one of Europe's most ecologically valuable rivers. It is the natural breeding site of Danube salmon and holds within its basin Austria's second largest alluvial forest, which is one of Europe's most species-rich habitats. The application of the Cordon approach to this river, dating back to the late nineteenth century, had separated the river's loops, branches, and floodplain forests, which are important for the health of its natural systems. Ecological conditions for flood forest tree species in its floodplains were deteriorating due to water shortages in oxbows, side arm channels, and soils. Many dams had also been constructed in the Upper Mur, affecting both river morphology and basin ecology. Restoration activities have focused on remedying the adverse effects of dams and embankments, to the extent possible, and preventing further damage.

In Slovenia, the River Mur corridor is up to 1 km wide and contains a high variety of typical plant and animal communities ranging from pioneer to mature stages. The river basin has large floodplain forests and side arm systems. Due to its high biodiversity, a large part of the Mur corridor in Slovenia has been designated a Natura 2000 site.[25]

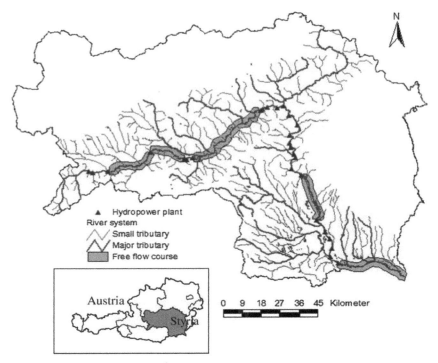

Figure 10.1 Restoration and preservation of the River Mur
Source: M. Getzner, M. Jungmeier, T. Kostl, and S. Weiglhofer (2011).

The mouth of the River Mur represents one of the last remaining preserved lowland rivers in Europe. Both the Drava and Mur are unrestrained here and continually create new habitats and restore existing ones with high ecological diversity. Restoring and recovering natural river habitats by reconnecting them with the dynamic river system has been the focus of River Mur management. These measures are offering environmental benefits, better flood-protection, and new natural recreation areas for residents. To preserve the free flowing parts of the river, there are plans to ban construction of dams in them. Management of the Mur offers a good example of policy integration and stakeholder dialogue, both of which are vital for successful river basin management (EEA 2016, p. 35).

10.4.5 The Netherlands' "Room for River" project

The Netherlands' "Room for River" project provides another example of moving away from the Cordon approach and toward the Open approach. For such a change to occur in the Netherlands is remarkable, because it is renowned for dykes and polders. As noted earlier, about one-third of the country comprises land reclaimed from the sea through construction of polders. Many of these polders therefore face the sea. Dykes have also been constructed extensively along the country's rivers, and channelization and canalization of rivers have been widespread.

However, a new thinking process started in the 1990s. Dutch confidence in the Cordon approach was particularly shaken by the near disaster in 1995, when the Meuse and distributaries of the Rhine almost burst their banks, and about a quarter of million people had to be evacuated. Dutch policymakers realized that the strategy of confining rivers to channels may not work for ever, particularly in view of climate change, which is altering calculation parameters and leading to unpredictable situations. To cope with changing conditions and an uncertain future, Dutch policymakers were forced to realize that "raising dykes' height" is no longer a panacea. Instead, they undertook a major policy review and came to see the merit of the Open approach. According to W. van Leussen of the Department of Public Works and Water Management, "the preferred solution is now the restoration of natural processes," and especially opportunities to create more space for the river. This means that the floodplains should only be used for necessary river-related activities, while measures should be taken to give the river more room to expand. Following this change in approach in 2006, the Dutch policymakers initiated the "Spatial Planning Key Decision," popularly known as the "Room for Rivers" project, the basic goal of which is to loosen the confines of river channels and allow more space for river water to spillover, when necessary.

To achieve this goal, the Dutch are doing several things. The first is to set dykes back further so as to open up more of the floodplains to river overflows. Second, they are creating flood bypasses, called green channels, allowing high stage river water to move out to floodplains beyond the dykes, including one around Veessen-Wapenveld. To minimize any damage such overflows may cause, the Dutch are relocating a large number of villages and townships from the floodplains to higher ground. Third, they are increasing the depth of floodplains in some places, so that these can hold more water and increasing the depth of side channels so more water can flow out through them. Fourth, the Dutch are removing obstacles to river flows, lowering, for example, the height of groins and removing the hydraulic bridge at Oosterbeek.[26] The Dutch have been an important promoter of the Cordon approach and polder construction in many developing countries. It may be hoped that their policy advice to these countries will reflect the change in thinking and practice in their own country.

10.4.6 Open approach in France

Significant moves toward the Open approach can be seen in France, where much more attention is now given to keeping floodplains open to rivers. The French National Strategy for Flood Management (FNSFM) notes that preventing construction on floodplains is the most effective way to limit damage from floods. Proceeding from this realization, the FNSFM does not allow new urban development on floodplains, even when these are protected against some level of risk (EEA 2016, p. 44). The general principles adopted by the French strategy include the following:

- Preserving floodplains in non-built-up areas, wet areas, and dune areas along the coastline;
- Banning building in areas that are extremely vulnerable to flooding;
- Limiting the presence of sensitive equipment and reducing its vulnerability in flood risk areas to avoid excessively complicated crisis management;
- Adapting all new construction projects, if they can be built, to the risk in question, in areas vulnerable to flooding;
- Ensuring that building is not allowed behind levees except where this is justified in urban areas or in areas of strategic importance;
- Identifying areas that are dangerous to human life and studying how to make existing populations safe in these areas, through not only monitoring, forecasting, warning, and evacuation measures but also through relocation projects, by reinforcing protection, or retention works (EEA 2016, p. 44).

10.4.7 Necessity for further progress in Europe

Despite the progress described above, halting and reversing the general degradation of river-related ecosystems remain formidable challenges even for many European countries. For example, local communities in France, under a law passed in 1995, are required to formulate "Plans for the Prevention of Flood Risk" to restrain development on floodplains. However, RFREV informs that progress has been slow and by 2004 only a third of the 10,000 communities at risk had prepared their plans. An important problem faced in this regard is path dependence. As RFREV notes, it is almost impossible to go back to a complete natural state in areas that are highly developed or that have been developed for long periods of time EEA (2016, p. 33). Thus, much work remains to be done.[27]

10.5 Open approach in the United States

There is an increasing recognition of the merits of the Open approach in the United States too. In particular, the experience of recent Mississippi floods has prompted many to look for a "new approach" to rivers and floods. The ideas and measures associated with this new approach agree, in many respects, with the ideas of the Open approach, presented in this book.

10.5.1 Change in approach following the Great Mississippi Flood of 1993

While the Great Mississippi Flood of 1927 served as a major impetus toward the Cordon approach and construction of the Mississippi levee system, the Great Mississippi Flood of 1993 has set off a train of thought and line of activities which go in the opposite direction. Though not the dominant one, this line of thinking and activities is in conformity with the Open approach, to some degree.

The Great Mississippi Flood of 1993, as noted in chapter 7, affected about one-third of the United States and caused damages of an estimated value of $18 billion, the highest in the history of the country until Hurricane Katrina.[28] Alarmed by these developments, the government set up the Interagency Floodplain Management Review Committee (IFMRC) to study the situation and offer recommendations. In its report, the IFMRC (1994) called for a "new approach" to the management of floodplains and river watersheds, reflecting the realization that levee construction, channelization, and other such measures were not going to be enough to protect floodplains from similar disasters in the future. Instead, floodplains and wetlands need to be allowed to perform their "natural functions."[29] Proceeding from this realization, the committee's report recommended that the optimum strategy for limiting flood damage was to restrain development on floodplains. Its recommendations triggered a range of activities that reflect the ideas of the Open approach. With some overlap among them, these activities may be classified into the following types:[30]

Reclaiming wetlands and floodplains: A process was begun to reclaim wetlands along portions of the Mississippi and Missouri rivers so that these could serve as natural sponges able to soak up a part of the river overflow. It was also decided "to return" parts of the floodplains to rivers.

Relocating people from floodplains: Reclaiming floodplains required relocating people or at least readjusting their settlement pattern. The Hazard Mitigation and Relocation Act (HMRA), passed in December 1993, helped this process. This act made an important change in federal-state-local cost-sharing, allowing federal funding of up to 75 percent of eligible costs. Total federal funds ultimately awarded for flood relief were $6.2 billion, part of which was used for buyouts of flood damaged homes and businesses from willing sellers. The buyouts represented the complete removal of flood-prone structures and the dedication of the purchased land "in perpetuity for a use that is compatible with open space, recreational, or wetland management practices." Four entire communities, two in Illinois (Valmeyer and Grafton) and two in Missouri (Rhineland and Pattonsburg), were relocated. Altogether about 7,700 properties were bought, and the land was made open for river overflow. (Changnon 1996, p. 314; Pinter 2005). [31]

10.5.1.1 Restrictions on further building on floodplains

The IFMRC also recommended that the optimum strategy for limiting flood damage was to restrain development on floodplains. As a result, some restrictions were placed on construction of new structures on floodplains.

10.5.1.2 Restraints on channelization and other measures

One important change in policy that followed the IFMRC report was restraint on channelization. On-going and future plans for channelization were curtailed and in some cases channelization already implemented was partially reversed. Earlier in 1989, during the administration of George H. W. Bush, a policy of "no net loss" of wetlands was adopted. According to this policy, loss in wetlands due to stream channelization in one place has to be offset by the creation of new wetlands in another, a process known as "mitigation." The Great Mississippi Flood of 1993 reinforced this policy.

10.5.1.3 Modifications of the Mississippi levee system

Most importantly, the 1993 flood shook the confidence in the levees as a foolproof bulwark against flooding and prompted efforts toward loosening the levee framework in order to avoid catastrophic flooding in future. In 1990, the US Congress had already given the USACE a specific mandate to include environmental protection in its mission and to undertake restoration projects. As noted in chapter 7, several floodways and spillways were already included in the Mississippi levee system.[32] It was recognized that more such floodways and spillways are needed in order to cope with likely future floods.[33]

10.5.2 Pushback and reversal

However, not everybody was happy with this new approach. Many communities and townships, used to the protection of the levees, did not want this (albeit deceptive) protection to go away. They would rather see more public money spent on making the levees taller and stronger. Also, Congress gave responsibility for the restoration projects to the USACE—the very organization that had built the levees. It would not be surprising if some members of this organization either did not agree with the new approach or did not feel that enthusiastic about its implementation. Partly as a result of this pushback, measures toward reclaiming floodplains, relocating settlers, and restricting further development on floodplains slowed down and, in some cases, were reversed. As a result, by 2005 more than 28,000 new homes had been built on land flooded in 1993, most of them apparently protected by new and strengthened levees (Pinter 2005; McCully 2007, p. 11).

10.5.3 Impact of Hurricane Katrina on approach to rivers

The experience of Hurricane Katrina, as noted in chapter 7, led to diverse responses. One was to demand strengthening of the New Orleans levees and floodwalls to make these capable of withstanding a Category 5 hurricane (up from the current Category 3 standard). One estimate of the cost of such upgrading is $32 billion. As also noted in chapter 7, some steps have indeed been taken in that direction, though the full project embodying this idea is yet to be funded and implemented.

At the same time, doubts were voiced about the eventual efficacy of this line of effort, particularly in view of the well-known prediction, also noted in chapter 7, that the main course of the Mississippi may shift toward a shorter course to the sea either down the Atchafalaya River, breaking out of the Old River Control structures, or through Lake Pontchartrain, bursting through the Bonnet Carre Spillway. Such a shift would undermine the viability of New Orleans as a port and undercut the rationale for spending huge sums of money to upgrade the levees.[34]

In view of this, many observers have recommended "working with the forces of nature rather than against them" and accordingly build a new port and a new city of New Orleans at the mouth of Atchafalaya River and abandon the current port of New Orleans. In this scenario, the higher parts of the current New Orleans—along the bank of the Mississippi where the city originally developed—will remain as a historical and cultural treasure and its lower parts will be converted into natural areas and parks, open to periodical flooding.[35]

While the long-term future of the city of New Orleans remains unsettled, the new approach advocated by IFMRC (following the 1993 flood) finds reflection in the inclusion of efforts to reverse the loss of coastal wetlands as part of the post-Katrina strategy for New Orleans.[36]

10.5.4 Need for further progress of the Open approach in the United States

Despite the encouraging moves just described, the progress toward the Open approach in the United States is still limited.[37] As noted earlier, there are powerful forces opposed to it and especially to the policy regarding "net wetland loss" that partly reflects this approach. First of all, as noted earlier, many farmers, businesses, and dwellers neither want to move from nor adjust to open floodplains.[38] The levees protect them in most years and when the levees fail, residents receive compensation for damages from various public sources. Existing infrastructure and policy frameworks serve their material interests well, and these interests outweigh ecological concerns. Path dependence thus creates a major barrier to the switch to the Open approach.

It may be noted here that advocacy for the Cordon approach by the USACE may come from two sources. The first is conviction, the genuine belief that the Cordon approach is the only or the superior way of dealing with floods (or managing rivers). The second is material interests. Observers note that the desire for bigger budgets plays an important role for the USACE in deciding which projects to push for. Even during the time of President Franklin D. Roosevelt, the interior secretary blamed the USACE for "reckless and wasteful behavior" and termed it "insubordinate and self-seeking."[39] Jimmy Carter, while governor of Georgia in 1974, rebuked the USACE for their "false justifications" and "grossly distorted" analyses of costs and benefits of their proposed dams.[40] More recently, the USACE has again been accused of "dysfunctional and dishonest activities," of continuing to inflate benefits and hide costs, misrepresent environmental impacts, and justify its projects no matter what in order to "keep its employees busy" and "its Congressional patrons happy."[41] As McCully (2007, p. 4) notes, "the Corps—encouraged by its Congressional allies who relish bringing federal dollars back to their home districts via big water projects—continues to pour billions of dollars into old-style dam expansion, dredging, pumping, and levee-raising."

It is difficult to be sure which of these two factors drives the USACE position. However, it is clear that no matter which of these dominates, it is time to change. In this regard, it may be noted that the expertise of the USACE will not become redundant if the Open approach is adopted as policy. Earlier, I noted that hard/structural measures, even embankment construction, can be part of the Open approach tool-kit. It may therefore be possible to direct USACE activities toward flood-proofing and flood-regulation measures that are in conformity with the Open approach. Perhaps, some evidence of this may be seen in the fact that the same USACE is undoing some of its past errors in some places (such as the Kissimmee River in Florida and Sun Valley and Napa in California) and trying new management techniques.[42] This may suggest that with a radically changed incentive structure, the USACE may play an important role in implementing the Open approach to rivers too.

10.5.5 Growing support for the Open approach in US academia and civil society

While politicians, bureaucrats, technocrats, and the direct material beneficiaries of Cordon approach projects may continue to oppose the Open approach, there is now widespread support for this approach among US academia and civil society. In the academia there is an increasing consensus that "working with nature" is a better way than "working against nature," even to serve human interests (especially in the long run). More concretely, there is a growing consensus among academia and civil society that Cordon structures—levees, channelization, etc.—have done more harm than good, and the best way to deal with floods is to open up floodplains to rivers, allowing the latter to claim what they need.[43]

This movement can be seen in the recent influential article (Munoz et al., 2018) published in *Nature*, which shows scientifically that river engineering has contributed to the increasing frequency and intensity of Mississippi flooding. The authors note that one of the difficulties reaching conclusions in this regard is the short longitudinal span of data. To overcome this, they use innovative techniques to reconstruct 500 years of data to conclude that "the interaction of human alterations to the Mississippi River system with dynamical modes of climate variability has elevated the current flood hazard to levels that are unprecedented within the past five centuries" (p. 95).[44]

Within civil society there is now greater support and activism for the Ecological approach. The important role that Sierra Club played in thwarting several dams in the western part of the United States was noted earlier, as was noted the crucial role that the IRN, together with the NBA, played in the setting up of the WCD. Recently, organizations such as Waterkeeper Alliance and Riverine People have been spreading the river movement, advocating the Ecological approach, to the entire world. There is a growing realization that it is possible to live successfully on floodplains in a way that does not require elimination of floods through construction of levees. It is a way that relies on the dig-elevate-dwell principle and that tries to make use of the benefits that regular flooding brings to floodplains.[45]

10.6 Open approach in other parts of the world

The Open approach is gradually spreading to other parts of the world, including developing countries. For example, in the Indian state of Bihar, houses and sometimes whole villages are now being built on mounds in order to make them flood proof. Similarly, hand pumps are installed on raised platforms so as to ensure safe drinking water during floods. As noted earlier, these measures signify a return to the dig-elevate-dwell principle that pre-industrial societies followed in settling and living on floodplains.[46]

However, progress toward the Open approach in the developing world is still quite limited. One reason for this is external influence. River policies in developing countries—like development policies in general—are influenced heavily by the ideas and practices of developed countries, as noted earlier; this will be discussed further in chapter 13. To the extent that powerful financing agencies, such as the World Bank, Asian Development Bank, and other such multilateral banks, and eminent implementing organizations, such as the USACE or the BuRec, have been—and, by and large, are still—championing the Cordon approach, it is difficult for developing countries to break out from the mold of the Cordon approach and adopt the Open approach.[47]

Also, while resource rich developed countries, such as the United States, can think of reclaiming floodplains in the sense of clearing them altogether of settlements through relocation, it is difficult for resource-scarce developing countries to follow this strategy. This is particularly the case for densely populated countries where arable and habitable lands are scarce.[48] In these countries, the Open approach requires more adjusting to living in floodplains than abandoning them altogether. Despite these difficulties, many efforts are going on toward implementing the Open approach. In chapter 10, I will discuss the experience of applying the Open approach in deltaic countries.

10.7 Concluding remarks

The Open approach—the opposite of the Cordon approach—argues for keeping floodplains open to rivers by preserving natural connections between river channels and floodplains, instead of severing these connections through construction of continuous embankments, as is done under the Cordon approach. It recognizes that a river channel and its floodplains constitute an organic whole and separating them forcibly harms both. It draws attention to the necessity of allowing the floodplain nurturing functions of regular river inundations to continue. The Open approach does not require switching from natural, gravity-flow-based to pump-based irrigation and drainage. It discourages below-flood level settlement and reduces the possibility of catastrophic floods. The Open approach therefore proves beneficial for rivers, floodplains, and the people living in them.

Pre-industrial societies practiced the Open approach, in part due to their lack of technological capacity to build embankments along large rivers, and in part because they understood that, given their conditions, it was better to make use of regular river inundations than to prevent them. For example, they depended on the silt deposited during inundation for renewal of soil, instead of chemical fertilizers, which they did not have.

The post-industrial Open approach is not due to the lack of technological capacity to build embankments along major rivers and carry out their channelization. Instead, this approach emerged from the realization that trying to confine rivers to their channels for all time is counter-productive in the long run. The Cordon approach may help prevent regular inundation but it also increases the possibility of catastrophic floods causing extensive damage. The effects of climate change are aggravating the dangers of the Cordon approach.

Contrary to popular perceptions, the Open approach is not a passive approach. Instead, it implies sustained and vigorous activities aimed at both flood proofing and flood management. The Open approach is more conducive to sustainable development and, apart from protecting the environment, can help to achieve sustainable agriculture, urbanization, and industrialization and to confront climate change.

In view of this, countries in different parts of the world are gradually switching from the Cordon to the Open approach. In the European Union, the Flood Directive of 2007 calls for the principle of living with floods rather than trying to prevent them. It asks Member States to reclaim floodplains for rivers, as much as possible. There is a widespread movement in the continent for restoration of rivers, with the River Mur providing a good example. In the Netherlands, more floodplains have been opened to rivers under the "Room for Rivers" project. In France, the FNSFM notes that preventing construction on floodplains is the most effective way to limit damage from floods and has, accordingly, prohibited new development on many floodplains.

In the United States, a "new approach," containing elements of the Open approach, has been called for, in the wake of the Great Mississippi Flood of 1993. The ideas of the Open approach received further impetus following the New Orleans disaster during Hurricane Katrina in 2005 and also subsequent Mississippi floods, including the flood of 2011. The belief that the Mississippi levee system can provide fool-proof protection against inundation has been shaken, and the idea of returning floodplains to rivers has gained ground. Similarly, there are strong doubts regarding the efficacy of the strategy of trying to strengthen the New Orleans levees, as compared to the strategy of returning the below-flood-level parts of the city to the river and, in the long run, even relocating the city, particularly in view of the possibility of the Mississippi River itself changing its course and bypassing the current city of New Orleans. However, these alternative views regarding how to deal with rivers are yet to become dominant in the United States, despite considerable support for them in the academia and the civil society.

The ideas and measures of the Open approach are spreading in other parts of the world too, including developing countries, which ironically were following the (pre-industrial) Open approach until recently. However, switching from the Cordon to the Open approach is complicated by the further fact that river strategies of developing countries are heavily influenced by policy prescriptions of multilateral and bilateral aid agencies, which still, by and large, cling to the Cordon approach.

Notes

1. In urban areas, it is possible to encourage depavement to allow more river flow percolate to groundwater tables and thereby reduce flood depth.

2. Sometimes, embankments under the Cordon approach, at the time of their inception, are also motivated by the need for and portrayed and justified as flood control projects. However, in reality, they generally turn out to be flood prevention or elimination projects. This transmutation happens because of the underlying philosophy of the Cordon approach that sees river overflows as basically a nuisance that is unwelcome and hence needs to be prevented. In other words, the attitude to river overflows is negative and based on a lack of appreciation of their positive role for floodplains. As a result, first, the number and size of openings in the very design of embankments are generally inadequate. Second, in actual construction of embankments, these openings are neglected further, making them narrower and placing them at inappropriate points. Third, proper operation and maintenance are neglected. In other words, what were supposed to be discontinuous embankments become continuous embankments. The difference in the underlying philosophy makes embankments under the Open approach less likely to get converted into flood prevention embankments than it is the case with embankments under the Cordon approach. For more discussion of various activities under the Open approach, see Islam (1990, 1999, and 2001). See also Rashid (2001) and Rashid and Quayes (2000) for consolidation of settlements as a flood proofing measure.

3. In pre-industrial times, temporary dams were prevalent, but these used to be made with mud and soil. Industrial era technologies help achieve basically the same purpose through the use of modern rubber dams. These rubber dams provide an example of "negation of negation," i.e., a return to a previous position but at a higher technological level.

4. The EEA (2016) Report on Flood Risks and Environmental Vulnerability (RFREV) describes the process of disconnecting floodplains from river channels and the consequences as follows: "Large parts (up to 80 or 90 percent) of the previously intermittently inundated land adjacent to rivers are now disconnected from the riverbed or river channel and do not function as active floodplains any longer. Main pressures are economic developments, the regulation of water levels, and loss of connectivity as a result of flood-protection measures. Land use changes from natural (in central and southern Europe mainly forested) vegetation into agriculture, housing development and industries turn irregular but rather frequent inundations into *undesired phenomena* because of the economic damage caused. When at the same time the water level is regulated for navigation or hydropower and areas are protected by hard flood-protection measures, *the remaining active floodplains are inundated with higher water levels*. Nevertheless, remaining floodplains are biodiversity hot spots and they have a key role in sustainable flood risk management, as these are the locations in which NWRMs can be implemented most efficiently and effectively" (EEA 2016, p. 61, added emphasis). Figure 1.1 (p. 11) of RFREV shows a clear tendency of increase in the "number of reported flood phenomena." The same has been the case with flood damages. See EEA (2016, pp. 23 and 29) for relevant discussion.

5. See EEA (2016, pp. 13 and 41). Also see Feyen and Watkiss (2011) on economic costs of river flooding in Europe.

6. See EU (2007, p. 16). However, it is recognized that planning level recognition of and attention to the beneficial relationship between floods and ecosystem services is a "rather recent development" and "recent policy approach." (See Sayers, Galloway, Penning-Rowsell, Yuanyuan, Fuxin, Yiwei, Kang, Quesne, Wang, and Guan 2015).

7. By "flood hazard maps," the EU Flood Directive meant maps showing geographical areas which could be flooded (with different probabilities), showing the extent, depth, and flow velocity (EU 2007, p. 2).

8. By "flood risk maps," the EU Flood Directive meant maps showing "potential adverse consequences associated with flood scenarios" (EU 2007, p. 5), indicating (i) number of people affected; (ii) type of economic activity affected; (iii) installations (which may accidentally create pollution) affected; and (iv) information about areas where floods "with a high content of transported sediments and debris" can occur, and information about other significant sources of pollution (pp. 2 and 5).

9. According to the EU Flood Directive, "Flood risk management plans should focus on prevention, protection, and preparedness. With a view to giving rivers more space, they should consider where possible maintenance and/or restoration of floodplains, as well as measures to prevent and reduce damage to human health, the environment, cultural heritage, and economic activity" (EU 2007, p. 2). The EU Flood Directive asks for "an assessment of the potential adverse consequences of future floods for human health, the environment, cultural heritage, and economic activity, taking into account as far as possible issues such as the topography, the position of watercourses, and their general hydrological and geomorphological characteristics, including floodplains as natural retention areas, the effectiveness of existing man-made flood defence infrastructures, the position of populated areas, areas of economic activity

and long-term developments including impacts of climate change on the occurrence of floods" (p. 4).

10. "The elements of flood risk management plans should be periodically reviewed and if necessary updated, taking into account the likely impacts of climate change on the occurrence of floods" (EU 2007, p. 2).

11. It advocates the solidarity principle: "In the interests of solidarity, FRMP established in one Member State shall not include measures which, by their extent and impact, significantly increase flood risks upstream or downstream of other countries in the same river basin or sub-basin, unless these measures have been coordinated and an agreed solution has been found among the Member States" (EU 2007, p. 5).

12. As per the Flood Directive, EU member countries were required to report on four impact categories, namely (i) human health, (ii) environment, (iii) cultural heritage, and (iv) economic activities (EU 2007, p. 9).

13. The RFREV distinguishes two types of floodplain, namely "genetic floodplain," by which it refers to "the alluvial landform adjacent to a river and built of its sediments." The other it calls "hydraulic floodplain," meaning that part of the floodplain "which is the area inundated with a certain frequency regardless of land use, soil, etc." (EEA 2016, p. 13). It further observes that there is no comprehensive classification of floodplains. (See also Nanson and Croke 1992.) Following Ward (1989), RFREV distinguishes four types of water flow in a floodplain. These are longitudinal (upstream-downstream); lateral (between river channels and floodplains); vertical (between surface water and groundwater tables); and temporal (over time) (EEA 2016, p. 14). It notices that though lateral spillovers from river channels are the main cause, high groundwater levels can also contribute to floodplain inundation. See also Tockner and Stanford (2002, p. 40).

14. However, elsewhere RFREV offers a more nuanced discussion of "grey" and "green" structures, noting that the distinction between them is "somewhat arbitrary." See for details EEA (2016, pp. 13, 34–35, 61, 63).

15. The RFREV goes on to observe that, implementing the Flood Directive, it needs to be taken into consideration that flow requirements for the conservation of certain species or habitats may go beyond those to reach 'Good Ecological Status,' as set out in that Directive. In this context, RFREV further notes that floodplain ecosystems do not always need high flows in every year. For example, floodplain forests require only occasional flood events for their continued regeneration. This allows considerable flexibility to long-term water-quantity management for these forested ecosystems in which environmental flows are used as water-quantity management approach (EEA 2016, pp. 12-13). See also Hughes, Moss, and Richards (2008).

16. RFREV notes that the Sendai Framework aims for substantial reduction of disaster risk and losses in lives, livelihoods, and health and in the economic, physical, social, cultural, and environmental assets of persons, businesses, communities, and countries. It sets four specific priorities for action: (i) understanding disaster risk; (ii) strengthening disaster risk governance; (iii) investing in disaster risk reduction for resilience and enhancing disaster preparedness; (iv) "Build Back Better" in recovery, rehabilitation, and reconstruction. To support the assessment of global progress in

achieving the outcomes and goals of the Sendai Framework, seven global targets have been agreed, which aim to reduce the impacts of disasters, enhance preparedness, enhance international cooperation and develop and improve access to early warning systems. The framework pays attention to social vulnerability and recognizes social processes and weak institutional arrangements as drivers of risk. It pays ample attention to environmental aspects. RFREV further notes that many of the new elements of the Sendai Framework are already included in the Flood Directive, such as stakeholder involvement, the importance of governance, ecosystems and eco-based solutions, and growing importance of climate change. However, it points out that, unlike the EU Flood Directive, the Sendai Framework does not provide enough guidance to governments to link Disaster Risk Reduction with actual planning and funding mechanisms (EEA 2016, p. 44).

17. The Ramsar Convention uses a very broad definition whereby wetlands include "all lakes and rivers, underground aquifers, swamps and marshes, wet grasslands, peatlands, oases, estuaries, deltas and tidal flats, mangroves and other coastal areas, coral reefs, and all human-made sites, such as fish ponds, rice paddies, reservoirs and salt pans" In total, 42 different wetland types are distinguished in the Ramsar multinational classification system; 12 types of marine and coastal wetlands, 20 inland wetland types and 10 man-made types of wetland. Floodplains are not listed as a specific type of wetland, but overlap partly or in full with several wetland types. See EEA (2016) and Ramsar Convention Secretariat (2016).

18. The RFREV also draws attention to the necessity for greater utilization of the synergies that exist among water, environment, and sectoral policies, such as the European Union's Common Agriculture Policy (CAP). It notes that early cooperation, negotiation, and flexibility can help avoid counteracting activities emanating from these different policies (EEA 2016, p. 7). In this context, it draws attention to the fact that the EU LIFE program advocates for coordinated implementation of the Water Framework Directive (WFD), the Marine Strategy Framework Directive, the Birds and Habitat Directives, the Biodiversity 2020 strategy, and the Flood Directive (EEA 2016, p. 7).

19. The RFREV states that the first generation RBMPs mention "over-abstraction" as the second most common reported pressure on rivers (EEA 2016, p. 12).

20. The RFREV however notes that in advocating for restoration of floodplains, SFRM recognizes the importance of protection of high value assets (EEA 2016, p. 31).

21. See EEA (2016, pp. 17–18 and Box 2.4 in p. 34 in EU 2011). According to RFREV, "SFRM can be described as a section of the wider integrated water management and planning approach for river basins and coastal areas. It focuses on reducing flood risks and promoting environmental, societal, and economic opportunities both at present and in the longer term." In particular, it notes that SFRM, "contrary to the still widespread misconception, is *much more than maintaining the integrity of flood control structures* now and in the long term" (EEA 2016, pp. 50-51, emphasis added).

22. See (EEA 2016, p. 56). The RFREV also pointed out that lack of data prevented it from "a proper status and trend assessment" of Europe's floodplains. Data gaps also existed

with regard to land use in floodplains and flood protection measures adopted by different countries and sub-national units within particular countries (p. 57).

23. "Many restoration projects in rivers and floodplains across Europe were also initiated with the help of EU's financial instrument supporting environmental, nature conservation, and climate action projects (LIFE)" (EEA 2016, p. 35).

24. https://ascelibrary.org/doi/abs/10.1061/%28ASCE%29WR.1943-5452.0000442.

25. The Natura protected habitats are alluvial forests and hydrophilous tall herb fringe communities of floodplains.

26. For more on the Room for Rivers project, see van Leussen, Kater, and van Meel (2000). See also McCully (2007, p. 12). For more on evolution of Dutch thinking regarding rivers, see GoN (2014), Hendriks and Buntsma (2009), Hooijer, Klijn, Perdroli, and van Os (2004), Klijn, van Buuren, and van Rooji (2004), Vis, Klijn, De Bruijn, and van Buuren (2003) and Wiering and Arts (2006).

27. According to RFREV, "Restoration activities have taken place in Member States (see Boxes 2.5 and 2.6) but have not yet halted the trend of degradation of ecosystems and services. Outside the Natura 2000 network, in particular, much remains to be done to halt the loss of ordinary biodiversity EEA (2016, p. 33)." For more on wetland solutions, restoration, flood management, and water improvement activities in Europe, see Johannessen and Granit (2015), Johannessen and Hahn (2013), Nienhuis and Leuven (2001), Pearce (2014), Petrow, Thieken, Kreibich, Bahlburg, and Merz (2006), Tunstall, Johnson, and Penning-Rowsell (2004), van Alphen and van Beek (2006), and White and Howe (2003).

28. As Changnon (1996, p. 264) notes, "The flood's magnitude and damages *raised fundamental questions about the nation's floodplain management approach* and the utility of the flood insurance program. Many questions about restoring the area centered on the future land use in the floodplains and the flood control system. Severely questioned were the benefits and effects of the development of the lock and dam system by the Corps of Engineers and the levee system (italic added)."

29. There were earlier stirrings of this kind of policy. For example, the Federal Interagency Floodplain Management Task Force report of 1986 noted the necessity of bringing together "concerns for mounting flood losses with increasing interests in maintaining important natural functions of floodplains and wetlands (Changnon 1996, p. 258). However, this thinking gained full force following the 1993 flood. See Kusler and Larson (1993).

30. "As we review the Great Flood of 1993 and our policies for managing this nation's floodplains, there is a call for a new approach to their management and their related watersheds" (IFMRC 1994). Changnon (1996) states that in the first blush of post-flood shock, some local and federal officials decided that trying to hold back the Mississippi River was likely to be a costly and never-ending enterprise. Instead of depending on levees and other structures for protection, some thought, it was time to move people's homes and workplaces off the floodplain and cede ground to the river. "We must and can work to design and build our communities better and, to the extent possible, out of harm's way," then-director of the Federal Emergency Management Agency (FEMA) James Lee Witt told Congress later that year. "Mitigation must become a priority throughout all levels of our government. We must be proactive on mitigation and not reactive." And FEMA acted on this

notion: In the nine states flooded in 1993, the agency ultimately moved more than 300 homes and bought and razed nearly 12,000, at a cost of over $150 million; the lands were turned to flood-friendlier uses like parks and wildlife habitat. The village of Valmeyer, Ilinois, just downriver of St. Louis, became the buyout poster child: devastated when floodwaters overtopped its levee (an event that likely helped save St. Louis itself from a major flood), the entire town packed it in, selling out its bottomland location for a new site 2 miles away—and 400 feet above the Mississippi floodplain. It may have been the greatest exodus of Americans from floodplain homes and businesses in the nation's history. (See Changnon 1996.)

31. "The Administration established buyouts of flood-damaged properties as the first priority for mitigation funds available for Midwest flood. As of October 1994, the federal government had approved 160 projects for elevation, acquisition, or relocation of 7,500 buildings. It is anticipated that as many as 8,000 buildings in 140 different communities will eventually be elevated, acquired or relocated. Projects range in size and complexity from elevations of one or two homes in a neighborhood to whole communities relocating to new locations (Valmeyer and Grafton, IL and Rhineland and Pattonsburg, MO). This initiative represents a turning point in flood recovery policy; it is the first time that buyouts have been attempted at such as large scale. Buyouts are viewed by many as an appropriate governmental response to the 1993 flood and future floods like it. Under the right circumstances, buyouts will not only reduce flood damages and protect people and property, but will also achieve other objectives, such as improving the quality of affordable housing, increasing recreational opportunities and wildlife values, and general betterment of the community" (Changnon 1996, p. 256). Since 1993, the government has bought out some 7,700 riverside properties in the two worst affected states, Illinois and Missouri (Pinter 2005).

32. Among these are the Birds Point-New Madrid Floodway in Missouri (to protect the city of Cairo) and the West Atchafalaya Floodway, Morganza Floodway, and Bonnet Carré Spillway to protect the cities of Baton Rouge and New Orleans, respectively.

33. The major agency involved in the enforcement of this policy is the same USACE which for so long was the primary promoter of wide-scale channelization. In 1990 the US Congress gave the USACE a specific mandate to include environmental protection in its mission, and in 1996 it authorized the Corps to undertake restoration projects. The US Clean Water Act regulates certain aspects of channelization by requiring non-Federal entities (i.e., state and local governments, private parties) to obtain permits for dredging and filling operations. Permits are issued by the USACE with Environmental Protection Agency (EPA) participation.

34. As McCully (2007, p. 12) explains, "Hurricane Katrina has, not surprisingly, sparked much debate over how to best protect New Orleans. One positive outcome of this debate may be significant investments in efforts to reverse the loss of coastal wetlands. The USACE is committed to completing its Category 3 protection system for the city. Meanwhile, many New Orleanians are, for good reason, dismissive of only receiving Category 3 protection and are pushing for protection from a Category 5 storm, which estimates say would cost around $32 billion and take decades to build. Yet the key economic reason for the existence of New Orleans, its major deep-water port, may

not survive for many decades. The port depends on the Mississippi to keep it supplied with water and reasonably free of sediment. But at a time of the river's choosing, the Mississippi will, for geomorphological reasons, shift course upstream of New Orleans and take a shorter, steeper route to the sea, most likely down the Atchafalaya River."

35. "Richard E. Sparks, director of research at the National Great Rivers Research and Education Center in Illinois, argues that working with forces of nature rather than against them implies building over a period of decades a new port and city of New Orleans on higher ground near the mouth of Atchafalaya. The natural process of sediment deposition around the "lobe" where the Atchafalaya meets the Gulf would be encouraged, offering protection against subsidence and storm-surges. The higher parts of 'Old' New Orleans would remain as a cultural treasure and a tourist and convention city, but the lowest lying sections would be converted to natural areas and parks" (McCully 2007, p. 12).

36. It is true that the USACE Louisiana Coastal Protection and Restoration (LACPR) Final Technical Report (USACE 2009) identifies areas to not be rebuilt and areas where buildings need to be elevated. These measures would represent deviation from the Cordon approach. However, the focus was on strengthening the levees, expanding the pumping capacity, and ensuring flood free conditions for areas which are clearly below-flood-level. The Final Technical Report includes locations of possible new levees; suggested existing levee modifications; "Inundation Zones"; "Water depths less than 14 feet, Raise-In-Place of Structures"; "Water depths greater than 14 feet, Buyout of Structures"; "Velocity Zones"; and "Buyout of Structures" areas for five different scenarios. In January 2007, the USACE, after having visited the extensive "Delta Works" levee system in the Netherlands, awarded a $150 million contract to a group of Dutch engineering companies for the evaluation, design and construction management of levees and floodwalls, special closure structures for protection of the communities adjacent to the Inner Harbor Navigation Canal, major pumping facilities and planning studies for improved levels of flood protection for New Orleans and southern Louisiana. It may be recalled here that the Delta Works are a series of structures built between 1953 and 1997 in the southwest of the Netherlands to protect a large area of land around the Rhine-Meuse-Scheldt delta from the sea. The works consist of dams, sluices, locks, levees, and storm surge barriers. The works were initiated after the North Sea flood of 1953 in which 2,170 people were killed.

37. As Changnon (1996, p. 264), notes, "Though these were good beginnings, many were not satisfied." Referring to Faber and Hunt (1994), Changnon further notes that "the Midwest floods of 1993 may have shaken people's faith in levees, but the vast majority of federal and state relief dollars will be spent to rebuild and put people in harm's way. Perhaps $500 million will be spent to help relocate river towns. In contrast, $2.35 billion will be delivered in the form of one-time crop-loss payments to farmers. If we are to avoid disaster bills of this magnitude in the future, our funding priorities must change" (Changnon 1996, p. 264).

38. This policy (regarding "net wetland loss") also has its critics, however. Farmers who are losing land to soil erosion as channelized streams cease to be maintained feel

particularly aggrieved. Not only are they losing their valuable property to erosion, the erosion ends up in the stream or river, and contributes to decreased water quality. They also point out that if such policies had been in place in the United States at the time of white settlement, the country probably would never have become a leader in world agriculture and a net exporter of food. Channelization critics however respond that this is immaterial, as we are no longer living in the era of initial settlement.

39. See McCully (2007, p. 4).

40. See Clymer (1977).

41. See Grunwald (2006) and a series of articles he wrote in the *The Washington Post* in 2006.

42. See Pottinger (2003).

43. Surveying the views, McCully concludes that "flood-risk management has reached the status of 'conventional wisdom' among most contemporary analysts on floods" (McCully 2001).

44. In view of the importance of its findings, the article's abstract is quoted here:

> Over the past century, many of the world's major rivers have been modified for the purposes of flood mitigation, power generation and commercial navigation. Engineering modifications to the Mississippi River system have altered the river's sediment levels and channel morphology, but the influence of these modifications on flood hazard is debated. Detecting and attributing changes in river discharge is challenging because instrumental streamflow records are often too short to evaluate the range of natural hydrological variability before the establishment of flood mitigation infrastructure. Here we show that multi-decadal trends of flood hazard on the lower Mississippi River are strongly modulated by dynamical modes of climate variability, particularly the El Niño–Southern Oscillation and the Atlantic Multidecadal Oscillation, but that the artificial channelization (confinement to a straightened channel) has greatly amplified flood magnitudes over the past century. Our results, based on a multi-proxy reconstruction of flood frequency and magnitude spanning the past 500 years, reveal that the magnitude of the 100-year flood (a flood with a 1 per cent chance of being exceeded in any year) has increased by 20 per cent over those five centuries, with about 75 per cent of this increase attributed to river engineering. We conclude that the interaction of human alterations to the Mississippi River system with dynamical modes of climate variability has elevated the current flood hazard to levels that are unprecedented within the past five centuries. (Munoz, Giosan, Therrell, Remo, Shen, Sullivan, Wiman, O'Donnell, and Donnelly 2018, p. 95) (See also Marris 2018.)

45. As Changnon (1996, p. 293) notes, "Even though we cannot 'turn back the clock,' it is interesting to reflect on how the Native Americans who originally occupied the area behaved more than 1,000 years ago, thriving in a culture centered on floodplain residence and use. They farmed on the floodplains, fished the rivers, hunted bison, and built earthen mounds on the floodplains for their homes, for their food storage buildings (to keep their corn and dried meats when floods occurred), and to bury their dead. They had adapted to floodplain living with its intermittent flooding. And within a few decades after the Native Americans had been driven from the Mississippi-Illinois River floodplains, the white man's levees, dams, and drainage systems had changed the river and its floodplains forever."

46. "In Bihar, pakka houses built of brick and cement survive floods much better than the majority of houses that are built of mud and other local materials. The importance of a pakka house is not just that the house itself can withstand swirling flood waters, but that if it has a sturdy, flat roof, this can serve as a place of refuge and storage during flood. Another flood-proofing measure used in Bihar is to build houses, and some-times whole villages, on mounds. Raising the height of hand pumps is also vital to ensure access to drinking water during floods. Earth mounds used to be built in the Dutch province of Zeeland as refuges in the event of embankment failures" (McCully 2007, p. 12).

47. As McCully (2007, p. 5) notes, "Perhaps partly because there are no agencies with the clout of the World Bank or the Corps to promote internationally the soft path of flood management, in much of the world the old flood-control mentality still appears to dominate among government planners and politicians. The standard response to a flood disaster in many parts of the world is to call for more levees and dams—regardless of any role that these technologies may have played in failing to prevent, or actually worsening, the latest disaster." He notes further, "Yet the Corps' sullied reputation in the US has not stopped the agency's concept of flood control from being promoted as a model for the rest of the world. A Corps of Engineers official is one of the office holders on the board of the influential Global Water Partnership (GWP), and Corps generals and civilian employees are regular presenters at major international water conferences. The World Bank's position paper for the World Water Forum in Mexico in 2006 not only applauds the economic benefits of the Corps' flood control efforts but lauds the performance of their infrastructure during the 2004 hurricanes in Florida—while, astonishingly, not even mentioning the dev-astating failure of one of their grandest schemes, the hurricane protection 'system' around New Orleans".

48. "Trying to impose floodplain management is particularly difficult in developing countries where large number of people live in informal settlements—often in ex-tremely flood-prone areas. Colin Green from Middlesex University is sadly cor-rect when he says that 'land use control is most likely to be effective when it is least needed: it will fail where the development pressures are great'. Still, planning controls are a vital part of the flood manager's toolkit" (McCully 2007, p. 11).

11

Cordon approach in a delta

11.1 Introduction

Nowhere is the contrast between the Cordon and Open approaches to rivers starker than in deltas, which generally comprise almost entirely floodplains and tidal plains. Among deltas of the world, the Bengal delta is the largest and most active. Created jointly by three of the world's mighty river systems, the Ganges, the Brahmaputra, and the Meghna, the Bengal delta currently spreads across Bangladesh and India, with the former containing its main and active part. In this and the following chapter, I use the experience of the Bengal delta to illustrate the contrast between the Cordon and Open approaches. This analysis can be useful for understanding the experience of other deltas too.[1]

The Cordon approach was applied to both the Bangladeshi and the Indian parts of the delta. The process began during British colonial rule and gathered momentum after 1947 when the subcontinent was partitioned into India and Pakistan, and Bangladesh became part of the latter as East Pakistan. Application of the Cordon approach was part of the application of the broader Commercial approach, under which both India and Pakistan embarked on a vigorous program of construction of dams, barrages, and embankments, as noted in chapter 3.

In the Indian part of the Bengal delta, the embankments constructed along the Ajay River, an important tributary of the Hoogli River, are a prominent example of the application of the Cordon approach. The experience of these embankments confirms the general points about the Cordon approach, discussed in chapter 9. The frontal and lateral versions of the Commercial approach were also applied to different stretches of the same river in the Indian part of the Bengal Delta. This was most prominent in the case of the Damodor River, another prominent tributary of the Hoogli River.

However, the Cordon approach found its most extensive application in the Bangladesh part of the Bengal Delta. The process gave rise to many different types of cordons, such as rural, urban, full, partial, inland, and coastal. In many cases, cordons in Bangladesh assumed the character of polders, with embankments constructed on all sides of tracts of flood and tidal plains. This has been the case particularly with coastal cordons, though many inland cordons of Bangladesh also have polder characteristics.

Rivers and Sustainable Development. S. Nazrul Islam, Oxford University Press (2020). © Oxford University Press.
DOI: 10.1093/oso/9780190079024.001.0001

While experiences vary across different types of cordons and across individual cordons of a particular type in a delta, many of their consequences are common albeit more acute than for cordons in non-deltaic floodplains, discussed in chapter 9. For example, in all types of cordons, natural gravity irrigation and drainage systems were substituted by pumps, accompanied by new systems of irrigation and drainage canals. Most often these systems did not work well, so that over time irrigation became more dependent on groundwater, and lack of drainage led to the new problem of waterlogging. In some sense, waterlogging has become a more serious problem than the older problem of floods. While unusual floods were infrequent events, waterlogging has become annual and, in some places, a permanent state of things. Numerous regulators and other structures built under the Cordon approach have disrupted the river system of Bangladesh. Together with large-scale upstream diversion of water by India, application of the Cordon approach has led to a general decay of rivers in Bangladesh and the deltaic character of the country.

The discussion of this chapter is organized as follows. Section 11.2 provides an overview of the delta formation process in general, noting the stages through which it progresses. Section 11.3 presents the basic facts regarding the Bengal delta. Section 11.4 examines the experience of the Cordon approach in the Indian part of the Bengal delta. Sections from 11.5 to 11.11 review the experience of the Cordon approach in the Bangladesh part of the delta. Of these, Section 11.5 shows how the Cordon approach in Bangladesh began in the 1950s and 1960s. Section 11.6 explains the classification of these cordons into different types. Section 11.7 reviews the experience of partial rural cordons. Section 11.8 reviews the experience of full rural cordons. Section 11.9 examines the experience of coastal polders, the most numerous among the different types of cordons in Bangladesh. Section 11.10 reviews the experience of urban cordons. Section 11.11 discusses the overall experience of the Cordon approach in Bangladesh. Section 11.12 concludes.

11.2 Different stages of life of land in a delta

From the geologic viewpoint, land in a delta passes through broadly four stages of life. In the first stage, deposited silt remains under the sea, though rising gradually toward the surface. In this stage, land remains, so to speak, permanently inundated. According to a 1968 marine seismic study, the undersea fan of sediment from the Himalayas deposited in the Bay of Bengal by rivers is 1,000 km wide, over 12 km in depth, and 3,000 km long. It extends as far south as and beyond Sri Lanka. The sheer size of this undersea fan testifies to the scale and vigor of the Bengal delta. This huge undersea fan represents the first stage of land in a delta.

As rivers deposit more silt, a second stage is reached. The land now emerges from under the sea, however only periodically: it rises above sea level during low tides but remains otherwise submerged. This is the situation with many tracts of land lying further south of Noakhali and Bhola districts in Bangladesh, at the mouth of the combined flow of the Ganges, Brahmaputra, and Meghna rivers.

As more silt is deposited, elevation of the land increases further, and it gets out of the range of tidal inundation. However, it still remains subject to river inundations during peak seasons. This is the third stage, when the new land becomes part of the floodplain, which then matures with time.

With continued siltation, the elevation of the floodplain increases, and after a certain point of time, parts of it get out of the reach of river overflow even during the peak season. These lands then become part of the old, mature, or moribund delta, which generally no longer experience inundation and sedimentation.[2]

In the case of the Bengal delta, the Ganges River shifted gradually eastward, abandoning, for example, the Bhagirathi-Hoogli channel (of India), and making the Padma River of Bangladesh its main channel. As a result, most of the Indian part of the Bengal delta became moribund. Active delta formation is now occurring mainly in the Bangladesh (i.e., eastern) part, where the sediment brought by the Ganges River is joined by the sediment brought by the Brahmaputra and Meghna river systems. It is at the mouth of the combined channel of these three rivers (south of Chandpur in Bangladesh) where the most vigorous delta formation process is currently going on.

However, Bangladesh contains some of the moribund parts of the delta too, mostly in the greater Kushtia and Jessore districts of western and some parts of northern Bangladesh. It is estimated that about two-thirds (63 percent) of Bangladesh's cultivated area falls under the active delta, while the remaining 37 percent falls under the moribund part.[3] In addition, Bangladesh contains the Barind, Madhupur, and Lalmai Tracts, which are generally referred to as the Old Alluvium Tracts, and are mostly out of the range of river inundation.

11.3 Basic facts regarding the Bengal Delta

The Bengal Delta extends from the Rajmahal Hills of India in the west to the Chittagong Hill Tracts of Bangladesh in the east, and from the hills of the Sikkim and Meghalaya states in the north to the Bay of Bengal in the south. Among the important characteristics of the Bengal delta are its gigantic scale, extreme seasonality of river flow, bank erosion and shifting channels, political fragmentation of the river basins, and the small size of the part of the basins lying in Bangladesh relative to that lying outside its borders.

11.3.1 Gigantic scale

While most deltas are the creation of single rivers, the Bengal delta—as noted earlier—is the creation of three mighty rivers, the Ganges, Brahmaputra, and Meghna, and their distributaries. As a result, the dimensions of the Bengal delta are enormous.[4] To have some comparative perspective, the following facts may be noted:

- The amount of rainfall in the catchment basins of the Bengal rivers is more than four times the rainfall in the Mississippi basin, although in terms of area the former is less than half of the latter.
- Under average conditions, from June to September, 775 cubic km of water flow into Bangladesh through the main rivers and an additional 184 cubic km of streamflow is generated by rainfall in Bangladesh. This may be compared with the annual flow of only 12 cubic km of the Colorado River at Yuma, Arizona.
- The combined channel of the Ganges, Brahmaputra, and Meghna is about three times the size of the Mississippi.
- The amount of sediment carried annually by the rivers of the Bengal delta was, until recently, about two billion tons. This is far more than any other river system in the world.

Often the Netherlands—much of which lies in the Rhine Delta—is cited as the model to be followed by the Bengal Delta with regard to land and water development. In making this recommendation, however, it is often not taken into sufficient consideration that the dimensions and characteristics of the Bengal and Rhine Deltas are very different. Figure 11.1 helps to illustrate this point by comparing the water flows of the Bengal rivers with the Rhine River of Europe.[5] It shows that the Rhine's flow is only about 7 percent of the combined Ganges-Brahmaputra flow. This ratio will decrease further once the Meghna flow (about 200 cubic km) is taking into consideration. The situation regarding the volume of sediment is even more different – about 2 billion ton vs. 1.25 million ton (Frings et al. 2019). Thus, in terms of dimensions of water and sediment flows, the Rhine Delta does not compare with the Bengal Delta.

11.3.2 Extreme seasonality of rainfall and river flow

The Bengal Delta also differs from most other deltas in terms of the seasonality of flows. About 85 percent of the precipitation in the catchment basin of the Bengal river system occurs in just four months of the year, between June and September. The Bengal and Rhine deltas differ strikingly in this respect too. Figure 11.2 compares the seasonal variation of rainfall in Bangladesh with that in

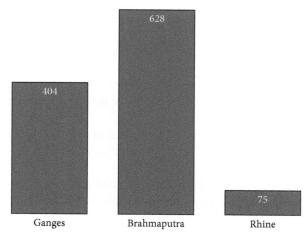

Figure 11.1 Comparison of annual flows (cubic km) of rivers of Bengal and Dutch deltas

Source: Author, based on data available in Dai and Trenberth (2002)

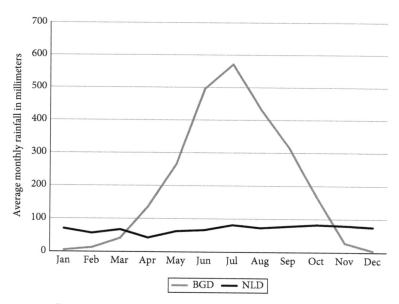

Figure 11.2 Comparison of seasonal variation of rainfall in Bangladesh and the Netherlands (average monthly rainfall in millimeters)

Source: Author, based on average monthly rainfall data for Bangladesh and the Netherlands.

the Netherlands. It shows that while rainfall in the latter varies very little across seasons, in Bangladesh, it is concentrated in just a few months. This extreme seasonality is another important factor that needs to be kept in mind when thinking about interventions in the rivers of the Bengal delta.

11.3.3 Bank erosion and frequent course changes

Bank erosion and frequent course changes are an important feature of the Bengal Delta. Avulsion of river channels in deltas are not uncommon. However, bank erosion is more serious in the Bengal Delta, partly because of the alluvial character of the soil and partly because of the extreme seasonality of flow, noted earlier. The peak season flows of the Bengal rivers are so huge that riverbanks come under enormous pressure and—given the alluvial character of the soil—give way easily. Minor lateral shifts of river courses are therefore a regular phenomenon in the Bengal Delta. However, they may accumulate over time to cause major shifts too. As mentioned earlier, the Ganges River shifted from the Bhagirathi-Hoogli channel to the Padma channel in the sixteenth century through such a cumulative process.[6]

However, the changes are sometimes sudden and are often triggered by seismic events. For example, the Brahmaputra River shifted its main course, during the late eighteenth century, from the previous south-easterly direction—along the north and east of the Madhupur Tract—to a more directly southern course represented by the current Jamuna River, flowing along the west of the Madhupur Tract. This shift is considered to have been more sudden, triggered by either the earthquake of 1787 or a shift of the course of the Teesta River from a more southerly direction (reaching the Ganges) to an easterly direction (reaching the Brahmaputra) around that time.[7]

Though such radical course shifts are rare, bank erosion and gradual lateral shifts of channels, are, as noted earlier, a regular phenomenon for rivers of the Bengal delta. Experience shows that when a combination of circumstances set a major river to change its course, no amount of earthwork can prevent that from happening. Even nationally concentrated efforts have so far failed to prevent such course changes.[8]

11.3.4 Political fragmentation and the miniscule nature of river basins' parts located inside Bangladesh

Another feature of the Bengal Delta is its political fragmentation, which has several aspects. First, as noted earlier, the delta itself is spread across two sovereign countries, namely Bangladesh and India. Second, the basins of the rivers of the Bengal delta are divided across several sovereign countries, including Bangladesh, Bhutan, China, India, and Nepal. Third, more than 90 percent of the catchment area of the Bengal Rivers lies outside of Bangladesh. As a result, there is an asymmetry in the situation, and Bangladesh has little control over the river flows that ultimately pass through it.[9] By contrast, for example, both the Yangtze Delta and the Yangtze River Basin are contained entirely within China.

This political fragmentation makes management of the Bengal Delta difficult, because it requires inter-country cooperation, which is not always easy to achieve. Second, the political fragmentation and the smallness of the parts of the river basins contained in Bangladesh, compared to those lying outside, together with its lower riparian location, create a challenging situation for Bangladesh. As noticed in chapter 3, India, using its upper riparian location, has been diverting water away from Bangladesh rivers, and plans to divert more. The situation is made even more difficult for Bangladesh by the fact that most of it comprises the delta, so that it has little space left for maneuvering (such as creating reservoirs) regarding its management—a situation very different from that of other countries in which deltas form only a small part of their total area. For example, the Mississippi Delta is a very small part of the total area of the United States.[10] The high population density in Bangladesh—with more than 1,000 people (more accurately, 1,116 in 2019) per sq km[11]—further constrains Bangladesh's capacity to maneuver.

11.3.5 Beginnings of intervention in the Bengal Delta

The people of the Bengal Delta followed the Open approach in settling on it. They practiced the dig-elevate-dwell principle, dug ponds, and used the soil to raise the elevation of the ground on which to build houses. Other than that, they left floodplains mostly unobstructed. They did not build many roads, relying instead more on water transportation during the rainy season. However, they undertook a lot of canal digging—organized often by local rulers rather than by the central government—extending river flows deep inside the floodplains.

In this general setting of unobstructed floodplains, the first major intervention came in the form of railways, constructed by the British, begining in the nineteenth century. Often these railways were not aligned with rivers and thus created obstructions to both longitudinal and cross-sectional flows of water on the floodplains.[12] In fact, railways became the first set of "unintended" embankments on the Bengal Delta. However, soon the era of "intended" embankments began. The process of abandoning the pre-industrial Open approach accelerated with political independence in 1947, when governments in both parts of the Bengal Delta started to adopt and implement the Cordon approach more vigorously.

11.4 Cordon approach in the Indian part of the Bengal Delta

The Indian part of the Bengal Delta falls entirely in the state of West Bengal, which used to be part of undivided Bengal before 1947.

11.4.1 River basins of West Bengal

India's West Bengal state government distinguishes three distinct river basins. These are: (i) the Ganges basin; (i) the Brahmaputra basin; and (iii) the Subarnarekha basin. While the first two are well known, the third is a relatively small river, which however, has its own catchment area and reaches the Bay of Bengal independently of the other two.[13]

Several tributaries of the Brahmaputra River, such as the Sankosh, Teesta, Torsa, Jaldhaka, and Raidak, pass through West Bengal's northern districts of Koch Bihar, Alipurduar, Jalpaiguri, and Darjiling, before flowing into Bangladesh. The part of the Brahmaputra basin lying in these districts extends over 10,584 sq km and comprises 12 percent of the geographical area of the state.

Most of the districts of West Bengal however lie in the Ganges Basin. These include the northern districts of West and South Dinajapur; central districts of Maldah, Murshidabad, and Nadia; western districts of Purulia, Birbhum, Bardhaman, and Bolpur; and the southern districts of Medinipur and 24-Parganas. The total area of this basin is 74,575 sq km and comprises about 84 percent area of the state. The most important river in this basin is the Bhagirathi-Hoogli River, which, as noted earlier, used to be the main channel of the Ganges, before it shifted eastward to become the Padma River of Bangladesh in the sixteenth century. A large number of tributaries flow into the Bhagirathi-Hoogli River, before their combined flow reaches the Bay of Bengal south of the city of Kolkata. The most important among them are the Ajay, Damodar, Silai, and Kasai Rivers, all of which originate in the hills of Jharkhand and flow from the west. Several rivers flowing from the north and east also join the Bhagirathi-Hoogly River. They include the Mahananda, Filhar, Atreyee, Punarbhaba, and Ichamati.

The River Subarnarekha, originating in Jharkhand flows through only the Purulia and Medinipur districts of West Bengal before reaching the Bay of Bengal. Its basin extends to 3,595 sq km, comprising 4 percent of the area of the state.

11.4.2 Application of the Cordon approach to the Ajay River

The Ajay River experienced one of the early applications of the Cordon approach in West Bengal.[14] Flowing from west to east, it forms the boundary between the Bardhaman district in the south and Birbhum district in the north. It is served by the Hinglo and Kunur Rivers as its major right-hand side tributaries (Figure 11.3).

As a river in a delta, the Ajay obviously overflowed during high seasons and inundated adjoining lands as part of the process of floodplain creation and nourishment. Also, the river itself shifted its course over the floodplain, widening its basin.[15] In some cases, local landlords (Zamindars) constructed "bundhs" (embankments) of local dimensions to prevent the flooding. However, embankment construction scaled up to a new level after independence when the government took over the responsibility, particularly after the flood of 1959.[16] Over time, 136.16 km of embankment was built, with 80.97 km on the right bank and 55.19 km on the left, apparently to protect from flooding 37,040 hectares on the right side and 29,785 hectares on the left side.[17]

As expected, the embankments caused more sediment to settle on the riverbed, increasing its elevation. As a result, the water holding capacity of the river decreased, and the water level started to cross the danger level too soon following rainfall, and after some time, the Ajay embankments started to fail to contain the water.[18] Meanwhile, sudden releases of water from upstream dams and barrages increased the intensity of flooding.[19]

The problem was aggravated by the fact that embankments do not continue along the entire stretch of the Ajay river on both sides. As a result, water spills out with more ferocity and causes deeper flooding in areas where there are no

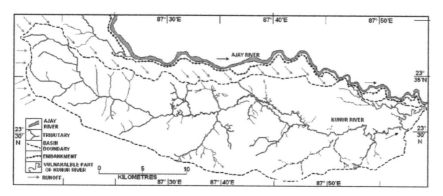

Figure 11.3 Map of Ajay River embankments
Source: Roy and Dutta (2012).

embankments. For example, on the right-hand side of the river, the embankment does not continue between Kogram and Joykrishnapur, so that Mangalkote Gram Panchayate and the adjoining areas witness severe flooding.[20]

The Ajay embankments illustrate another pitfall of the Cordon approach, namely that maintenance becomes a problem, reflecting the vicious cycle between embankment repair demand and budget constraints (Figure 9.2). As a result, breaches become common, leading to catastrophic flooding. For example, the number of breaches on the right-hand side embankment of the Ajay River increased from 12 in 1978 to 22 in 1999 and 25 in 2000.[21] As a combined effect of all of this, the frequency of flooding by the Ajay River has increased over time.[22]

The Cordon approach applied to other rivers of West Bengal had similar consequences. For example, embankments on the Fulhaar River in Malda district gave way in July 2013, inundating 15 villages, 3,000–4,000 hectares of farmland, and requiring the evacuation of 30,000 people.[23]

11.4.3 Combination of frontal and lateral versions of the Commercial approach in West Bengal

Rivers in the Indian part of the Bengal Delta also illustrate how the effects of the frontal and lateral versions of the Commercial approach can combine to produce devastating consequences. This is particularly the case with the Damodar River, for which the Indian government set up in 1948 the Damodar Valley Corporation (DVC), modeled on the TVA in the United States. Though portrayed as multipurpose, DVC focused mostly on power generation and built several dams, including the Panchet Dam on the Damodar River; Tilaiya and Maithon dams on the Barakar River (a tributary of the Damodar); and Konar Dam on the Konar River (another tributary of the Damodar). It also built the Durgapur Barrage on the Damodar River. In addition to the DVC dams, the state government of Bihar built the Tenughat Dam on the Damodar River in 1978. The DVC dams are claimed to have created a total reservoir capacity of 1,299 million cubic m, which could be used to moderate a peak flood of 18,395 cumec to a safe carrying capacity of 7,076 cumec. (See Box 11.1 for more information about the Damodar Valley Corporation.)

However, instead of moderating, these dams on upper reaches of the Damodar River and its tributaries have often ended up aggravating floods. An example of this could be seen in July 2017 when untimely releases from the DVC dams and barrages added to the unusually heavy rainfall to cause serious flood, killing at least 50 people.[24] The flooding affected 14 districts, including 24-Parganas Hoogli and West Midnapore, with the latter two as the worst affected. In Hoogli district, embankments along the Darakeswar River were washed away causing inundation of the Arambagh town. Referring to the releases from DVC dam reservoirs, the West Bengal Chief Minister characterized the floods as "man-made."[25]

Box 11.1 Damodar Valley Corporation of India

The Damodar River, an important river in the Indian part of the Bengal Delta, provides an example of the application of the frontal and lateral versions of the Commercial approach, inspired in part by the western model of river basin development. The Damodar River was well known for frequent floods.[26] The flood of 1943 prompted the government to institute the Damodar Flood Inquiry Commission,[27] which recommended formation of an authority similar to the US TVA with the goal of multipurpose development in the valley area. Inspired by the Commercial approach, it recommended construction of dams and reservoirs with a total capacity of 1,850 million cubic m.[28] To follow up on the commission's recommendation, the Government of India formed the Central Technical Power Board and appointed Mr. W. L. Voorduin, a TVA senior engineer, to make recommendations for comprehensive development of the Damodar Valley. Voorduin's plan, submitted in 1944, was evaluated and accepted by the Indian government. Following the agreement reached by the

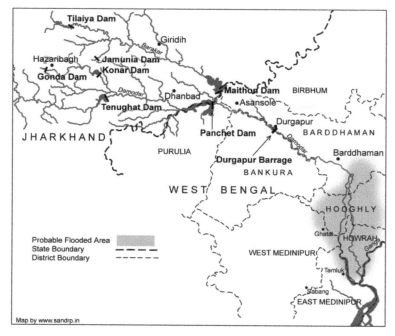

Figure 11.4 Map of Damodar Valley Project of India
Source: Thakkar (2015).

Indian central and West Bengal and Bihar state governments, the Damodar Valley Corporation Act (Act No. XIV of 1948) was passed for the joint setting up of the Damodar Valley Corporation (DVC) in 1948, as the first "multipurpose river valley project" in India (Figure 11.4).[29]

Though portrayed as a multipurpose development undertaking, the thrust of DVC activities was on setting up power plants, as was also the case with TVA. Over time, DVC constructed 10 thermal power plants, totaling a capacity of 8,768 MW.[30] Of these, four were located in West Bengal, totaling 4750 MW. DMC also constructed several dams: Panchet Dam on the Damodar River; Tilaiya and Maithon Dams on Barakar River, a tributary of Damodar River; and Konar Dam on Konar River, another tributary of the Damodar River. Though close to the state border of West Bengal, these dams are located in Jharkhand State, making use of its hilly terrain. The total power generation capacity of these dams is 147 MW, making it clear that the dimensions of these dams and reservoirs are rather small. In addition, DVC constructed the Durgapur Barrage on the Damodar River and a canal network of 2,494 km that were handed over to the West Bengal government in 1964. In 1978, before the separate state of Jharkhand was created, the government of Bihar constructed Tenughat Dam across the Damodar River outside the control of DVC.[31]

It is claimed that with a large flood reserve capacity of 1,292 million cubic m, the reservoirs of the DVC dams can moderate a peak flood of 18,395 cumec to a safe carrying capacity of 7,076 cumec.[32] The reservoirs are also claimed to have created an irrigation potential of 3,640 sq km,[33] and they are also used to supply water to meet industrial, municipal, domestic requirements in West Bengal and Jharkhand.[34] Unfortunately, the DVC dams and barrages have also become a source of problem for the lower regions of the Indian part of the Bengal Delta. Impounding and abstraction of the Damodar water has led to less flow to the lower Hoogly-Bhagirathi River, causing more sedimentation through tidal processes. As noted in chapter 5, Kapil Bhattacharya, the Chief Engineer of West Bengal in the 1950s, pointed to the Damodar dams and barrages as the main cause of Kolkata port's siltation problem—a problem that could not be solved by the Farakka Barrage.

The above shows that the Cordon approach has not been able to solve the "flood problem" in the Indian part of the Bengal Delta. Instead, a combination of the frontal and lateral versions of the Commercial approach has often worsened floods. More importantly, the Cordon approach has left the Indian part of the Bengal Delta unprepared for the unusual precipitation patterns resulting from the climate change.

11.5 Introduction of the Cordon approach in Bangladesh

In the Bangladesh part of the Bengal Delta, too, the Cordon approach became a comprehensive official strategy only after the partition in 1947. Few river-related projects embodying the Cordon approach were quickly taken up. However, the adoption of the Cordon approach was accelerated soon by a process set off by the arrival and recommendations of the Krug Mission.

11.5.1 Krug Mission (1956–1957)

Following consecutive unusual floods in 1954, 1955, and 1956, the then government of Pakistan approached the UN Technical Assistance Administration (UNTAA) for help. Accordingly, a UN Water Control Mission (UNWCM) was sent to study and advise, with J. A. Krug, former US secretary of interior, as leader, and Walton Seymour, Consultant to the Puerto Rican Water Resources Authority, as deputy leader. Four of the remaining five members of the mission were water experts and the fifth was an economist.[35] In his letter to Mr. Krug on March 21, 1957, Huseyn Shaheed Suhrawardy, then prime minister of Pakistan noted, "the very existence of East Pakistan was at stake" if the problem of flood control and water resources development was not solved satisfactorily (UNTAA 1959, p. iv). He also noted the integrated nature of the water system of the country, including surface water, groundwater, and tidal flows in the estuaries. The issues of flood control, irrigation and drainage, channel maintenance, and power generation and use were therefore interrelated, he observed.[36]

Starting its work in November 1956, the Krug Mission took a comprehensive look at the water and power situation of Bangladesh (then East Pakistan). Within a short amount of time, it amassed and analyzed an impressive amount of data and information and submitted its report in May 1957. The report began by noting that "probably nowhere in the world is there such a confluence of great rivers as in East Pakistan. Probably nowhere else do flowing waters pose at once such a challenge and such an opportunity." The report covered both water and power development issues and both technical and institutional sides of the problems. With regard to water, it considered flood, irrigation, transportation, and pollution issues. With regard to the central issue of flood control, the Krug Mission Report noted that a variety of methods were ideally possible. These include: a) catchment conservancy; (b) reservoirs in the hills; (c) detention reservoirs in lower reaches; (d) diversions; (e) flood embankments; and (f) channel improvements. It considered that the first two were not available to Bangladesh because the catchment and hilly areas were mostly outside its borders.[37] The mission did not take a favorable view of the "diversions" option (such as diversion of the Jamuna River water along the Old Brahmaputra, treating the latter basically as a floodway). It was also skeptical of dredging as an option for improvement of channels of major rivers, because of the heavy sediment loads these rivers carry.

Regarding embankments as a method of flood control, the mission noticed both pros and cons. On the side of pros, it took note of the fact that: (i) the location of embankments coincided with the land that would be protected; (ii) the embankment option was relatively less costly; (iii) embankments could be built with local labor and materials; (iv) they could be used to protect "the richest and populous parts" of river basins; and (v) they were immediately effective. On the side of cons, the Krug Mission Report noted that embankments (i) raised the upstream flood level; (ii) increased the flood level downstream; (iii) were likely to raise the riverbed in the embanked reaches; (iv) deprived floodplains from the source of natural irrigation and fertilization; (v) were likely to create drainage problems and waterlogging; and (vi) created the problem of catastrophic flooding by giving a false sense of security and encouraging below-flood-level settlement. It noted however that embankments with sluice gates allowing "controlled flooding" and "natural drainage" might help to avert some of the problems noted above.

In view of these pros and cons, the Krug Mission recommended, as an immediate task, to investigate "the feasibility and effects, beneficial and adverse, of a system of flood embankments (along the major rivers) with sluices, flushing channels, and drainage channels." It recommended planning, design, and construction of such a system only "if the investigation . . . shows that there will be substantial net benefits" (UNTAP 1959, pp. xii-xiii).[38] Thus, first, the Krug Mission did not provide an unequivocal recommendation for embankment construction. Second, the embankments it conditionally recommended, did not entail complete separation of floodplains from river channels.

Two other aspects of the Krug Mission's report are important. First, it argued for setting up a powerful, considerably autonomous, public agency entrusted with the task of both water and power development.[39] Second, it recommended engaging one or more international management and engineering consulting firms to organize and train the necessary personnel of the newly formed agency. These consulting firms would also be responsible for executing the agency's work programs until the training of its personnel was completed. At the same time, the mission recommended that major works should be put to international tender until the new agency became capable of implementing such major works.

In response to the mission's recommendations, the government formed the East Pakistan Water and Power Development Authority (EPWAPDA) in 1959, tasked with both water and power development. It also engaged the San Francisco based International Engineering Company (IECO) as "general consultants" of EPWAPDA to suggest and design water development projects for the agency to undertake.

11.5.2 IECO Master Plan (1964)

IECO was already involved in the water and power sector of East Pakistan. Together with the Utah Construction Company, it was engaged in designing and implementing the Karnaphuli hydroelectric project, a major project of the time.[40] In its new role, IECO started preparing a master plan in December 1962 and completed it in December 1964.[41] The plan was for both water and power development and covered the twenty-year period of 1966–1985.[42] The plan's basic objective for agriculture was to achieve food self-sufficiency, taking into account the anticipated population growth.[43] For this purpose, it put together a portfolio of about 50 water development projects, requiring a huge amount of investment.[44] Table 11.1 presents the list of these projects, and Figure 11.5 shows their location as proposed by the Master Plan. IECO presented these projects as highly beneficial, with benefit exceeding cost for most projects by a factor of three or more.

As noted earlier, some of the projects included in the IECO Master Plan were already underway. Among these were the Ganges-Kobadak (G-K) project, Coastal Embankment Project (CEP), Brahmaputra Right Hand Embankment Project (BRHEP), Chandpur Irrigation Project (CIP), and Dhaka-Narayanganj-Demra (DND) Project. The financial and technical support for these projects came from either the UN Food and Agriculture Organization (FAO) or International Development Assistance (IDA), the concessional lending window of the World Bank.[45] However, the IECO Master Plan went much further and looked for the possibility of water development projects in all parts of the country. It had at least one project in every district and in some districts, all the land area was included in a project.[46] Since IECO was involved in both identification and implementation of projects, it was likely to have an incentive to make the list of suggested projects longer.

The IECO Master Plan was a "massive scheme for *empoldering* huge portions" of the country, covering about 8 million acres, more than one-third of its arable land.[47] It envisaged construction of about 2,000 miles of river embankments, including embankments on both sides of the three main rivers, the Brahmaputra-Jamuna, Ganges-Padma, and Lower Meghna. The latter embankments were to be completed by 1975. These embankments generally aimed at flood "protection" and were to be equipped with "regulators" for admitting floodwater inside the floodplain and for drainage. It was hoped that the controlled environment would allow for increasing the cropping intensity and cultivation of higher yielding varieties. Irrigation was generally to be added at the second stage, several years after the provision of flood protection and drainage. A total of 4 million acres of the flood protected areas were to receive irrigation through canals by gravity flow using regulators (at river high stage) and by pumping (at low stage). Relift

Table 11.1 Water projects proposed by the IECO Master Plan (1964)

Number	Project	Type	Gross area (thousand acres)	Cost and benefits (millions of rupees)			Annual benefit-cost rati\o	Present status
				Capital cost	Annual cost	Annual benefits		
Northwest (NW)								
1	Belkuchi	F, D, I	78	70.01	5.12	13.31	2.6	FSR
2	Bogra	F, D, I	1,220	731.96	56.39	189.42	3.4	FSR
3	Brahmaputra Right Hand Embankment	F	594	44.39	2.69	26.39	9.8	UC
4	Chalan Beel	F, D	333	132.60	9.29	20.74	2.2	FSR
5	Groundwater and pump irrigation—Northern districts	I	187	84.90	11.54	51.09	4.4	UC
6	Kurigram	F, D, I	345	157.43	12.74	64.04	5.0	FSR
7	Pabna	F, D, I	508	255.63	27.23	100.70	3.7	FSR
8	Rangpur	I	288	53.38	6.00	13.49	2.3	FSR
9	Southern Rajshahi	F, D, I	165	70.70	6.39	34.86	5.5	FSC
10	Teesta	I	1,850	302.40	17.52	92.51	5.3	FSC
11	Teesta South Embankment	F, D	150	14.73	0.88	3.80	4.3	FSR
12	Tentulia	I	181	19.16	2.44	7.50	3.1	FSR
Northeast								
1	Bara Haor	I	19	19.26	1.20	2.24	1.9	FSC
2	Brahmaputra Left Flood Embankment	F, D	939	77.04	4.53	34.23	7.6	FSR
3	Dhaka-Narayanganj-Demra	F, D, I	21	15.00	1.30	3.92	3.0	UC
4	Dhaka North Irrigation	F, D, I	158	101.84	8.01	33.38	4.2	FSR

Continued

Table 11.1 *Continued*

Number	Project	Type	Gross area (thousand acres)	Cost and benefits (millions of rupees)			Annual benefit-cost rati\o	Present status
				Capital cost	Annual cost	Annual benefits		
5	Dhaka Southwest	F, D, I	580	457.40	37.10	113.42	3.1	FSR
6	Habiganj	F, D	250	162.20	17.46	24.01	1.4	FSR
7	Hail Haor	I	45	52.00	3.35	8.00	2.4	FSR
8	Hakaluki Haor	I	151	18.92	2.65	6.35	2.4	FSC
9	Karangi	I	10	12.14	1.26	4.31	3.4	FSR
10	Meghna-Baulai Valley Channelization	F. D	641	84.08	5.51	9.00	1.6	FSR
11	Monu	F, D, I	56	25.59	2.26	8.08	3.6	UC
12	North Mymensingh Irrigation	I	276	109.90	10.98	33.94	3.1	FSR
13	Old Brahmaputra, Phase I	F, I, N, P	806	540.81	40.86	180.65	4.4	FSC
14	Old Brahmaputra, Phase II	I, D	610	356.04	23.28	75.62	3.3	FSR
15	Old Brahmaputra, Phase III	F, D, I	729	430.51	31.77	101.12	3.2	FSR
16	Shaistaganj	F, D	261	195.07	17.98	25.36	1.4	FSR
17	Surma-Kushiara Rivers	I	242	111.98	8.13	27.04	3.3	FSR
18	Titas-Salda	F, D	262	185.70	11.38	24.51	2.2	FSR
Southwest								
1	Coastal Embankment	F, D	3,300	680.00	41.89	117.00	2.8	UC
2	Faridpur-Barisal Barisal Unit, Phase I	F, D, I	630	283.30	20.26	92.62	4.6	FSR
3	Faridpur-Barisal Barisal Unit, Phase II	F, D	540	115.25	7.36	25.10	3.4	FSR
4	Faridpur-Barisal Faridpur Unit	F, D, I	882	433.11	30.82	113.02	3.7	UC

Table 11.1 *Continued*

Number	Project	Type	Gross area (thousand acres)	Cost and benefits (millions of rupees)			Annual benefit-cost rati\o	Present status
				Capital cost	Annual cost	Annual benefits		
5	GK-Jessore Unit, Phase I	I	223	107.14	7.93	36.04	4.6	FSC
6	GK-Jessore Unit, Phase II	F, D, I	126	57.25	3.91	19.26	4.9	FSR
7	GK-Jessore Unit, Phase III	F, D, I	345	125.08	9.87	46.73	4.7	FSR
8	GK-Jessore Unit, Phase IV	I	17	6.35	0.57	2.50	4.4	FSR
9	GK-Jessore Unit, Phase V	I	550	274.48	21.22	81.98	3.9	FSR
10	GK-Jessore Unit, Phase VI	F, D, I	406	172.80	12.35	61.25	5.0	FSR
11	GK-Kushtia Unit	F, D, I	420	193.63	15.08	61.70	4.1	UC
12	GK-Kushtia Unit, Phase I&II	F, D, I	488	203.13	14.98	70.86	4.7	FSR
13	Sureswar	F, D, I	250	150.40	10.73	34.42	3.2	
Southeast								
1	Chandpur	F, D, I	187	93.52	7.06	38.33	5.4	FSC
2	Comilla-Noakhali	F, D	830	328.68	19.10	58.23	3.1	FSR
3	Gumti River	F, D, I	61	56.94	4.07	21.11	5.2	FSC
4	Karnaphuli Irrigation	F, D, I	132	92.77	6.06	24.91	4.1	FSR
5	Little Feni Drainage	F, D	259	50.67	3.08	11.87	3.9	UC
6	Meghna-Dhonagoda	F, D, I	50	30.55	2.35	8.02	3.4	FSR
7	Sangu	F, I, P, N	125	293.82	17.85	24.82	1.4	FSR
8	Karnaphuli Multipurpose	F, P, N	–	488.80	32.26	51.93	1.6	Complete

Note: F = Flood control; D = Drainage; I = Irrigation; P = Power; N = Navigation
FSR = Feasibility study required; FSC = Feasibility study completed; UC = Under construction
Source: IECO (1964), Master Plan, Volume II, Table V-2, pp. 120–121

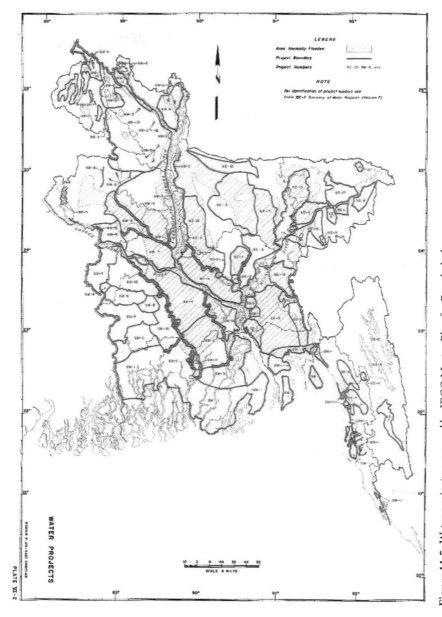

Figure 11.5 Water projects proposed by IECO Master Plan for Bangladesh
Source: IECO (1964a), p. 98.

pumps were to be used to transport water to fields not commanded by gravity canals. The plan envisaged year-round gravity flow irrigation when barrages were constructed on the Ganges and Brahmaputra rivers after 1985. Irrigation was also to be provided to areas not requiring flood protection—1.5 and 4 million acres by 1975 and 1985, respectively. In particular, it recommended tube-well irrigation for limited areas in Rangpur, Dinajpur, Mymensingh, and Comilla districts, where groundwater recharge was thought to be adequate. The Master Plan also envisaged extending coastal embankments by another 1,400 miles.

Thus, the main idea of the plan was to keep water flowing from outside Bangladesh's borders confined within river channels, to be conveyed to the Bay of Bengal without allowing it to spill over the floodplains. For this purpose, all the major rivers were to be double-embanked. The rain water falling within floodplains would be pumped out to rivers. Of course, this basic proposition was qualified with provision of regulators to be inserted in embankments and to be used to let in controlled amount of river water and also to let rain water accumulating inside to flow out to rivers by gravity. However, the main thrust was on cordoning off floodplains from river channels. This idea could be seen clearly from the IECO map showing how Bangladesh would look like after the master plan projects had been implemented by 1985, the final year of its planning period (Figure 11.6).

While focusing on flood protection and irrigation, the Master Plan also considered measures for stabilizing riverbanks and maintaining channel alignments to improve navigation. For the former, it recommended use of flexible fence, jetties with anchored jacks, groins and revetments, and planting of trees and shrubs. For the latter, it recommended maintaining navigation channels that were 250 feet wide and 12 feet deep during the dry season even after withdrawal of water for irrigation. For the same purpose, it included locks or transfer systems in some of the projects (as we shall see).

IECO clearly envisaged a substantial increase in the use of chemical inputs (fertilizers and pesticides) for the success of its proposed projects. It allowed for "development periods" of 3 to 10 years for training and adjustment of famers to change their cultivation practices and increase their cropping intensity.

IECO recognized the paucity of data and the necessity of studies before undertaking many of the proposed projects. It admitted that most of the projects required more detailed investigation than was possible in preparing the Master Plan. In fact, it noted that of the 47 projects included for implementation by 1985, 30 required feasibility studies. It pointed out that more knowledge was necessary regarding canal linings; river hydraulics; sedimentation; riverbed aggradation and degradation; saltwater encroachment; groundwater; agriculture; navigation; supply of material; and integration of the earth embankments in the highway system. Given the lack of local expertise, it recommended hiring external consultants for planning and designing of the projects.

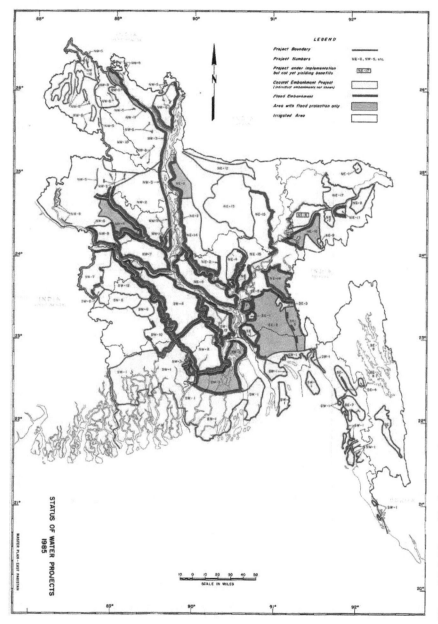

Figure 11.6 Master Plan's projected view of Bangladesh of 1985
Source: IECO (1964a), p. 126.

11.5.3 Thijsee's (1964, 1965) review of IECO Master Plan

The massive Master Plan obviously drew considerable attention. The first commentaries came from those who were already involved in water development efforts in the country. As noted earlier, some water development projects, including the Coastal Embankment Project, was underway through FAO assistance, and the Dutch professor J. Th. Thijsee, hydrology consultant at FAO Bangladesh, offered one of the first reviews.

Given that Thijsee was serving as a consultant of the Coastal Embankment Project, it is not a surprise that he approved the general Cordon approach embodied in the Master Plan. He too thought that all the major rivers should be cordoned (double embanked) so that the transboundary water could pass to the sea without overflowing onto floodplains. He wanted a completely controlled environment in the cordoned off areas. Excess water would be pumped-out to the river, and shortages would be met by pumping-in water from the river. The coastline would be shortened by closing off many channels and keeping only the mouths of large rivers open. Finally, a sea wall would be built to prevent flooding (from the sea). Clearly, Thijsee's proposal reflected the Dutch "Delta Works" plan that was undertaken at that time in the Netherlands. He basically imagined recreation of another Dutch landscape in the Bengal delta.[48]

However, Thijsee expressed concerns about the haste with which the Master Plan wanted to proceed.[49] Being a hydrologist, he focused on the hydrological issues and raised four sets of hydrological questions that needed to be addressed before detailed work on the proposed projects commenced. These were: (a) the high-flow problem of rivers; (b) the low-flow problem; (c) the stability problem; and (d) coastal problems.

Thijsee emphasized many uncertainties that existed at that time regarding the effects of the projects on water levels and discharges in the major rivers; the changes in their morphological and hydraulic behavior; and the salt-water intrusion problems associated with the tides and decreased upland flows. In general, Thijsee was of the view that with such large rivers, variable flows, and unstable riverbeds, it was almost impossible to predict the consequences of project works. For example, he pointed out that prevention of river overflow during high stages could increase river discharges and aggravate floods downstream. Similarly, he noted that additional diversions during the low-flow season could aggravate saline intrusion in coastal areas, affecting adversely drinking water supply and navigation. He also raised questions about the rate of recharge of groundwater in areas where it was to be used for irrigation under the Master Plan.[50]

In view of these concerns, Thijsee suggested proceeding slowly and by stages, allowing time to ascertain the effects at each stage. He therefore divided the Master Plan's projects into two categories. One category comprised those which could be regarded as safe (or harmless), in the sense of not likely to create serious adverse effects because of their nature, size, and location. In his view, 12 of the 25 projects suggested by the Master Plan for inclusion in the third Five Year Plan (1966–1970) and 7 of the 13 projects suggested by the Master Plan for inclusion in the fourth Five Year plan (1971–1975) or later were in the "harmless" category. Thijsee recommended going ahead with these projects and delaying the others until the full effects of implemented projects were assessed more accurately.[51]

11.5.4 World Bank's (1965) review of IECO Master Plan

More influential was the review of the Master Plan by the World Bank, which, as noted earlier, was already involved with several important water development projects in Bangladesh and thus could have a *team* of experts examine and assess the master plan.[52] In its review submitted in 1965, the World Bank appreciated the IECO effort to identify projects and prepare estimates of their costs and benefits, despite the very limited data available. It characterized the Master Plan as "a welcome and valuable addition to the available knowledge on prospective development in East Pakistan," because of its "comprehensive coverage of the water sector."[53]

Given that it was itself financing several cordon projects, the World Bank was obviously in general agreement with the Cordon approach embodied in the Master Plan. It thought that the Master Plan provided the orders of magnitude of investments required for a thorough application of the approach.[54] However, in the same vein as Thijsee, the World Bank thought that the Master Plan represented a too hasty attempt to implement the Cordon approach, so hasty that it might even create a "disaster."[55] While agreeing with its general objectives and approach, the World Bank concluded that the Master Plan did not represent a feasible program for development of the country's land and water resources for the next 20 years.[56] It expressed "serious reservations" with regard to its "specific proposals," and instead suggested a "modified" water development strategy, involving a more gradual and step-by-step implementation of the Cordon approach. More concretely, it envisaged the following four phases, with distinctive tasks for each:[57]

Phase 1: Focus on the 14 million (about two-thirds of the total 22 million) acres of arable land for which flooding was not a serious problem and increase agricultural production from this land through improved and enhanced use of known inputs (such as fertilizer, pesticides, improved seeds, improved practices, etc.) Encourage adoption by individuals or groups of farmers, with limited institutional assistance, of minor irrigation schemes (such as low lift pumps (LLPs) and small-scale water works, implemented through works programs, etc.).[58]

Phase 2: Pay attention during the initial years (1966–1970) of the Master Plan's reference period[59] to (i) institution building; (ii) data collection; (iii) completion of on-going projects;[60] (iv) preparation and initiation of small pilot projects aimed at finding solutions to problems likely to be encountered in implementing major embankment projects, including problems of land acquisition, land consolidation, and flood protection; (v) initiation of feasibility studies by outside consultants on projects that have high priority and are considered hydrologically safe. Limit embankments to only one side of major rivers.

Phase 3: Continue during 1971–1975 the work specified in phase 2 and start executing some priority projects, if the capacity of EPWAPDA permits. However, the program should be limited and focused on general agricultural development, flood protection, and drainage and not aspire to include irrigation, because more experience and evidence are needed before deciding on the role of large-scale gravity flow irrigation in the medium-term development of the country.

Phase 4: Use studies, pilot projects, data collection, and experience of the first decade (1966–1975) to develop a sounder technical basis and institutional capability for taking on the bigger and more demanding projects of the Master Plan. From that point onwards, the Master Plan's "general path" could be followed, with a more extended time frame and modifications in the light of experience.

Apart from recommending this more phased approach, the World Bank emphasized the need for trained human power for its implementation.[61] It also noted that its modified strategy would require some redistribution of resources from EPWAPDA to the Department of Agriculture. At the same time, it would require greater cooperation among various government agencies, various international and foreign aid-giving agencies engaged in the country, and consultants hired for data collection and studies (IBRD 1965, p. vii). In this connection, the World Bank encouraged the idea of relying on foreign consultants and advisers for "extensive technical assistance" regarding both long-range development planning and project formulation and execution. The World Bank also suggested setting up a permanent Land and Water Development Advisory Board (LWDAB) comprising international experts—such as water development engineers, agriculturists, and development economists—national representatives, and representatives of the bank. The board would advise, on a continuing basis by meeting several times during the year, on policy formulation and implementation in the area of water development. The LWDAB, proposed by the World Bank, was not formed. However, the World Bank later considered that IECO was failing to give EPWAPDA adequate advice and assistance, and therefore, together with UNDP, selected a new set of general consultants consisting of a team of 39 experts, with 21 for water, 13 for power, and the remaining 6 for management (of both water and power).

11.5.5 World Bank's Land and Water Resources Sector Study (LWRSS 1970–1972)

To strengthen the technical basis of its recommendations for a four-phased water development strategy, the World Bank conducted a massive Land and Water Resources Sector Study (LWRSS) that began in 1970 and was completed in 1972, resulting in a nine-volume report, based on 26 technical reports. The main purpose of LWRSS was to provide further support to the strategy of four-phase implementation of the Cordon approach, noted earlier. For this purpose, LWRSS divided Bangladesh into finer sub-units—than was done in the IECO Master Plan—and presented estimated water balances for them.

LWRSS came out with three main conclusions. First, most land units of Bangladesh had enough local water supply—in the form of local rivers, rivulets, ponds, lakes, and other surface waterbodies—for dry season irrigation to be carried out through single or multi-stage LLPs. Second, the underground water table that underlies much of Bangladesh's land is more or less fully rechargeable, so that irrigation can also be carried out using tube wells, if surface water sources were not adequate. Third, in view of the above, there is no urgent necessity for largescale schemes to import water through canals by gravity flow from faraway rivers for irrigation purposes.

However, adhering to the Cordon approach, the LWRSS maintained that flood protection was desirable, and embankments were necessary for that purpose. It continued to advocate double-embankment of the major rivers, and as a corollary, double-embankment of their (major) branches too. However, as before, it recommended postponing double-embankment of major rivers until more knowledge and experience were gathered.

11.5.6 From "flood control" to "flood elimination"

What is important in understanding the experience of the Cordon approach in Bangladesh is that, though IECO, World Bank, and other agencies talked about and used the terms flood control and flood protection, what they actually meant was flood elimination. To demonstrate that this was indeed the case, I quote below few statements from relevant IBRD documents. For example, explaining some modifications of the Chandpur Irrigation Project, one of the early cordon projects, the World Bank states,

Irrespective of whether a scheme is based on canals, on LLPs or on a combination of the two is adopted, *flood control* would remain as a part of the project. However, inasmuch as the sudden *elimination* of flooding would be too rapid a change, the Team proposes that *flooding be eliminated* gradually so as to make the problem of accommodating the farmers to the new conditions more manageable. The main *flood control* features involved are the peripheral embankment and two groups of structures: an inlet regulator, navigation lock, and two drainage pumping stations; and an outlet regulator and navigation lock. Owing to the revised conception involving gradual *elimination of flooding*, and for other, technical reasons, these structures would require review by new consultants (IBRD 1968, pp. 38-39, emphasis added).

Thus, from the viewpoint of IECO, World Bank, and other such agencies, the terms flood control and flood elimination were interchangeable, because by the former they actually meant the latter.[62] Yet, there is a world of difference between the two, as we shall soon see.

11.6 Implementation of Master Plan and creation of different types of cordons

Apparently, the water establishment of the country was not too impressed by the World Bank's four-phased approach and did not approve its LWRSS. Instead, it wanted to go for immediate execution of large-scale embankment construction projects. However, unlike IECO, World Bank was the main financing institution, making it difficult to ignore its advice. In the end, therefore, implementation of the Cordon approach in Bangladesh proceeded more or less according to the strategies suggested in the bank's review of the Master Plan and its LWRSS. The plan for double-embankment of major rivers was put on hold; instead numerous small- and medium-scale Cordon projects were implemented; and the work on the massive Coastal Embankment Project, now taken over by the World Bank, proceeded at full steam. Formally, these were called Flood Control (FC), Flood Control and Drainage (FCD), or Flood Control, Drainage, and Irrigation (FCDI) projects, depending on the purposes they were geared for.[63] Other aid agencies joined the World Bank in sponsoring these projects. Important among them was the Asian Development Bank (ADB) and bilateral aid agencies, such as Japan International Cooperation Agency (JICA) and the Kuwait Development Fund. As a result of these efforts, Bangladesh is now littered with cordon projects. There is hardly a district, where cordon projects have not been built; there is hardly a river on which regulators have not been constructed (Figure 11.7).

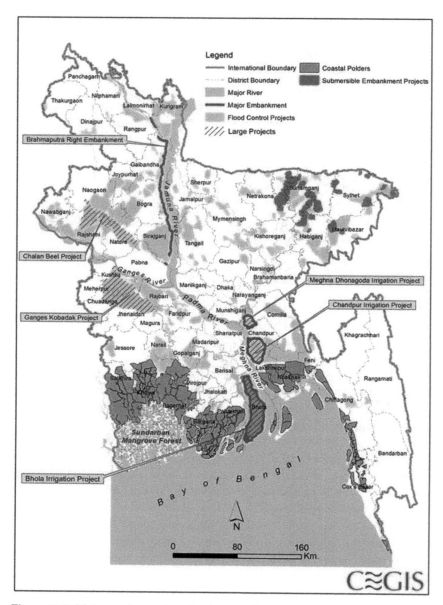

Figure 11.7 Major cordon projects implemented in Bangladesh

Source: Centre for Environmental and Geological Information Services (CEGIS), Dhaka, Bangladesh

Implementation of the Master Plan, however modified, led to the creation of many types of cordons in Bangladesh. Earlier, in chapter 9, I classified cordons into several types. As noted there, from the viewpoint of the land-use pattern, they may be classified into rural and urban. From the viewpoint of proximity to tidal flows, they can be either inland or coastal. From the viewpoint of the extent of sealing off, they can be either partial or full. These classifications are not mutually exclusive. For example, coastal cordons are also generally rural cordons.

The distinction between full and partial cordons may be considered from several viewpoints. For example, cordons may be considered partial, because they have not been embanked on all sides. However, even when embankments are erected on all sides, cordon projects may be deemed partial if there are enough openings by design to let river water in or rainwater accumulated in floodplains to get out (drainage), or both.

The nature of a cordon may change over time. What was designed as a partial cordon may be converted into a full cordon if the openings become dysfunctional and/or clogged up. On the other hand, a full cordon may become partial if—often under pressure from people within the project area—new openings are made or clogged up openings are cleared up, widened, and put into use. Similarly, what was initially conceived as a rural cordon may transmute into an urban cordon because of changes in land-use characteristics.

In the next section, I review briefly the experience of some of the cordon projects in the Bangladesh part of the Bengal Delta in order to help understand the overall impact of this approach. In particular, I focus on some cordon projects that can be regarded as representatives of the different cordon types, mentioned earlier.

11.7 Rural partial cordons in Bangladesh

In the early 1960s, when the IECO Master Plan was formulated, Bangladesh was overwhelmingly rural, and agriculture was the dominant sector of the economy. Rural cordons were therefore the focus of attention and raising crop production was the main goal. Many of these cordons were theoretically of the partial type. Prominent among them were the G-K project, the Brahmaputra Right Hand Embankment project, and the Pabna Irrigation Project. In the following, I provide a brief review of the experience of these projects.

11.7.1 Ganges-Kobadak Project

The G-K project is one of the initial cordon projects undertaken in Bangladesh, as noted earlier. It was motivated, in part, as a response to India's Farakka Barrage then being planned. The idea was to show that Bangladesh needed Ganges water for irrigation purposes (under the G-K project), so that it was inappropriate for India to divert Ganges water using the Farakka Barrage. The G-K project was therefore focused on irrigation. As it was located in the relatively mature, even moribund, part of the delta, flooding was not a major issue.

The initial survey for the G-K project was conducted in 1951, and the government of Pakistan approved the project proposal in 1954. Implementation of the project started during the fiscal year 1954–1955, with support from FAO. Thus, as noted earlier, the project was already underway before EPWAPDA was created, and it was later included in the IECO Master Plan.

The G-K project was originally conceived as having two units, an upper Kushtia Unit and a lower Jessore Unit, with the names referring to the greater districts of Kushtia and Jessore, in which these two units were to be largely located, respectively. There was also a Khulna unit, extending further south. Altogether, the project was to extend irrigation to 500,000 hectares. In the end, however, the project remained limited to the Kushtia Unit (Figure 11.8), and the Jessore and Khulna units were never implemented, for reasons that we will soon see.

Situated on the south (right-hand) side of the river Ganges, just after the Hardinge bridge, the project area is bounded by the Ganges on the north, the Garai and Modhumati on the east, Noboganga on the south, and Mathabhanga

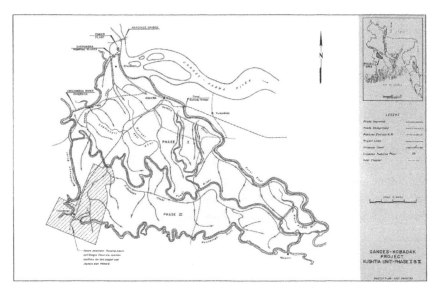

Figure 11.8 Map of the Ganges-Kobadak Project
Source: IECO (1964b), p. 321.

on the west. Embankments running 39 km were constructed to prevent river water from getting in, particularly from the Garai and Modhumati rivers in the east. No embankment was constructed on the west, along the Mathabhanga River, in part because of the absence of any serious threat of flooding from its side. Similarly, the southern edge was left open because, first, as per the original conception, the project was to extend south into Jessore and toward the Kobadak River (hence the name Ganges-Kobadak project). Second, there was no serious threat of flooding from the Nabaganga River either. Third, natural drainage in the project area flows generally from the north to the south. Thus, keeping the south side open was conducive to drainage too. In view of the lack of embankments on the west and south, the G-K project may be characterized as a partial cordon project.

The main objective of the G-K project was to introduce large-scale pump-cum-gravity-flow irrigation either to replace natural, river overflow-based irrigation or to complement irrigation by rainfall, which was less than the national average in this region and did not always take place at the right time. To pump water from the Ganges, an intake channel and two pump houses were constructed near Bheramara in Kushtia District.[64] A power plant was also constructed there to ensure electricity supply to these pumps. To carry the water to the fields, a network of canals was constructed comprising main canals, secondary canals, tertiary canals, and field channels.

The Kushtia Unit of the G-K project has a command area of 197,500 hectares, of which 142,000 hectares are irrigable. It is served by two main canals, namely the Kushtia Canal and the Ganges Canal. The area served by the Kushtia Canal is known as Phase I, which started functioning in 1969–1970. It has a command area of 85,020 hectares, of which 48,700 hectares are irrigable. The area served by the Ganges Canal is known as Phase II, which started functioning since 1982–1983. It has a command area of 117,814 hectares, of which 93,300 hectares are irrigable. The total irrigable area of 142,000 hectares is served by 1,655 km long network of irrigation canals.[65] The G-K Project also aimed at improved drainage and required construction of 971 km long drainage canals for that purpose.[66] Many additional infrastructure components, such as roads, bridges, and culverts were also built.

However, the G-K project was soon beset with problems; and one main problem was of external origin. The proposed G-K project could not prevent construction of India's Farakka Barrage, which went into operation in 1974. As a result, there was not enough water to be pumped into the G-K irrigation canal network, particularly during the lean season. Decrease of water flow in the Ganges due to the Farakka Barrage also caused the intake channel to become sedimented. Up to 1 million cubic m of silt has to be dredged annually from the intake canal. Rehabilitation work to increase the capacity of the water delivery system was initiated in 1984 and completed in 1993. However, in view of the continued fall in the Ganges water level, these measures proved to be of temporary effect. The paucity of the Ganges water was an important reason why the Jessore and Khulna Units of the G-K project were never implemented.

Other problems of the G-K project were internal. For example, as per its design, it was not enough to have water in the primary, secondary, and tertiary canals. Instead, the water level had to be high enough to flow into field channels by gravity. This design characteristic aggravated the water paucity problem created by the Farakka Barrage. Second, the flood protection embankments of the G-K project were constructed long before the irrigation system became operative. As a result, farmers were deprived of the natural river overflow irrigation before the alternative irrigation method was made available to them. This timing mismatch caused significant crop loss and generated resentment among farmers toward the project. As a result, farmers were not that enthusiastic about constructing field channels, which were their responsibility. Third, the success of the gravity-flow irrigation required considerable cooperation and coordination among farmers, not only to construct field channels but also to synchronize crop choice, agree on a schedule and quantum of water allocated, and so forth. Ensuring this cooperation proved to be more challenging than anticipated. Fourth, for early summer crops, farmers could also rely on rainfall, so irrigation by G-K project was not absolutely essential. During the initial years of the project, winter *boro* rice was yet to emerge as the main crop. Fifth, the pumps frequently malfunctioned, so that G-K irrigation was often neither adequate nor reliable. Sixth, long transportation through canals led to significant passage-loss of water, sometimes reaching up to 50 percent.[67] On the other hand, since G-K water was practically free, there was an incentive to use as much of it as possible, leading to wastage and thus constraining further expansion of G-K irrigation coverage. For example, while fields in the G-K project were using 120 acre inches of water, fields in the LLP project areas—another public irrigation project implemented by Bangladesh Agriculture Development Corporation (BADC)—were using only 24 acre inches. Seventh, and most important, the advent of shallow tube wells (STWs) undermined the usefulness of the G-K project. STWs provided an easy and efficient way for farmers to obtain irrigation as necessary, drawing upon groundwater. Due to their small command areas, STWs face fewer coordination problems and are portable and multipurpose. People can easily move them from field to field and also to their homes (if flooding is expected). They can also use the STW engine to motorize their boats and vehicles and to perform other functions. As a result, STWs proved to be a hugely popular and appropriate technology for Bangladesh in many respects. Over time, groundwater irrigation using STWs became the dominant method of irrigation in G-K command area, frustrating its very purpose. The same happened to most other surface water-based irrigation projects of Bangladesh, as we shall see.

The response of the authorities to the problems faced by the G-K project was to come up with rehabilitation projects. After 1983, when the plan for expansion (to Jessore Unit) of the G-K project was shelved, a project for rehabilitation of the existing part of the G-K project was undertaken. An important component of this project was to protect the pumping stations from erosion of the Ganges River's bank, which had reduced the length of the intake channel from 1,160 m to about 500 m.[68] The rehabilitation project however could not solve the basic problems of the G-K project created by the Farakka diversion and the irrigation design characteristics.

As a result of all this, the G-K project lost its usefulness over time, and the project could never be that successful in collecting charges from farmers for their use of the G-K irrigation water, as was originally planned. The project's redundancy was borne out further by the fact that the area that was supposed to be included in its Jessore Unit, which was not implemented, experienced a similar growth in agricultural output as the Kushita Unit, where G-K project was implemented.[69] Researchers who took a close look at the project therefore recommended against taking up projects similar to G-K.[70] Similarly, ADB, based on its review, concluded that a project like G-K would not have been accepted under current price conditions.[71] As noted earlier, IBRD too was skeptical about large-scale, surface water irrigation projects such as the G-K.

The G-K project thus proved to be largely a waste of investment. One particular aspect of the wastage was the loss of a large amount of land that was used up for construction of new sets of irrigation and drainage canals. The very idea of separate channels for irrigation and drainage is alien to Bangladesh and inappropriate for a delta. All rivers, rivulets, *khals*, etc. of Bangladesh have always served the dual purposes of both supplier of (river) water (for irrigation and other purposes) and receiver of (rain) water (i.e., drainage). Forcible separation and allocation of these two functions to two different sets of canals manifested a lack of understanding of the very nature of Bangladesh's deltaic environment. Apart from the loss of original investment and land resources, the G-K project continues to be a drain on the government budget, as the project involves a large amount of recurrent (annual) expenditure.[72]

To make things worse, the G-K project has also contributed to the waterlogging that has now emerged as a more serious problem than flooding, particularly in the southwest part of the country where the G-K project is located. As the ADB (1998) review revealed, serious design flaws led to waterlogging in the Agarkhali and Borabali Beel areas, instead of improving drainage. As we shall see, most of the cordon projects in Bangladesh ended up creating waterlogging. The G-K project probably has the distinction of initiating this process.[73]

11.7.2 Brahmaputra Right Hand Embankment project

Another prominent partial cordon project of Bangladesh is the Brahmaputra Right Hand Embankment. Probably the most important embankment of the country, it extends 216 km from Kaunia Bridge on the Teesta River, then proceeding south along the Jamuna River—the main channel of the Brahmaputra River—to end at Sirajganj. While the G-K project focused on irrigation, the BRHE was mainly for flood protection. It can be considered a partial cordon for two reasons. First, embankments were not constructed along all sides of the project area. However, while the BHRE covered the north and east side, the pre-existing Kaunia-Bogra railway track and a high ridge of land between Bogra and Ullapara served as the western barrier, and the Shirajganj-Ishurdi railway track served as the southern barrier and thus completed the cordon (Figure 11.9). The second reason why BRHE project may be considered a partial cordon is that the embankment itself contained eight openings, particularly at points where tributaries joined the Jamuna River from the west. Regulators were constructed to control the flow of these openings, which were meant mainly to let water flow out to the Jamuna River (i.e., for drainage) and not to flow in. The drainage of the area was to be achieved mainly through pre-existing rivers, flowing mostly south, and flowing out through the openings of the Shirajganj-Ishurdi railway. BRHE is the only major embankment along any of the main rivers of the country,

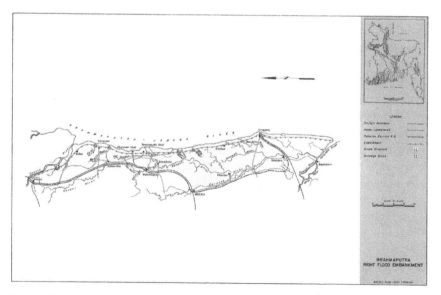

Figure 11.9 Map of Brahmaputra Right Hand Embankment Project
Source: IECO (1964b), p. 200.

and hence it may be considered the first step in the direction of double embankment of these rivers. The consequences of this embankment are therefore important for evaluating the idea of double embankments in general.

The important consequences of BRHE can be summarized as follows. First, it prevented water and sediment flow to the right-hand side floodplains. An important victim of this change was the Chalaan Beel, the vast wetland extending across greater Rajshahi and Pabna districts, that used to serve as a retention pond, balancing the stages of the Jamuna and Padma Rivers. Second, though BRHE protected the area from the normal, annual overflow of the Jamuna River, it couldn't protect it from unusual floods, such as that of 1988 when water from the Teesta and Jamuna rivers entered the project area through several points, particularly in the north. Third, unable to spread onto the right-hand side floodplains, more sediment settled on the Jamuna riverbed, increasing its elevation, and as a result, the Jamuna has become a shallower river than it used to be. Fourth, BRHE aggravated flooding in the districts lying on the left-hand side of the Jamuna River. These include the districts of Jamalpur, Tangail, and Manikganj. Fifth, BRHE aggravated flooding for a large number of people who continue to live on the riverside of the embankment and on the shoals of the river. Sixth, BRHE has not been able to prevent riverbank erosion even on the right-hand side, as the perpetual crisis with protection of the town of Shirajganj demonstrates. In fact, one of the important unintended consequences of the Cordon approach in alluvial deltas such as Bangladesh is the aggravation of bank erosion, because the banks come under more pressure, particularly during the peak season. Seventh, the main purpose of BHRE—to increase agricultural productivity by ensuring flood-free conditions—was frustrated when the project area, like most other parts of the country, switched to groundwater irrigation based *boro* rice, grown in winter, when no flood protection was needed. Eighth, the absence of sedimentation has led in places to subsidence, which together with riverbed aggradation has created a risky situation for many parts of floodplains. The 4.63 km Jamuna Bridge, with 50 pillars and completed in June 1998, has aggravated the riverbed aggradation problem.[74] The narrowing of the river through river training works approaching the bridge and the obstruction of its flow by the pillars have led to a funnel effect, whereby the Jamuna River is becoming shallower and wider upstream.

The consequences of BRHE thus validate the predictions for the Cordon approach made in chapter 9. Its experience therefore should serve as a caution against the idea of double embankments on all major rivers of Bangladesh.

11.7.3 Pabna Irrigation and Rural Development Project

The Pabna Irrigation Project (PIP), later renamed the Pabna Irrigation and Rural Development Project (PIRDP), is another partial cordon project located south of the BRHE project area. It is bounded by the Jamuna River on the east, Padma River on the south, Baral-Hoorsagar River on the north, and Abdulpur on the west (Figure 11.10). While the G-K project focused on irrigation and BHRE focused on flood protection, PIRDP aimed at both. The total project area is 196,680 hectares, of which 184,534 hectares were to have flood control. The net irrigable area is 145,263 hectares, but irrigation was to extend to 29,000 hectares (i.e., 12 percent of the command area) only, at the northeast corner of the project area. For flood control, about 200 km of embankment was constructed,[75] of which the eastern part—along the Jamuna River—was a continuation of the BRHE.

Natural systems of irrigation and drainage were replaced by pump-cum-gravity-flow systems. For that purpose, two dual pump stations were installed, with the main one at Bera near the area to be irrigated and other one at Koitola. An irrigation canal network with a total length of 255 km and a drainage canal system of 145 km were constructed, along with 23 drainage regulators.[76] PIRDP was designed to be a partial cordon, with 15 openings in its embankments, around the north, east, and southern edge of the project area. These openings were fitted with sluice gates, which were to be closed during the high season—so as not to allow excessive spillover of river water into the project area—and opened at the end of the high season to allow water accumulated inside to drain out to rivers.

PIRDP faced problems almost from the beginning. First, it had difficulties in reaching its irrigation target, even though this was modest. The problems with

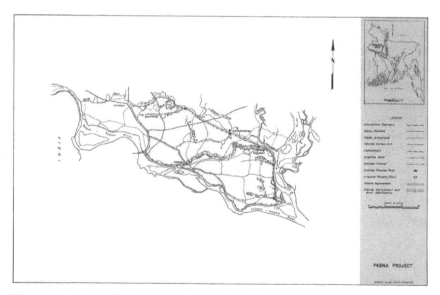

Figure 11.10 Map of Pabna Irrigation and Rural Development Project
Source: IECO (1964b), p. 210.

pump-cum-gravity flow irrigation noted earlier with the G-K project applied to PIRDP too. In addition, due to faulty design, irrigation canals were constructed far away from the area to be irrigated, so that much of the costly pumped water was lost on the way.[77] Second, as with BHRE, PIRDP embankments were able to prevent regular annual flooding, but they couldn't prevent unusual flooding, as during the 1998 flood. Third, as has been the case with other partial cordons in Bangladesh, PIRDP also tended to transmute into a full cordon, with inadequate and dysfunctional openings. However, such a transformation was less acceptable for the population of the PIRDP area, much of which lies in the Chalaan Beel. The people of this wetland were accustomed to deep flooding and had adjusted their cropping pattern and livelihood to it. Hence, flood elimination created more disruption in their lives and livelihood than was the case in other parts of the country. Fourth, the dissatisfaction with PIRDP embankments became more acute as STWs, based on groundwater, became the main method of irrigation, for which embankments were not necessary. In fact, as noticed earlier for the G-K project, the improvement in agriculture inside the project area was not markedly different from that observed outside the project.[78]

Fifth, PIRDP aggravated flooding in areas outside the embankments, in particular for the two townships that were left outside the project's embankments. Floodplains on the riverside of the embankments faced a greater pace of erosion than before, resulting in displacement of more people. In particular, river bends developed between the Bera and Koitola pump stations, threatening their integrity. Sixth, prevention of flooding by constructing embankments proved to be socially regressive. It is generally the landless and other asset-poor people who suffered more from the collapse of capture fisheries, as they depend more on it for both their livelihood and as a source of animal protein. Similarly, substitution of natural irrigation by either pump-cum-gravity flow or tube-well irrigation put the poor and smallholder more at a disadvantage.[79]

In view of the above, it is not surprising that the EIRR for PIRDP proved to be unacceptably low. The project was expected to yield an EIRR of 18 percent, whereas the Project Completion Report (PCR) estimate was 4.8 percent. Similarly, according to the sensitivity analysis carried out by Project Evaluation and Monitoring (PEM) team, the ex-post EIRR of the project could be between 2.6 and 3.7 percent. All these estimates were based on the assumption that flood embankments were a pre-condition for irrigation in the project area. The actual EIRR was therefore considerably lower, and most likely negative.[80]

As is generally the case, the response of the water authorities to the project failures was to chalk up another project. Thus, a Command Area Development Project (CADP) was taken up for PIRDP in 1994, only two years after the completion of the original project. This was in addition to the substantial amount per year that was spent on operation and maintenance of the PIRDP. The feasibility report for the CADP of April 1994 recognized that the tube-well irrigation that expanded in the command area did not require flood embankments. Furthermore, the ADB evaluation team was doubtful about the soundness of the project even after the CADP was implemented. Overall, it raised questions about the very rationale of cordon projects.[81]

11.7.4 Rural partial cordons—overall

Overall, rural partial cordons in Bangladesh were focused on either irrigation or flood protection or both. In all cases, substitution of natural gravity-flow irrigation by pump irrigation didn't work that well and in fact proved redundant when their command areas switched to a winter crop as the main crop based on tube-well irrigation using groundwater. The switch also made flood elimination unnecessary, so far as crop production was concerned. In all cases, partial cordons tended to become full cordons either through inadequacy of their openings or malfunctioning of the regulators constructed to control water flow at these openings. As a result, while floodplains within cordons were deprived of the nurturing functions of regular river overflows, there was increased flooding outside and more siltation of the riverbed. Also, the embankments could not protect the cordoned areas from floods when these were of unusual intensity.

11.8 Rural full cordons in Bangladesh

Many of Bangladesh's cordons were designed to be full cordons, involving a more radical substitution of natural irrigation and drainage by pumped irrigation and drainage. The following offers a brief review of the experience of some of them.

11.8.1 Dhaka-Narayanganj-Demra Project

The Dhaka-Narayanganj-Demra (DND) project was regarded as a show case of the full cordon idea and was already under construction before IECO's formulation of its Master Plan. Later, it was included in this plan, and EPWAPDA completed it during 1962–1968.

Conceived as a FCDI project, the DND focused on a triangular-shaped area, with Dhaka city on the north, Narayanganj city on the south, and Demra on the east. It has a total area of 8340 hectares and is bounded by the Buriganga River to the west and the Shityalakkha and Balu rivers to the east, and Dholai khal on the north (Figure 11.11). The purpose of the project was to insulate the area from

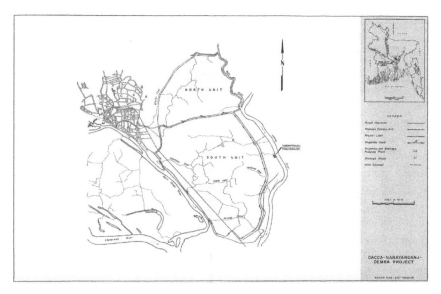

Figure 11.11 Map of Dhaka-Narayanganj-Demra Project
Source: IECO (1964b), p. 230.

regular river inundation, so as to raise food production by raising cropping intensity and productivity.

About 31 km of road-cum-embankment was constructed to seal off the area, though the railway running from Haziganj to Chashara also served as part of the cordon. DND was a full cordon, with embankments on all sides and no openings allowed in the embankments. So, both irrigation and drainage had to be conducted by pumping water across the embankment. For this purpose, a pump station was constructed at Simrail, close to the Shitalakhya River (with a 1 km long intake canal), with four pumps and a total discharge capacity of 14.5 cumec. Also, separate sets of irrigation and drainage canals were constructed to take irrigation water to the fields and to take drainage water to the pump head.[82]

Apart from the main area, called Area I, described above, the DND project formally included another part, called Area II, lying north of the Dhaka-Demra road (i.e., north of Area I), and hence closer to Dhaka city. Comprising 2,470 hectares, this area was not cordoned and hence did not have flood protection, though it enjoyed some irrigation facilities. Area II, therefore, could serve as a control for studying the impact of cordons on Area I.

Initially, the DND project appeared to be a success: both cropping intensity and output increased, and both irrigation and drainage canals were functioning. The pump station was new and efficient; it had the needed capacity to pump the rainwater out. However, this encouraging situation did not last for long. Unintended consequences soon overwhelmed the project and brought new miseries.

First, the artificial flood-free conditions led to expansion of below-flood-level settlement, encouraged in particular by the proximity of Dhaka city and the commercial and port city of Narayanganj. The process gained further momentum after independence in 1971, when Dhaka became the capital city of the new country, Bangladesh. As a result, what was intended to be a rural, agricultural cordon, metamorphosed into a semi-urban cordon. The amount of the project area under agriculture decreased from 80 percent in 1962 to 25 percent by 2008.[83] The construction of the Dhaka-Kanchpur road going from west to east to Shitalakhya bridge through the middle of the project area and the new diversion road that takes off from this road and goes south to Narayanjanj divided up the area into three compartments, disrupting the network of irrigation and drainage canals and accelerating urbanization and industrialization further. The population increased from 0.45 million in 1990 to 1.3 million in 2006, and land holding per family declined from 2.73 acres in 1972 to 0.071 acre in 2006. It is important to note that Area II, despite being closer to Dhaka city, witnessed a slower pace of urbanization than Area I, showing that the cordons were an important factor for the urbanization of DND.[84]

Second, the expansion of the below-flood-level settlement created the problem of catastrophic flooding. Each year, during the rainy season, saving the DND embankments from breaches and overtopping becomes a national challenge, often requiring deployment of the army. The DND embankments were overtopped during 1988 flood and breached in 1998. Despite the addition of floodwalls on top of the embankments, the danger of catastrophic flood looms large over the DND project area.

Third, drainage congestion and waterlogging, meanwhile, have become serious and persistent problems for the DND area. As noted above, both irrigation and drainage canals have become dysfunctional due to largescale, unplanned settlement. With agriculture largely out of the picture, the necessity of the irrigation canals has largely disappeared. However, the necessity of drainage remains. The problem has been compounded by loss of capacity of the pumping system, which is now used mostly for drainage (Figure 11.12). As a result, the DND area now suffers from serious waterlogging (Tawhid 2004). (See Figure 11.13.)

As is generally the case, instead of questioning the fundamental rationale of cordons, the response from the authorities to the problems faced by the DND project has been to adopt more projects. Thus, reflecting the "From lower to higher embankment!" syndrome, authorities have taken up and implemented projects to strengthen embankments and add floodwalls, as noted earlier. In addition, they have taken up various "command area development" and "rehabilitation"

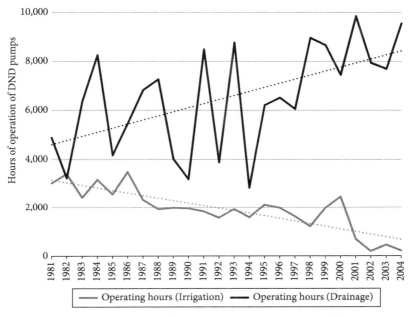

Figure 11.12 Hours of operation of DND pumps for irrigation and drainage
Source: Author based on BWDB data.

Figure 11.13 Waterlogging inside DND project
Source: Anisur Rahman, *The Daily Star* (January 1, 2016).

projects, including installation of additional pumps and restoration of drainage canals. Efforts have also been made to prevent any further construction in the area. However, as long as the cordons remain, the temptation for additional settlement and development is too strong to be resisted. As a result, the mutation of DND from a rural to an urban area continues unabated, with further aggravation of the problems of waterlogging and catastrophic flood risk.

11.8.2 Chandpur Irrigation Project

Chandpur Irrigation Project (CIP) is another pre-Master Plan project that was later included in the plan.[85] Located at the confluence of the Meghna and Dakatia rivers, it aimed initially at creating two polders. The northern one was to extend north of the Chandpur town and be bounded on the west by the Meghna River, on the north by the Dhonagoda River, on the east by Boaljuri Khal, and on the south by the Dakatia River. Embankments were to be constructed along the Meghna River, Dhonagoda River, and Boaljuri khal. On the south, the pre-existing Chandpur-Hajiganj railway line was to be reinforced to serve as the embankment protecting against the Dakatia River. The southern polder was to extend south of Chandpur town and be bounded on the west by the Meghna River, and on the north by the Dakatia River. Embankments were to be constructed along both the Meghna and Dakatia river to protect against their overflow. The South Dakatia River was to be enclosed within the polder (Figure 11.14).

Drainage to the south would be directed toward the Meghna River, where flap gates in the embankment would open when the river stage is lower than inside the floodplain. Being close to the sea, this part of the Meghna river experiences tidal fluctuations, allowing flap gates to open during low tides. For the northern unit, the drainage would be mostly pump-based, though openings with regulators would allow water accumulated inside to move out when river stages are lower. Pumps would be used to bring in water for irrigation from rivers to the southern unit. In the northern unit, the drainage pumps would also be used to bring in water for irrigation. In other words, both units of CIP would have pump-cum-gravity irrigation, as in other cordon projects reviewed earlier.

As noticed earlier, the World Bank had doubts about the efficacy of pump-cum-gravity irrigation, and the early experience with G-K project strengthened these doubts. The IDA therefore eliminated the irrigation component from the project in late 1965. The EPWAPDA however expressed doubts in January 1967 about the viability of the project without irrigation. As a result, a revised Chandpur Irrigation Project II was prepared. Under it, the north polder was dropped. The South polder was expanded, and instead of pump-cum-gravity, it was decided to switch to pump-gravity-pump irrigation. Under this strategy, LLPs

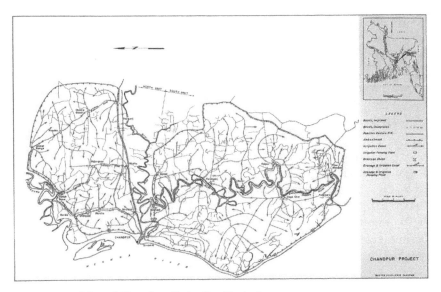

Figure 11.14 Map of Chandpur Irrigation Project
Source: IECO (1964b), p. 333.

would be used to lift water from both existing surface water sources and irrigation canals bringing in pumped water from outside rivers.[86]

Financed by IDA, the work on the modified project started in 1973 and was completed in 1978. It now extends across six upazillas in Chandpur and Laxmipur districts and covers an area of about 57,000 hectares. As with other cordon projects, CIP also had some immediate positive effects in terms of raising cropping intensity and expansion of cultivation of HYV of crops. However, with time the negative consequences of the cordon approach surfaced, including deprivation of the floodplain nurturing function, aggravation of flooding outside the project area, and aggradation of riverbeds.

In addition, being located in the most dynamic part of the Bengal Delta, where river flow patterns are yet to settle down, CIP faced a serious problem of river erosion, particularly from the Meghna River, which near Chandpur contains the combined flow of the Meghna, Padma, and Brahmaputra river systems. For example, during the 1988 flood, the CIP embankment along the Meghna River collapsed, and 5,000 hectares of land within the project and 3,000 hectares outside were devoured by the river. The flood embankment already had to be moved four times. However, the problem persists, and 72–82 km of the embankment is seriously threatened, even after recent construction of a 10 km embankment in Haimchar Upazilla.[87]

Instead of questioning the rationale of a cordon project in a most active part of the delta, the water authorities of Bangladesh came up with another project to mitigate the problems faced by CIP. Quite instructively, it is called the Chandpur Irrigation Project Protection Project with a price tag that exceeds the price of the original project!

11.8.3 Meghna-Dhonagoda Project

Meghna-Dhonagoda (M-D) Project is another, relatively new, full cordon project that was proposed by the IECO Master Plan and is located north of the CIP. Similar to the DND project in conception and design, the M-D project is also a FCDI project. Located near the confluence of the Meghna and Padma Rivers (19 km north of Chandpur and 40 km southeast of Dhaka city), it is bounded on the west and north by the Meghna River and on the east and south by the Dhonagoda River (Figure 11.15). The original feasibility study was conducted in 1964 and a revised one was carried out in 1977. The implementation of the project started in earnest in 1979–1980 and was completed in 1987–1988, with financing from the ADB. The project covers an area of 17,584 hectares falling under 13 unions and the *pourashava* (municipality) of Matlab Upazilla, Chandpur District.

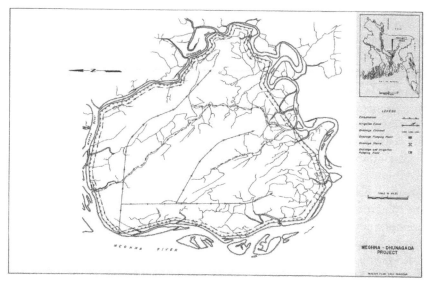

Figure 11.15 Map of Meghna-Dhonagoda Project
Source: IECO (1964b), p. 351.

Under the project, embankments of a total length of about 61 km were constructed, complemented by 218 km and 125.5 km long networks of irrigation and drainage canals, respectively.[88] To transfer irrigation and drainage water across embankments, four pump stations were constructed with a total capacity of 77.85 cumec.[89] ADB (1990) states that the original design allowed for seven drainage sluices, which were later deleted based on the assessment that water accumulation from local rainfall during the transition from the dry to rainy season—when the drainage sluices could operate—would not harm either the *boro* (winter) or *aus* (early summer) crops. Meanwhile, during the rainy season the sluices would not be useful for drainage because of high stages in the outside rivers. Also, the proposed navigation locks to be built next to the primary pumping stations were removed from the project, terminating connections between rivers and waterways inside the project with the outside rivers. This step was justified on the ground that existing roads as well as the embankment and irrigation canal dykes were adequate to meet transportation needs (ADB 1990, p. 8). Thus, despite conceived initially as a partial cordon project, M-D became a full cordon project.[90]

As with other cordon projects, the initial assessments of the M-D project were positive. However, the negative consequences of the project soon surfaced. First, the embankments could not protect the area from unusual floods. For example in both 1987 and 1988, the embankment on the eastern edge of the project, along the Dhonagoda River, breached, leading to submergence of a significant part of the project area with high velocity, sand-laden water, causing widespread disruption of the irrigation and drainage canal networks.[91] Second, in absence of the nourishment from annual flooding, the land is losing quality. According to Institute of Flood Control and Drainage Research (IFCDR) (1994), the organic nitrogen level by early 1990s stood at 0.75 percent, much below the critical level of 1.1 percent. Soil tests undertaken in collaboration with Soil Resources Development Institute (SRDI) laboratory show that the organic matter (OM) content of the soil has fallen to alarming levels. Levels of minerals, such as phosphorous, potassium, sulphur, and zinc are also dangerously low. Farmers are trying to compensate for these losses by using larger amounts of chemical fertilizers.[92] Third, prevention of sedimentation is also causing subsidence, a problem that will manifest more forcefully with time. Fourth, by deleting the navigation locks, which were included in the original design, the project has created the ironic situation whereby a piece of land that is completely surrounded by rivers cannot reach these rivers using waterways.

Fifth, the severance of links with the rivers has affected adversely the open-capture fisheries. The Meghna-Dhonagoda floodplain used to be a rich breeding area for various fish species, in particular, sweet water prawn. The cordon project has not only terminated river flooding, it has also led to marked shrinkage of surface waterbodies inside the project. The average capture fish catch per fisherman has declined from 4.4 kg per day to 2.7 kg, despite the fact that fishermen were trying to catch fish on more days than before (276 days vs. 252 days) (FAP 12, 1992, p. 25). The Rapid Rural Assessment (RRA) team reported a 75 percent loss of fish catch in the interior waterbodies (p. 26). Increased chemicalization of agriculture and resulting toxic runoff to waterbodies has caused further damage to open fisheries.[93] The collapse of the open fisheries in the M-D project area hurt more the landless and the poor, who depend on open fisheries both for employment and income and also their own protein intake. In general, minority groups suffered greater loss of occupations and employment from the project. Surveys show that in response to the question, "Are you better or worse off after the embankment?" positive responses came mostly from those who owned land, while the poor mostly answered in the negative (Vaughn 1997, p. 23; see also FAP 12, 1992).

Sixth, controlled irrigation, the main proclaimed benefit of the cordons, has also suffered. In fact, the area under irrigation has decreased from the original 13,602 hectares to about 5,620 hectares. Both technical problems—design of the irrigation mechanism—and institutional problems contributed to the decline. On the design side, the pump-cum gravity irrigation adopted by M-D faced many problems with the size, placement, and alignment of the turnouts, many of which were unauthorized and constructed by farmers.[94] Also, a large fraction of water was lost due to seepage and evaporation during its passage to the fields. Institutionally, M-D also faced the familiar problems of cooperation and coordination among famers, as noted for other Cordon projects such as G-K. Partly in response to these technical and institutional problems, more farmers resorted to irrigation using small-scale irrigation technologies, particularly STWs. In the early 1990s, just a few years after completion of the project, a significant part of the land was already irrigated by STWs and LLPs, and over time their importance increased, undercutting the very rationale of the project.[95]

Seventh, despite the elaborate network of drainage canals, some parts of the project area are experiencing waterlogging. This is partly due to technical reasons, such as lack of efficiency in pump operation, and also partly due to behavioral and institutional reasons. For example, in many cases, local influential people converted drainage canals into series of ponds for farm fishing. Unplanned construction of roads, culverts, and other infrastructures and household amenities have also contributed to the drainage congestion.[96] Eighth, though not as rampant as in the DND project, the M-D project area is also becoming urbanized. As noted earlier, the area under irrigation has shrunk by about a half. Thus, it is likely that the M-D project will soon lose its rural character, frustrating the project's irrigation goal further.

Finally, situated in the same active part of the delta as CIP, the M-D project also faces the serious problem of erosion, particularly from the Meghna River. For example, during the 1988 flood, the Meghna River cut a deep channel devouring more than 1,000 hectares of land.[97] During the 1998 flood, the river moved about 170 m inward threatening a 3 km stretch of the embankment (near Mohanpur and Dashami). The embankment was also threatened on the east and south, along the Dhonagoda River. Around 1.5 to 2 km of the embankment along this river was in peril at various points near Gazipur, Shibpur, Amirabad, and Torki.[98] Extraordinary efforts, consuming large amount of resources, have been needed to protect the embankments from river erosion so far. However, given the dynamic nature of the river landscape in this part of the Bengal Delta, it is doubtful whether the long-term physical sustainability of the M-D cordon can be ensured.[99]

As with other troubled cordon projects, the water authorities of Bangladesh have taken up a CADP, with a large budget, for M-D too. However, this will only serve as another example of the "From a troubled project to another project!" syndrome.[100]

11.8.4 Rural full cordons—overall

The above review shows that full cordons accentuate the effects of the Cordon approach. By depriving floodplains from the nurturing functions of regular river overflow, they trigger multidimensional decay of floodplain ecology. Below-flood-level settlement proceeds more vigorously and often transforms full cordons from rural to urban, particularly when these are close to large cities. The perennial risk of catastrophic flooding becomes more serious for them, and they also aggravate flooding in areas outside the cordons. Both irrigation and drainage depend more completely on pumps and networks of new canals. With urbanization of the cordons, the necessity for irrigation decreases but that for drainage increases. However, inadequacy and inefficiency of both pumps and canals lead to serious problems of waterlogging. Full cordons located in the un-settled river landscape of the most dynamic part of the Bengal delta also face the serious problem of riverbank erosion.

11.9 Coastal polders in Bangladesh

11.9.1 Bangladesh's coastal zone

A major area of application of the Cordon approach in the Bangladesh part of the Bengal Delta has been along its coast. Bangladesh has a relatively long coast line, estimated to have a length of 710 km. Officially, the coastal zone

covers 153 upazillas of 19 coastal districts and accounts for 32 percent of the
area and 28 percent of the population of the country. The elevation of 86 percent of the coastal zone is less than 5 m above sea level, and for 62 percent it is
less than 3 m.

The coastal zone of Bangladesh has been classified from different viewpoints.
From the viewpoint of exposure to the sea, it can be divided into two parts,
namely the exposed coast and the non-exposed coastal zone. The exposed coast
is closer to the sea and comprises 12 of the 19 districts (51 of the 153 upazillas).
The non-exposed zone comprises more inland areas of the coastal zone. While
the coastal zone as a whole is within 195 km of the sea shore, the exposed coast is
within 75 km (Figure 11.16).[101]

From the viewpoint of delta formation characteristics, the coastal zone may
be divided into three parts: western, central, and eastern. The western part
comprises the Gangetic tidal floodplain and lies in the Khulna and Barisal
divisions. Its lower part in Khulna Division is occupied by the *Sundarbans*. The
central part comprises the Meghna estuary, through which the combined flow of
the Ganges, the Brahmaputra, and the Meghna Rivers reaches the sea. As noted
earlier, this is where the most active delta formation process is going on. The
eastern part comprises the Chittagong coastal plain, which is narrow and runs in
a north-south direction. It ends at Teknaf, where hills reach close to the sea.

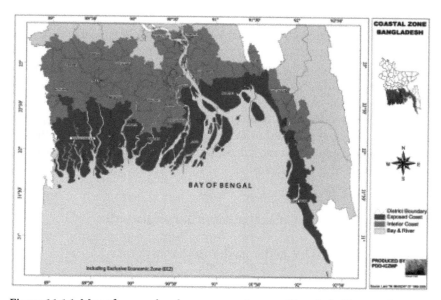

Figure 11.16 Map of exposed and non-exposed parts of Bangladesh's coastal zone
Source: Islam, Rafiqul (2004).

11.9.2   Coastal Embankment Project

Unlike the inland part of the delta, the coastal part is affected by both river and tidal flows, and salinity is therefore an important issue. Expansion of agriculture through prevention of salinity has been an on-going effort for the coastal people. Given its location at the apex of the Bay of Bengal, another problem for Bangladesh's coast is cyclones, which, together with tidal bores, hit it frequently.

In response to the above problem, some coastal embankment building activities were undertaken even during colonial and pre-colonial times (i.e., before 1947), generally by Zamindars, the local landlords. However, these were generally rudimentary earthen embankments of partial scope, and they generally fell into disuse after the abolition of the Zamindari system following adoption of the East Bengal State Acquisition and Tenancy Act (EBSATA) of 1950.[102] The idea of expanding crop cultivation by preventing salinity through construction of embankments gathered strength following independence from British rule in 1947. As noted earlier, already in the 1950s, a Coastal Embankment Project (CEP) was taken up with assistance from FAO. Later on, it was included in the IECO Master Plan, which aimed at constructing 73 polders, varying in size from about 480 to 80,000 hectares (Figure 11.17).[103] Over time, however, the scope of CEP was expanded, and by the 1980s, financed mostly by the World Bank, a total of 5,107 km of embankments were constructed to create 123 polders, with the goal of protecting an area of 1.5 million hectares.[104] It is considered to be the largest earthen embankment construction project in the entire world.

Most of the polders lie in the costal districts; however some lie in adjacent districts. From the viewpoint of the proximity to the sea, CEP polders can be classified into two broad groups: sea-facing and not sea-facing. The former comprises polders that have at least one side facing the sea, while the latter comprises polders that do not face the sea directly on any side. It is estimated that 957 km of CEP embankments are sea dykes.[105]

Coastal embankments in Bangladesh tend to play multiple roles. The main role is, of course, to prevent saline water from getting inside, so that more and high yielding crops can be cultivated. Second, coastal embankments also serve as roads in an area where transportation otherwise depended mostly on waterways. Third, coastal embankments serve as shelters during cyclones and tidal bores.[106]

During the initial 10 to 15 years following their construction, the coastal embankments appeared to be successful, helping to raise agricultural production. The total output in certain areas increased significantly.[107] They also performed well in their role as a means of communications and as places for shelter. The embankments were also used for coastal afforestation. However, over time, a host of problems caught up with the coastal polders, plunging many of them into a general crisis.

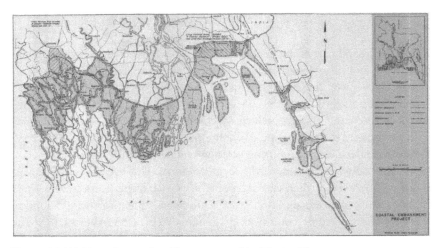

Figure 11.17 Map of coastal polders proposed by Master Plan
Source: IECO (1964b), p. 277.

11.9.3 Coastal polders in crisis

The coastal embankments, like the inland embankments, led to many unwarranted processes. First, they obstructed the deposition on coastal floodplains and tidal plains of sediment brought down by rivers and redistributed by high tides. As a result, and due to the additional processes of soil compaction and withdrawal of water from underground aquifers (for irrigation by tube wells), the land inside polders in many cases faced subsidence.[108] Meanwhile, more sediment settled on the riverbeds, causing riverbed aggradation. The polders were usually equipped with flap gates to allow drainage at low tides. These gates were mounted in sluice structures, usually located where the embankments crossed the pre-existing natural drainage channels (rivulets and canals). However, due to subsidence inside and aggradation of riverbeds outside, the gradient necessary for gravity drainage decreased or even got reversed. Also, the aggraded riverbed often clogged the flap gates making them immovable. The result is waterlogging, which in many polders has now become perennial, causing immense miseries for the people living in them, and often prompting them to cut the polder to create passage for the water (Figure 11.18).[109]

Second, with export opportunities opening up, shrimp cultivation has become a lucrative enterprise. Since shrimp cultivation requires brackish water, shrimp entrepreneurs have cut down the embankments of many polders to let saltwater in, thus undermining the very purpose of the polders. As a result, many polders have now become saltwater shrimp farms, either partially or wholly. The

Figure 11.18 Local people cut polders to restore water flows
Source: Mustafa Alam of Institute of Asian Research (IAR) of University of British Columbia.

area devoted to shrimp farming has increased from 51,812 hectares in 1983 to 137,996 hectares in 1994 and 218,649 hectares in 2004.[110] Since intrusion of salt-water affects paddy cultivation, the expansion of shrimp cultivation has often led to violent conflicts with rice cultivators (Karim and Stellwagen 1998; Deb 1998; Firoze 2003) and also caused environmental pollution (Chowdhury, Khairun, Salequzzaman, and Rahman 2011).

Third, coastal embankments suffer particularly from the vicious cycle of demand for repair and demand for budget, portrayed in Figure 9.2. Being far away from the capital and lacking local resource mobilization, demand for necessary repair of the coastal embankments often remains unmet. Furthermore, corruption and inefficiency lead to waste of whatever money is allocated by the national government. As a result, the condition of the embankments deteriorates further. The increase in the frequency, scope, and intensity of cyclones and tidal bores—due to climate change—has aggravated this cycle. As noted earlier, tidal bores and torrential rainfall—brought by cyclones—fill up the clogged polders quickly and damage the embankments, and by the time the water dissipates, another cyclone often hits the area. Thus, climate change is making the race between repair and damage harder to win. All these processes together have created for many coastal polders a crisis that has been termed by some observers as a man-made disaster.[111]

11.9.4 Inadequate response to the polder crisis

The water establishment of Bangladesh responded to the polder crisis in the same way as it did to the problems encountered by inland partial and full cordons reviewed earlier. Instead of trying to go deep into the problem and questioning the rationale of polders in the specific conditions of Bangladesh's coast, they developed, as is their usual practice, various rehabilitation projects, aimed at repairing and strengthening the embankments, raising their heights, and improving drainage by restoring the flap gates, etc. Unfortunately, these efforts proved to be of temporary relief, because, absent the restoration of the natural sediment distribution process, drainage outlets became clogged again soon. The authorities also took up various CADPs to overcome the perceived lack of cooperation among farmers and their lack of enthusiasm for the project. Recognizing that the polder projects were imposed on the people with little or no input from them, the water establishment, particularly their international financiers, wanted to give a greater role to local people's participation in these CADPs. This recognition and shift in attention was also in part a reflection of the "participatory approach to development" that became influential in the 1980s and 1990s. Thus, more importance was given to issues of "group formation," "local participation in management," "cost recovery," etc.

However, neither the rehabilitation projects nor the CADPs could be that successful when the underlying questions were not addressed. It was difficult to expect participation in solving issues that arose from the imposition of projects that were conceived and designed without any participation of the local people. The brief review below of the Khulna-Jessore Drainage Rehabilitation Project (KJDRP)—one of the rehabilitation and command area projects that were undertaken in response to the polder crisis in Bangladesh—can help to illustrate this outcome.

11.9.5 Khulna-Jessore Drainage Rehabilitation Project (KJDRP)

The KJDRP, financed by ADB, was initiated to ameliorate some of the problems that polders in the greater Khulna and Jessore districts were facing.[112] Discarding the previous practice of formulating the objectives of cordon projects in terms of increase in agricultural output, KJDRP defined its principle objective more broadly, in terms of poverty reduction (to below 60 percent). This would however require an increase in agricultural production and expansion of on-farm employment. To achieve this objective, the project intended to rehabilitate the polder's original physical infrastructure (embankments, sluice gates, etc.). In addition, it aimed at mobilizing the participation of local people (i.e., beneficiaries) in the design, implementation, and operation and maintenance (O&M) of the project; promoting agricultural extension services; and improving fisheries management "to ensure a continuing supply of noncommercial fishes caught and consumed primarily by the poor."[113] The latter two aspects of the project reflected some departure from the traditional formulation of water sector project objectives.

However, subsequent evaluation by ADB itself showed that KJDRP didn't prove to be successful. The plan to generate funds for O&M through leasing of government land to Water Management Associations (WMA) proved to be unrealistic. Lacking necessary funds, the O&M carried out was not adequate to confront the heavy silt deposit that filled up active rivers and drainage canals and clogged regulators. Second, the fisheries and environment development plans of the project were apparently not accepted by BWDB, the main government executing agency in the water sector. The technical assistance provided to BWDB for institutional strengthening did not create added capacity to engage with local communities and hence did not prove effective. The membership of the WMA's dropped from 35 percent at project completion in 2004 to 15 percent by 2007. The agriculture extension and fisheries management components of the project also lacked resources, coordination, and efficient implementation, with local people not seeing much benefit from them. Both these components were therefore dropped, causing further loss of interest among people in the project and weakening of group cohesiveness. Meanwhile, about 111 km of roads were constructed, when none was envisaged, aggravating the drainage problem. The ADB evaluation study assessed the project as overall "unsuccessful."[114]

11.9.6 Coastal polders—overall

The problems of the coastal polders created by the CEP in Bangladesh are rooted in the disruption of natural sediment distribution and redistribution processes that prevailed across the flood and tidal plains in coastal Bangladesh through a complex interplay between river flows from the north and tidal flows from the south. Hence, the coastal polder crisis cannot be resolved unless this disruption is stopped. Restoration of the sediment flow and its natural distribution across the flood and tidal plains has become more urgent in view of climate change. As noted earlier, this restoration can help to ensure the annual increase of coastal elevation by about 2 mm that can be of considerable help in countervailing sea level rise (Bhattacharya 2009).

11.10 Urban cordons in Bangladesh

Arguments for cordons acquire more force for cities and towns, because these are considered high-value areas deserving greater protection from floods. For drainage, pumps are to be used. Urban cordons are relatively recent additions to the water development portfolio of Bangladesh, pointing to the fact that originally urbanization took place on higher grounds while cordons were directed toward augmentation of agricultural output. However, urban cordons have now been built for a number of cities and towns, and these are now poised to be replicated on a wider scale. It is therefore useful to review the experience of urban cordons, and the best way to do so is to examine the lessons of the Dhaka Integrated Flood Protection Project, aimed at protecting the capital city of Dhaka from floods.

11.10.1 Dhaka Integrated Flood Protection Project (DIFPP)

The historic flood of 1988 flood triggered a process called Flood Action Plan (FAP), to be discussed in more detail in chapter 12. Under this plan, many studies and several action projects were formulated. Not all of these were ultimately implemented. However, the prevailing quasi-military government felt the urgency of moving ahead with one of the projects, namely the cordon project to prevent flooding of Dhaka city. The reason for this urgency is fairly clear. Being the capital, where the main airport and all foreign embassies and missions are located, protection of Dhaka city was seen as of utmost political importance.

In view of the above, the government appointed a committee to prepare a flood control plan for the Greater Dhaka Metropolitan Area. In January 1989, the committee submitted a detailed scheme for phased investment in flood protection and drainage. The scheme divided the metropolitan area into two parts, east and west, with the Tongi-Sayedabad Road as the dividing line. The western part was bounded by the Buringana River on the south and west and the Turag River on the north. The eastern part was bounded by Tongi Khal on the north, Balu River on the east, and DND embankment on the south. Protection of the western part was considered Phase I, while that of the eastern part Phase II (Figure 11.19).

Giving high priority to the scheme, the government did not wait for any detailed study and foreign financing and instead went ahead with implementation of Phase I using its own resources, under a crash program.[115] The project, covering 136 sq km, had three components: (i) flood protection, (ii) drainage, and (iii) improvement of the environment;[116] it was implemented

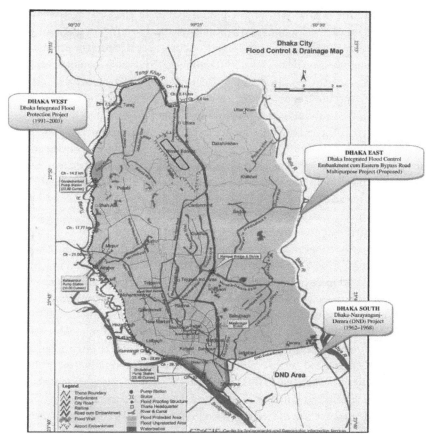

Figure 11.19 Map of Dhaka Integrated Flood Protection Project
Source: UN-HABITAT (2008)

during 1991–2003. The main physical components of Phase I included 30 km of embankment (from Tongi to Kellar Moar), generally called the Western Embankment, and 3.8 km of floodwall (from Kellar Moar to Buriganga friendship bridge), ending with the DND western embankment. Apparently, the DIFPP was conceived as a partial cordon, with the Western Embankment having 32 pipe outlets (for drainage). However, the emphasis seems to have been on pump-drainage, with pump houses constructed at Goranchatbari, Kallyanpur, and Dholaikhal. The project also included road and bridge construction, embankment slope protection, construction of water reservoirs as retention ponds, etc. Another important component was re-excavation of 30 km of canal, which were to bring rainwater to the pipe outlets and pump stations. The main proposed physical component of the eastern part of the scheme (Phase II) is the embankment, called the Eastern Bypass, along the Balu River from the junction with Tongi Khal to where it meets the Shitaylakha River at the northeastern tip of the DND project. This embankment is yet to be constructed, so that the experience of the eastern part can serve as a control area in assessing the impact of the Western Embankment. The experience of the Phase I of DIFPP can be summarized as follows.

Waterlogging: The major consequence of DIFPP has been waterlogging. It has converted Dhaka city into a "city of waterlogging." There are several dimensions of this process. First, the project disrupted the city's natural drainage system. Dhaka city used to have about 50 canals crisscrossing through it and connecting the bordering rivers with each other.[117] These canals used to bring the rivers to the city and take away the rain water that fell within the city to the rivers. Unfortunately, most of these canals and other waterbodies, that used to play the role of retention ponds, have now either disappeared or are on the way to disappearing. The high density of population was already providing strong incentives for encroachment of the canals and waterbodies. However, the DIFPP aggravated and hastened this process. It disconnected the city from the Buriganga and Turag Rivers. A direct result of this disconnection was marginalization of the role of the canals, which provided this connection. The pipe-outlets of the Western Embankment were poor substitutes for the links that the canals used to provide. The marginalization of the canals promoted further encroachment, neglect, and pollution of the canals. Particularly destructive in this regard has been the decision to convert many of these canals into box culverts, thus hiding them from sight and letting people forget their existence! Needless to say, keeping built-over canals unclogged is a much harder task than is the case for open canals—and less amenable to supervision. As a result, converting canals into box culverts sounded the death knell for many of these canals.

Second, while disrupting the pre-existing natural system of drainage, DIFPP failed to replace it with an effective new one. The pumped drainage that it seemed to have proposed was destined to fail. The difficulties with pump drainage in a country with an average of more than 200 cm of annual rainfall concentrated in four months was already illustrated by the DND project, reviewed earlier. With an area that is about 10 times larger, and a population of more than 20 million (as compared to about 1.3 million for DND), the magnitude of the drainage problem posed by Dhaka city is simply huge. Furthermore, even if the pumping capacity was adequate, it couldn't prove effective because the canals needed to bring the rainwater to the pump heads have either disappeared or decayed.[118] Also, pumping out water requires time, but the retention areas for rainwater to collect and wait until it is pumped out have also disappeared.

As a result of these, waterlogging has become a horrendous problem for Dhaka city. Even little amount of rainfall now leads to severe waterlogging. Whereas floods such as that of 1988—that triggered the DIFPP—are considered to be once-in-a-hundred-years events, waterlogging submerging streets and neighborhoods to the same depth as in the 1988 flood has now become frequent. Thus, the Cordon approach has created, in a way, a much more serious problem than the one it aimed at preventing.

Finally, as noted earlier, the eastern part of DIFPP hasn't been implemented yet. The centerpiece of this part is construction of the Eastern Bypass, aimed at separating the city from the Balu River. Once this cordon is constructed, Dhaka city's waterlogging problem will become more severe. Even without the Eastern Bypass, the city's connections to the Balu River are getting increasingly restricted due to dense settlement, encroachment, and clogging of the remaining canals, etc. As a result, even the eastern part of the Dhaka city is witnessing waterlogging, a problem that will certainly become worse once the Eastern Bypass is constructed.

Decay of the city environment: The DIFPP, has also led to a general decay in the city environment. Disconnected from rivers, the few public waterbodies that remain—such as the Dhanmondi Lake, Hatirjheel Lake, Banani Lake, and Gulshan Lake—are no longer recharged annually by rivers. They have thus lost their dynamic character and instead have become pools of stagnant water, facing long-term decay. The annual cleansing function of river overflow is now lost. With the disappearance of wetlands—canals, lakes, ponds, etc.—open spaces and green areas have also shrunk. Loss of waterbodies has also contributed to the rise of dust and other pollutants in the city air and decreased the availability of water to be used for clean-up efforts. The temperature tempering effect of waterbodies is also no longer there. The city has become a less pleasant area for habitation.

Risk of subsidence: With decay and neglect of rivers and canals, promoted by DIFPP and the Cordon approach in general, Dhaka city has been mining groundwater aggressively for the supply of water. This has led to a fall in the groundwater level to an alarming extent, aggravating the risk of subsidence.

Risk of catastrophic flooding: The Western Embankment has increased the risk of catastrophic floods by encouraging below-flood-level settlement. The influence of embankments on the pattern of settlement can be seen by comparing the experience of east Dhaka with that of the west. In the absence of the Eastern Bypass, settlement in east Dhaka has generally proceeded following the dig-elevate-dwell principle. Thus, land elevation was raised through earth filling before construction of buildings and structures.[119] By contrast, encouraged by the artificial dryness created by the Western Embankment, below-flood-level settlement has expanded in west Dhaka, particularly along the city side of the embankment, and also further inside. As a result, flooding will assume a more catastrophic character when, during years of unusual flood levels, river water gets in due to either overtopping or breaches. An indication of this was seen during the 1998 flood when water entered the city through the space between Kellar Moar and Buriganga Bridge in the Western Embankment.[120]

Transfer flooding: DIFPP has also caused "transfer flooding" and aggravated flooding in areas outside the embankments, especially in Basila and other areas lying to the west of the Buriganga River.

Overall, DIFPP, as an urban cordon, has boomeranged in several ways. By trying to prevent flooding, it has brought the more nagging problem of waterlogging. By promoting below-flood-level settlement, it has given rise to the risk of catastrophic flooding. By disconnecting the city from its neighboring rivers, it has contributed to the decay, degradation, and disappearance of water-bodies and worsened the general environment of the city. Completion of DIFPP Phase II, in particular, construction of the Eastern Bypass along the Balu River, will effectively seal off the city from all its bordering rivers and aggravate all the adverse consequences that DFIPP Phase I has already produced.

11.10.2 Secondary towns integrated flood protection project

One would think that the experience of DIFPP would deter the authorities from going forward with other urban cordon projects. Unfortunately, that is not the case. Backed by foreign lending agencies, they have moved ahead with similar cordon projects for many other cities and towns of the country. One such effort is represented by the Secondary Towns Integrated Flood Protection Project (STIFPP), through which 15 towns and cities are to be brought under FCD, beginning with six in the project's first stage.[121] Financed by the ADB, the project was implemented during the period from 1992–2000.[122]

As usual, the project had three components, namely flood protection, drainage, and improvement of the environment, with BWDB in charge of implementing the first, and Local Government Engineering Division (LGED) in charge of the other two. The flood protection component of the project involved construction of 55.06 km of embankment, 14.8 km of riverbank protection, 26 drainage regulators, and 2 groynes. The drainage component involved excavation of 144 km of main drains, 123 km of tertiary drains, and 633 culvert and road crossings.

According to BWDB, the project resulted in significant benefits with regard to flood protection, drainage improvement, sanitation improvement, solid waste disposal, slum improvement, and gender development activities. However, objective analysis shows that the STIFPP reproduced all the negative consequences observed earlier in the case of DIFPP for Dhaka city. In each of the cities and towns included in STIFPP, the embankments have led to decay, deterioration, and disappearance of waterbodies, expansion of below-flood-level settlement, and creation of the new problem of waterlogging. To the extent that these cities and towns are much smaller in size than the mega city of Dhaka, the scale and intensity of the negative consequences are of a lesser degree, for now. However, these problems are sure to become more acute with time.

Of particular note is the horrendous problem of waterlogging that now has emerged in the port city of Chittagong, the second largest city of Bangladesh. Lying at the mouth of the Karnaphuli River, where it meets the Bay of Bengal, the city has a steeper average gradient toward the river and sea. As a result, there are no natural reasons for this city to be waterlogged. Yet, due to negligence in keeping the city open to the river and the sea, Chittagong now faces a severe waterlogging problem. The cities and towns of Bangladesh that are building cordons to disconnect themselves from neighboring rivers are certain to face similar consequences.

In several cities and towns in Bangladesh, the Cordon approach has led to river aggradation to such an extent that the elevation of the riverbeds is now higher than that of the cities. For example, in Habiganj, the Cordon approach has led the Khoai River bed to have an elevation higher than that of the city, creating a risky situation. The problem has been aggravated by significant channelization, particularly in the form of disconnecting the Old Khoai River, which is now getting encroached and filled up. A similar problem can be seen in Comilla, where the elevation of the Gumti River bed, because of the Gumti embankments, is now higher than that of the adjoining plains.

11.11 Overall experience of Cordon approach in Bangladesh

The review presented here shows that Bangladesh has a rich experience of application of the Cordon approach that led to the creation of many different types of cordons, including rural, urban, full, partial, inland, and coastal. In general, we see that the Cordon approach has led to more problems than it intended to solve. The overall experience can be summarized under the following heads.

Irrigation: One of the main purposes of rural cordon projects was to promote surface water irrigation. However, the substitution of natural irrigation by pump-gravity irrigation did not work well. Modification from pump-gravity to the pump-gravity-pump method, such as in the CIP, did not solve the problem either. The entire irrigation goal of cordon projects was subverted when Bangladesh gradually switched to winter *boro* as its main crop, irrigated by tube-wells using groundwater.

Flood control: The flood control role was undermined by the fact that flood control was generally interpreted as flood elimination, which went against the very nature of a delta. Elimination of regular inundation led to decay of the floodplain ecology. Meanwhile, by encouraging below-flood-level settlement, and causing riverbed aggradation, the Cordon approach increased the risk of catastrophic flooding. In some cases, the problem of sedimentation differential (between inside and outside cordons) is so serious that elevation of riverbeds now exceeds that of adjoining floodplains, creating a permanently risky situation.

Transfer flooding: The Cordon approach has aggravated flooding outside the cordons, and cordons have pitted one rural area against another, and rural areas against cordoned off cities and towns. The strategy of fitting a FCDI in every nook and corner of the country was destined to fail due to the *fallacy of composition*. Meanwhile, the solidarity and cohesion that the nation needs to confront flooding and manage its rivers optimally have been fractured.

Bank erosion: The Cordon approach has aggravated the problem of bank erosion by inviting river onslaught on the banks by shutting off openings into floodplains that had previously taken off some of the pressure on the bank by allowing river water to flow into the floodplains freely through the canals and other natural openings.

Drainage: The plan to substitute natural drainage by either the gravity-cum-pump method or the gravity-cum-sluice method did not prove successful and instead led to waterlogging. The very concept of two separate systems of canals—one for irrigation and another for drainage—was alien to the Bengal Delta, where rivers, rivulets, creeks, and canals have performed, from time immemorial, the twin roles of bringing in and carrying away water. It is therefore not surprising that this concept did not work well in Bangladesh. Meanwhile, unable to spread over the cordoned off areas, more sediment settled on riverbeds, causing riverbed aggradation, clogging sluice gates and drainage pipes, and making drainage systems inoperative. The urban cordons did not have to worry about irrigation, but drainage congestion has become a severe problem for them. Disconnection with rivers has led to decay, encroachment, and disappearance of waterbodies, aggravating waterlogging. Meanwhile, expansion of below-flood-level settlement has increased the probability of catastrophic flooding.

Unwarranted metamorphoses: Most partial cordons with drainage openings morphed into full cordons. The same has been the case with partial cordons with openings for letting in controlled amounts of river water. Since, as noted earlier, water authorities and the foreign lending agencies behind them interpreted flood control as flood prevention, the metamorphosis of partial into full cordons wasn't a surprise. Similarly, many rural cordons metamorphosed into urban cordons, subverting their initial agricultural purposes, and meanwhile becoming waterlogged.

Subsidence, submergence, and salinity. Along the coast, the construction of more than a hundred polders has totally disrupted the natural process of sediment distribution and redistribution that used to take place through the interplay of river and tidal flows. Coastal embankments have prevented water and sediment from entering the polders, causing land subsidence inside and riverbed aggradation outside. As a result, drainage sluices have become clogged, causing waterlogging. Meanwhile, shrimp entrepreneurs have created breaches in the embankments of many polders, letting the saline water in and thus undermining the polders' original purpose. Many polders now face a crisis, and some of them are afflicted by permanent waterlogging. Absence of proper sediment distribution is making coastal Bangladesh more vulnerable to submergence.

Overall, cordons have disfigured the deltaic landscape of the Bengal Delta, disrupted the river system, and undermined the natural processes of delta formation, drainage, and sustenance. Their contribution to crop production increase has been modest, and their role in flood protection has been perverse.

11.12 Conclusions

The contrast between the Cordon and Open approaches assumes particular significance in the context of deltas, where rivers flow through floodplains and tidal plains. The Bengal Delta, being the largest and most active delta in the world, illustrates this contrast particularly well. Studying this rich experience can be helpful for other deltaic countries.

The review of the cordon projects in both the Indian and Bangladesh parts of this delta show their various flaws, particularly in the delta's more active parts. The review shows how the Cordon approach harms floodplain ecology, creates the risk of catastrophic flooding, and gives rise to the new problem of waterlogging. Though the Indian part of the Bengal Delta is less active, illustrations of the above can be seen from the experience of the application of the Cordon approach to, for example, the Ajay River, a major tributary of the Bhagirathi-Hoogly River.

The Bangladesh part of the Bengal Delta, the larger and more active part, provides a richer experience of the Cordon approach, which led to the creation of many types of cordons. The specific nature of problems created by the Cordon approach varies across different types of cordons, but there are some commonalities. In all cases, we find disruption of the natural drainage system, increase in the risk of catastrophic flooding, and decay of the environment. Coastal polders face a general crisis, with many witnessing permanent waterlogging. Moreover, the disruption of the natural process of coastal elevation increase through sedimentation has also made the coastal region more vulnerable to the sea level rise caused by climate change.

Overall, cordon projects have proved to be counterproductive in many respects. The wastage of resources they entail continues in the name of various rehabilitation and CAD projects. Natural systems of irrigation and drainage are a boon in a deltaic landscape. The Cordon approach has converted this boon into a bane.

Notes

1. The case studies of the Cordon approach reviewed in chapter 9 focus on floodplains that are generally *not* part of deltas. For example, the Mississippi floodplain, for the most part, is not a part of the Mississippi delta. The same is the case with the Huang He River floodplain. The floodplains of the Kosi and other rivers in the Indian state of Bihar are also not part of a delta.
2. There is a lack of unanimity among scholars about the use of the terms "old," "mature," and "moribund" parts of the delta. In this discussion, I use them interchangeably.
3. In the discussion of deltas, other terminologies are also used sometimes to refer to their different stages. For example, some authors use the term "hydraulic floodplains" when referring to those that are occasionally subject to river inundation. For further details on the dimensions of the Bengal delta, see Rogers, Lydon, Seckler (1989) and Islam (1999, pp. 3–5).
4. On average, 1,121.6 km³ of water crosses the borders of Bangladesh annually, of which 85 percent comes between June and October. Around 48 percent (537.2 km³) is contributed by the Brahmaputra, 47 percent (525.0 km³) by the Ganges, 4 percent (48.4 km³) by the Meghna/Barak and nearly 1 percent (11 km³) by other minor rivers to Chittagong in the southeast. http://www.fao.org/nr/water/aquastat/countries_regions/Profile_segments/BGD-WR_eng.stm.
5. While the average annual flows of the Ganges and the Brahmaputra Rivers amount to 404 and 628 cubic km, respectively, the volume of the Rhine annual flow is only 75 cubic km.
6. https://www.britannica.com/place/Bhagirathi-River.
7. The exact reason for the avulsion of the Brahmaputra in the eighteenth century is not known. According to some, the earthquake of 1787 played an important role. Others think that unusually large volumes of the Teesta River during that period pushed the Brahmaputra flow in a southerly direction. Some scholars think that the Brahmaputra alternated to flow along east and west of the Modhupur Tract several times during earlier geologic periods.
8. One recent example is the collapse of the Chandpur Irrigation Project's protective embankment in face of the Meghna River's onslaught during the 1988 flood. The river moved 550 m eastward and cut a 45 m deep new channel. Similarly, all efforts at stopping erosion by the Jamuna River near the town of Shirajganj have met with limited success. In 1966, the combined flow of the Ganges and Brahmaputra, downstream of

Faridpur, moved 1,500 m laterally and dug a 30 m deep new channel. As Rogers et al. (1988) notes in the Eastern Waters Study, there is no force on earth that can confront such raw power of nature. River experts are particularly worried about the Brahmaputra River, which they regard as one the world's most turbulent and dynamic rivers. They think that the way it is positioned on a fan of its own silt in northern Bangladesh is indicative of the possibility of another historic shift. As some of the areas through which Bengal rivers pass are seismically active, especially the Brahmaputra and the tributaries of the Meghna, shifts conduced by seismic events also cannot be ruled out.

9. The combined catchment basin of the Ganges, Brahmaputra, and Meghna measures 1,758,000 sq km, which is more than 12 times the size of Bangladesh. Thus, the situation of Bangladesh in this regard contrasts with the situation of most of other countries with important deltas. For example, the entire catchment area of the Mississippi lies in the United States. The catchment areas of the Yangtze, Ho, and Yellow rivers lie in China, as do their deltas. Similarly, Brazil contains both the Amazon Delta and much of the Amazon River's catchment basin.

10. The same is the case with Yangtze delta in China. Even in Egypt, the Nile delta constitutes a small part of the country.

11. This is the estimated figure for 2019, according to http://worldpopulationreview.com/countries/bangladesh-population/.

12. Road construction also started, but on a limited scale. Most communication was based on either the waterways or railways, and canals were dug to let river water reach deep inside the floodplains. People enjoyed the benefits of inundation. They also bore the risks associated with untimely and unusually high levels of flooding.

13. It is therefore considered to be a separate basin.

14. As noted in chapter 6, India is a vigorous follower of the Cordon approach and has built river embankments extensively in other states, including Bihar. For discussion of the Brahmaputra Basin in India, see Goswami (1985) and Pukhan (2005).

15. As Roy and Dutta (2012, pp. 3–4) informs, "During the 18th and 19th centuries, normal flood had occurred which resulted in the formation of floodplain as there was no embankment in those days. A huge volume of water was flowing through the river of sufficient depth. The river had tendency to shift its course and as a result flood plain had extended to a large extent of area." Needless to say, the floods were sometimes untimely and unusually deep, doing damage to crops and dwellings.

16. "To protect the fertile agricultural land and number of prospectus settlement Zamindari bundhs (embankments) were constructed and later these were transferred to Irrigation and Waterways Department, Govt of West Bengal, after the flood of 1959 for maintenance and reconstruction" (Roy and Dutta 2012, p. 4).

17. See Roy and Dutta (2012, p.1).

18. "Due to these high embankments, the main channel of Ajay gradually decreasing its water holding capacity due to the rising up the riverbed for huge amount of siltation. As a result, the river crosses its danger level quickly with a short span of rainfall and water level temporarily increases during rainy season at Natunhat Gauge station (Fig. 3). For this, flood is a common phenomenon in the study area" (Roy and Dutta 2012, p. 4).

19. "Along with massive downpour, sudden release of water from Hinglo barrage upstream triggers the breaching of embankment" (Roy and Dutta 2012, p. 4).

20. "In the lower Ajay River basin, embankments on both sides of the river were mainly constructed from Pandabaswar to Katwa. Particularly, in the right bank side, embankment is started from the confluence point of the Tumun River or Satkahonia to Kogram Village in a continuous way. But after Kogram Village there is no embankment up to the Joykrishnapur village. In between Kogram and Jaykrishnapur, river water of the Ajay River gets a free entrance to the surrounding floodplain and generate flood over the Mangalkote Gram Panchayate (G. P.) and others" (Roy and Dutta 2012, p. 5).

21. See Roy and Dutta (2012, p. 6).

22. Roy and Dutta (2012, p. 3) also provide data regarding decadal frequency of flooding in lower Ajay valley along with a graph showing that the frequency of flooding has increased after 1956 when the embankments were constructed. In their words, "From this frequency analysis, it is clear that *most of the devastating floods ever experienced by the existing generation of people are the floods of after 1956*. It is also clearly evident that the normal and cumulative frequency has been progressively increasing over time in this basin area (Fig 2)" (emphasis added). For more discussion on Ajay embankments, see Bandyopadhyay and Jana (2010), Chakraborty (2010), Ghosh, Mandal, and Banerjee (2015). For information on flooding in West Bengal in recent years, see GoWB (2015, 2016, and n.d.)

23. http://floodlist.com/asia/west-bengal-flood-defences-collapse.

24. http://floodlist.com/asia/india-floods-west-bengal-july-august-2017.

25. https://www.financialexpress.com/india-news/flood-woes-in-west-bengal-man-made-says-cm-mamata-banerjee/799408/.

26. The Damodar River was sometimes called as the "Sorrow of Bengal."

27. The Maharaja of Burdwan and the noted physicist Meghnad Saha were members of this Commission.

28. 1.5 million acre feet.

29. For information on DVC and its activities, see its official website at https://www.dvc.gov.in/dvcwebsite_new1/

30. Of these, Maithon (1,050 MW) is a joint venture with Tata Power, and Bokaro Power Supply Corporation Limited (BPSCL) is a joint venture with Bokaro Steel Limited. The rest are fully DVC owned.

31. It also proposed constructing a dam across the Barakar River at Belpahari in Jharkhand.

32. This suggests reducing the river flow from 18.4 thousand cumec to 7.1 thousand cumec.

33. DVC dams have an irrigation command area (gross) of 569,000 hectares (5,690 sq km), with an irrigation potential of 364,000 hectares (3,640 sq km). The project has canals running 2,494 km.

34. Of the water stored in the four DVC reservoirs, about 419 mcm is used to supply 19.3 cumec of water to meet industrial, municipal, domestic requirements in West Bengal and Jharkhand. For more discussion of the DVC and the impact of its structures,

see Bhattacharya (2011), Chandra (2012), Ghosh (2013), Ghosh and Mistri (2015), Nandy (2001), Thakkar (2015).

35. Other members of the mission were (i) Dr. W. J. van Blommestein, land and water development expert, and leader of the Food and Agriculture Organization team in Dhaka; (ii) Mr. Anthony R. Thomas, consulting hydraulic engineer from the United Kingdom, who served for about 20 years in the Indian Service of Engineers; (iii) Dr. Fred C. Shorter, assistant professor of economics at the California Institute of Technology, serving at the time as adviser to the Pakistan Planning Board and the Government of East Pakistan; (iv) Mr. James Shaw, consulting hydraulic engineer from the United Kingdom, who served in India for about 26 years having retired as chief engineer of the State of Orissa in 1952; and (v) Mr. Charles W. Okey, consultant to the TVA, who served with that organization as director of flood control for many years (UNTAP 1957, p. 2).

36. This letter was part of the report submitted by the Krug Commission, suggesting that it was not written before the mission's work. Instead, the mission's work and report might have influenced it.

37. This recognition reflects Bangladesh's limited scope of maneuverability regarding river management, as noted earlier.

38. Thus, we notice two salient aspects of the Krug Mission's recommendations. First, it recommended studies of feasibility and cost-benefit ratios of a system of flood embankments, and to proceed to their construction only if these studies confirm that there are significant net benefits. Second, the mission was recommending embankments with sluices, flushing channels, and drainage channels. In other words, it did not recommend embankments that entailed complete separation of floodplains from river channels.

39. At the time of this mission's work, both the Karnaphuli hydroelectric project and the Bheramara power station (to serve the Ganges-Kobadak irrigation project) were under construction. Both these power stations were constructed by government water related agencies even though utilization of their output required the participation of the power related agencies. This lack of cooperation made a strong impression on the mission, and it therefore made putting both water and power development tasks under one agency an important part of its recommendation.

40. This pre-existing connection probably partly explains why it was chosen as general consultant.

41. It may be noted that there was some ambiguity in the Krug Mission's recommendation regarding the tasks of the international companies to be hired. On the one hand, it suggested that the firm(s) hired would help to "organize and train the necessary personnel" of the newly formed agency and would also be responsible for "*executing* work programs of the agency until the training of its personnel is completed." On the other hand, the mission recommended that "major" works should be put to international tender until the new agency becomes capable of implementing such major works. Several questions arise. Who will identify the "major works"? Who will design them? If the consulting firm is to "execute work programmes," why would it then be necessary to go for international tender to select companies to implement the

major works? It seems that IECO assumed the role of selecting, designing, as well as implementing major water projects. This multiplicity of IECO's role had some effects on Bangladesh's water development trajectory, as we shall see later.

42. IECO proposed an investment of $3.3 billion, with $1.9 billion for water and $1.4 billion for power. The power generation capacity was to increase from 279 MW in December 1966 to 2,352 MW in December 1985 (IECO 1964a).

43. IECO projected that without water development projects (i.e., under the baseline scenario), rice production would increase at the rate of 2 percent per annum, resulting in a deficit of 5.8 million tons by 1985. With the projects implemented, the country would not only meet this deficit, but also have a modest surplus. The Master Plan envisaged an increase in gross crop value at constant prices from Rs 3,059 million (US$644 million equivalent) in 1964 to Rs 6,694 million (US$ 1,400 million equivalent) in 1985. Direct agricultural benefits arising from projects, measured in terms of the increase in net annual value or production, would increase from about Rs 58 million (US$12 million equivalent) in 1965 to Rs 1,934 million (US$ 407 million equivalent) in 1985 (IECO 1964a).

44. To implement this plan, IECO proposed an investment of US$210 million per year during the third (1966–1970) and fourth (1971–1975) of the Five Year Plans. Nearly 75 percent of the total (1966–1985) expenditure would be made during the first half of the period (IECO 1964a).

45. The Ganges-Kobadak project and the Coastal Embankment project were implemented with assistance from the United Nations Food and Agriculture Organization (FAO). On the other hand, the Brahmaputra Right Hand Embankment project, the Dhaka-Narayanganj-Demra (DND) project, and the Chandpur project were financed largely by credit from the International Development Assistance (IDA). As noted above, G-K started in 1955, and the credit agreements on BRHE and Chandpur were signed in June and July of 1963, respectively (IECO 1964b).

46. See IBRD (1965, p. 2).

47. As IBRD (1965, p. 2) put it, *"The Master Plan is basically a massive scheme for empoldering huge portions of East Pakistan—about 8 million acres of agricultural land by 1985.* The Plan envisages extending the coastal embankments along the Bay of Bengal by a further 1,400 miles and building river embankments of about 2,000 miles. Embankments along both sides of the three main rivers—Brahmaputra-Jamuna, Ganges-Padma, and Lower Meghna—would be completed by 1975" (emphasis added). The protecting embankments would be designed with regulators for admitting floodwater to the project areas when required and for evacuating excess quantities of water by gravity drainage or pumping. See IBRD (1965, pp. 2– 6) for more details.

48. According to Thijsee, "the river flow is confined to a stable and fixed bed at all stages of discharge, allowing for efficient inland navigation. The water in the land between the rivers as well as between the rivers on the one hand and the sea on the other, is completely controlled. It is kept at the most favorable level. When there is too much, the surplus is evacuated; when the crops and the raising of cattle need more, the rivers supply the shortage. The coast line is continuous and as short as possible. It

is interrupted only by a small number of mouths of the principal rivers and by the entrances for navigation to the main harbors. A sea wall prevents flooding, even in exceptional conditions" (IBRD, 1965).

49. As IBRD (1965) notes, "The 1964 and 1965 reports of Professor Thijsse do not disagree with the overall views of either Netherlands Engineering Consultants (NEDECO) or IECO, but they emphasize, correctly, the many uncertainties at that time regarding the effects of the development proposals on the water levels and discharges in the major rivers, the changes in their morphological and hydraulic behaviour and the salt-water intrusion problems associated with the tides and decreased upland flows."

50. IBRD (1965, Annex 1, p. 2).

51. IBRD (1965, Annex 1, p. 3).

52. Already in November 1963 (i.e., before the Master Plan was finalized), the government of Pakistan asked IBRD to "participate in studying East Pakistan's water and power resources." In response to this request and the subsequent elaboration in December 1963, IBRD decided in March 1964 to review the IECO master plan, when completed. It was also decided that IBRD would subsequently hold discussions with the government "to see what more should be done to develop a comprehensive plan for the use of the province's water resources (IBRD 1965, p. 1)." IBRD received the Master Plan in May 1965 and submitted its report on November 8, 1965. The review was led by A. R. Sadove and T. C. Creyke, assisted by A. J. Carmichael, G. F. Darnell, J. Dumolin, P. Z. Kirpich, J. J. Stewart, S. Takahashi, C. R. Willoughby, and G Wyatt— bank staff members—and four consultants, W. H. Cummings, W. A. Dexheimer, J. B. Hendry, and Charles Senour (IBRD 1965, p. 1).

53. IBRD (1965, pp. 6–7).

54. "The Bank review agrees on the necessity, in the long run, to harness the rivers of East Pakistan in order to minimize the destruction caused by floods and to increase the agricultural capacity of the Province. Flood control, drainage, and irrigation projects constitute the means to accomplish this, and at this level of generality, the kinds of projects contained in the Master Plan would be consistent with a rational use of East Pakistan's resources. The Master Plan therefore constitutes a guide on the orders of magnitude such an effort will require, as determined on the basis of a series of specific proposals" (IBRD 1965, p. 6).

55. "The effects of many of the schemes that have been included in the IECO Plan are only partially foreseeable and could prove disastrous" (IBRD 1965, Annex 1, p. 1).

56. IBRD (1965, p. iii). One reason for IBRD's reservations was future uncertainty regarding water availability in Bangladesh due to its downstream location. It noted that Bangladesh was "extremely susceptible to any action taken upstream by India" (IBRD 1965, p. viii). It noted further, "The successful planning, execution, and operation of any large program of development of East Pakistan's land and water resources depends on a solution of the international water problem. Without some international agreement, the sound programming of some Master Plan projects is impossible, particularly those dependent upon the Ganges-Padma system. To make intelligent planning possible, an agreement should be reached to coordinate river

basin planning between India and Pakistan. This may not be achieved easily or soon" (IBRD 1965, p. viii).

57. IBRD (1965, pp. v-vii).

58. Works programs were relief programs aimed at creation of job opportunities and income for the rural poor. See Jones (1985) for discussion of ground water availability in Bangladesh.

59. In particular, during the third Five Year Plan (1966–1970) period.

60. Such as the Ganges-Kobadak; Coastal embankment; Dhaka-Demra-Naraynaganj; Chandpur; Brahmaputra Right Bank Embankment projects, as noted earlier.

61. In particular, it suggested conducting a survey to determine (i) existing availability of human power in the various agencies concerned; (ii) capacity of existing training facilities to meet the shortfall in the time available; and (iii) additional training facilities needed to fill the gap, particularly regarding technicians.

62. Hence, regulators included in Flood Control and Drainage (FCD) and Flood Control, Drainage, and Irrigation (FCDI) embankments recommended by the World Bank were mostly for drainage and not for controlled flooding. Furthermore, soon these regulators often failed in their drainage function too.

63. The idea of double-embankment as an immediate task surfaced again in the wake of the 1988 floods. However, as we shall see in chapter 12, it was put off again in order to gather more knowledge and experience.

64. The project has two pump houses with pumps having various capacities. The main pump house had three large pumps with a total designed capacity of 111 cumec. An auxiliary pump house was also constructed with 12 medium-sized pumps with a total capacity of about 42 cumec. For information on the Ganges-Kobadak project, see BWDB (n.d.).

65. The combined length of the two main canals is 193 km; of 17 secondary canals is 467 km; and of 154 tertiary canals is 995 km. The field channels number 1,772 and have a total length of 2,771 km.

66. See Mohammad (1966) for details.

67. For example, of the 240 acre-inches water pumped, only 120 acre-inches could reach the fields. See Mohammad (1966).

68. See ADB (1998, p. 2).

69. See ADB (1998, p. 11), which found, "The areas proposed for inclusion in the subsequent phases of the development, which have not yet started, have achieved remarkable production growth even without the intended surface irrigation."

70. For example, based on his review, Golam Mohammad recommended against similar projects. See for details Mohammad (1966).

71. "Large irrigation schemes like the G-K system would probably not be viable if undertaken anew considering current rice prices" (ADB 1998, p. iv).

72. As many as 1,402 personnel are employed by this project, of whom 57 are officers and technical experts, while the other 1,345 are supporting and field staff.

73. See ADB (1998, p. 3) for details.

74. At the time of its completion, the Jamuna bridge of Bangladesh was the eleventh longest bridge in the world. For information on the BRHE project, see BWDB (n.d.).

75. According to Hoggarth-Vicki, Cowan, Aeron-Thomas, McGregor, Garaway, Payne, and Welcomme (1996), 200 km.

76. The project also involved 220 km of "inspections roads" and other components of infrastructure. For information on PIRDP, see BWDB (n.d.).

77. The area irrigated by STWs in PIRDP increased from about 30,000 hectares before the project in 1985 to 55,000 hectares after the project was completed in 1992 (ADB 1996, pp. 8–9). As a result of the pre-existing STW-based irrigation, the project actually led to little incremental irrigation. As ADB (1996, p. 12) put it, "The Bank-supported Command Area Development Project (CADP) is being implemented to address this issue. The CADP is resulting in the replacement of private tube-wells by large-scale public pumping irrigation, a technology little developed and tried in Bangladesh" (ADB 1996, p. 12).

78. As ADB evaluation noted, "As regards crop benefits, the Post Evaluation Mission (PEM) estimated the situation using district/sub-district and other information. *Crop output growth in the project area shows no discernible difference from that of overall Bangladesh or nearby districts*, raising questions about the benefits of flood protection" (ADB 1996, p. 12, emphasis added).

79. ADB (1996, pp. 12–13).

80. See ADB (1996, p. 11).

81. See ADB (1996, p. 12).

82. The total length of the irrigation canals was 55.2 km and that of drainage canals was 45.4 km. In addition, there were 216 minor water control structures. For information on DND project, see BWDB (n.d.).

83. Shams and Hossain (2013, p.2). See also Saha, Tareq, and Saha (2012).

84. See Munshi (2008) .

85. See IBRD/IDA (1963).

86. See IBRD (1968, p. 38–39) for details. See also BWDB (n.d.).

87. In 1988, the CIP's embankment collapsed in face of the Meghna's onslaught. The river moved 550 m eastward and cut a 45 m deep, new channel.

88. Of these, 34 km were main channel, 64 km secondary channel, and 120 km tertiary channel. The project also included many outlets, 72 footbridges, and 811 turnouts. It cost Tk 175.04 crore (of which Tk 48.28 crore was in foreign currency). ADB (1990) further states, "The existing creeks in the project area were interconnected under the project. They were then directed towards the main drainage ways and were finally conducted to the primary pumping stations."

89. Of these, the main two pump stations were located at Kalipur and Uddhamdi, while the two, considered auxiliary, are located at Dubgi and Eklaspur. Kalipur pumping station can drain 40 percent of the project area and irrigate 6,426 hectares, while Uddhamdi pumping station can pump out excess water from 60 percent of the area and supply irrigation water to a command area of 8,046 hectares. For information on Meghna-Dhonagoda project, see BWDB (n.d.).

90. The project also included 72 footbridges and 811 turnouts. See ADB (1990) for details.

91. In 1987, a portion of the embankment along the Dhonagoda River on the east breached near Durgapur, and in 1988 a similar breach developed at Rishikandi. The high velocity, sand-carrying water that entered and submerged several thousand acres of the project area disrupted the networks of irrigation and drainage channels.

The breaches were caused not by overtopping but embankment or subsoil failure (see Islam AZMM 1992, p. 21). Observers found that the Dhonagoda embankment between Nandalalpur and Durgapur was riven with gullies and riddled with incipient piping and apprehended that another breach might occur if there was an attack by the river on the embankment in future (pp. 29-30).

92. See Islam, AZMM (1992, pp. 32–33). See also FAP 12 (1992).

93. Farmers were bound to apply high doses of chemical fertilizers to make up for the nutrient deficit for reasonable crop productions, leading in turn to greater toxification of open water bodies and damages to capture fisheries (Islam, AZMM, 1992, pp. 32–33).

94. See FAP 12 (1992) and Islam AZMM (1992, p. 35).

95. See FAP 12 (1992) and Islam AZMM (1992, p. 22).

96. See Islam AZMM (1992, p. 32).

97. See Islam AZMM (1992, p. 29) and FAP12 (1992).

98. See Islam AZMM (1992, p. 31).

99. See Islam AZMM (1992, p. 30). The budget of this CAD is Tk 1.1 billion.

100. See Vaughn (1997, pp. 29–30) for further discussion of other environmental effects of the M-D project. See also, FAP 12 (1992); Thompson (1990); Thompson and Sultana (1996); and Sultana, Thompson, and Daplyn (1995).

101. See Islam, R. (ed.) (2004) and Islam R. (2006, 2007).

102. See World Bank (2005, p. 1).

103. See IECO (1964b, p. 275).

104. See World Bank (2005, p.1). See also BWDB (n.d.).

105. Islam, R. (ed.) (2004, p. 11) and Islam R. (2006, and 2007)

106. During tidal bores, seawater often breaches or overtops the embankments and submerges the polders. Rain falling inside the polders aggravates the situation further. In these conditions, parts of the embankments that remain above water become the main shelter for the submerged polders' inhabitants. For discussion of the impact of coastal embankments on flash flood, see Choudhury, Paul, and Paul (2004).

107. See Islam, R. (2006, and 2007).

108. See, for example, Brammer (2014, p. 56), who explains the process as follows: "An additional dynamic factor in Region E is land subsidence. Under natural conditions, such subsidence was counteracted by sedimentation from the rivers at high tide. The construction of embankments cut off this natural sediment accretion within polders. Few studies have been made to-date to measure actual subsidence rates or determine its causes, and the situation is complicated by the breaching of embankments in some areas for shrimp farming and by storm surges which periodically enable new (but probably irregular) sedimentation on the land."

109. As ADB (2007, p. v) notes, "The shortcomings of the coastal embankment project (CEP) in southwest Bangladesh worsened drainage congestion and caused prolonged inundation of farmlands, household lots, and the internal communication networks. The results were declining agricultural production, fewer employment opportunities, and deteriorating salinity conditions, which collectively led to lower living standards, reflected by 75 percent of the population living below the poverty line at the time of project formulation (i.e. 1993)."

110. In the whole coastal zone, 52,906 brackish water shrimp farms, with a total area of 172,833 hectares, have been established. About 93 percent of all shrimp farms inside polders had an average size of less than 10 hectares. Bangladesh now accounts for 2.5 percent of the global shrimp trade (DoF 2013). See also Kashem, Siddiqui, and Rahman (2017).

111. See Rahman (1995).

112. See for details ADB (2007). For alternative methods of coastal embankment erosion control, see Islam N. Md. (2000). See World Bank (2005) for an assessment of the Coastal Embankment Rehabilitation Project.

113. See ADB (2007, p. v).

114. See ADB (2007, p. vi).

115. However, a feasibility study of the Dhaka Integrated Flood Protection Project (DIFPP) was conducted later, in 1991.

116. BWDB, Dhaka Water and Sewerage Authority (WASA), and Dhaka City Corporation (DCC), respectively, were put in charge of implementation of these components costing Tk 523 crores, Tk 267 crores, and Tk 47 crores, respectively, with a total of Tk 837 crores. For information on DIFPP, see also BWDB (n.d.).

117. Of these 50 canals, the three major canal systems were as follows: (i) Degun–Ibrahimur–Kallyanpur canal that drains out to the Turag River; (ii) the Dhanmondi–Paribagh–Gulshan–Banani–Mohakhali–Begunbari canal that drains out to the Balu River, and (iii) the Segunbagicha–Gerani–Dholaikhal canal that drains out to the Balu and Buriganga rivers. See https://ejatlas.org/conflict/disappearing-of-canals-of-dhaka-city-bangladesh, which also provides the list of the 50 canals. For example, the Buriganga River on the west was connected with Dhanmondi Lake, which was in turn connected with Hatirjheel Lake area through what is now the Panthapath; and Hatirjheel Lake area was connected with the Balu River through Begunbari Canal.

118. The Cordon approach's reliance on pumped drainage led to under-appreciation of the role of the canals for Dhaka city's drainage.

119. This has been the general practice, for example, for housing construction in the vast Bashundhara Housing Project, stretching from the Bishwa Road to the bank of the Balu River and beyond. For details see, http://www.bvsde.paho.org/bvsacd/cd46/cap9-flood.pdf.

120. In that year, water from the Balu River spread over the eastern part of the city and entered even the western part though unblocked culverts and open regulators. Water entered Mahakhali, Tejgaon, Gulshan, Banani, Badda, and Baridhara areas through Begunbari Khal regulator, Shahjadpur Bridge on Pragati Sarani, and Khilkhet pipe culvert at Nikunja. Other crossings on the Bishwa Road allowed water to move to Rajarbag, Gopibag, and Fakirapul area. All these openings were closed, using regulators and temporary earthen bunds, and 32 pumps were installed at Segunbagicha khal to pump out the flood water, for details see http://www.bvsde.paho.org/bvsacd/cd46/cap9-flood.pdf. For more discussion of Dhaka city's flood experience and vulnerability and the new problem of waterlogging, see Huq and Alam (2003), Nishat, Reazuddin, and others (eds.) (1999), Mowla and Islam (2013) and Saleh, Ahmed, Rahman, Salehin, and Mondal (2002). For increase in flood frequency in Bangladesh as a whole, see Khalequzzaman (1994).

121. These are Khulna, Dinajpur, Kurigram, Panchagarh, Habiganj, and Moulavi bazar. For information on STIFPR, see also BDWB (n.d.).

122. Feasibility studies were conducted during 1991–1992 by Sir William Halcrow and partners of the United Kingdom, in association with Engineering and Planning Consultants Ltd. of Bangladesh.

12

Open approach to rivers in a delta

12.1 Introduction

The crisis of the Cordon approach has led more people to the Open approach in both parts of the Bengal Delta. Though the Cordon approach still dominates government policymaking, the influence of the Open approach is spreading among various social groups, including river activists and scholars; non-governmental and civil society organizations; media; judiciary and local government; and the general public. Sometimes, they even have been successful in influencing government policies and projects.

Chapter 11 reviewed criticisms of the embankments constructed along the Ajay and other rivers in the Indian part of the Bengal Delta. The debate is more acute in Bangladesh, where the approach adopted toward rivers is of fundamental importance for the ecology and economy of the entire country. After many decades of experience with the Cordon approach, a popular resistance has developed against it, as was evidenced during the movement against the FAP of 1989, to be discussed later. Scholars, environmentalists, non-governmental organizations (NGOs), and the affected local people put up a stiff resistance against re-imposition of the ideas of the Cordon approach. Instead, they called for the Open approach.

In some cases, people came up with their own, indigenous, Open approach-conforming solutions to the problems created by the Cordon approach. For example, the people of southwestern Bangladesh, fed up with the waterlogging created by coastal polders, came up with their own solutions, which were later accepted by the water establishment and came to be known as the Tidal River Management (TRM) system. Under this system, openings were made in the cordons to let river water in during high tides, deposit the sediment in depressions within the polders, become sediment free, and scour the riverbed on the way back during low tides. Needless to say, TRM can be regarded as a partial application of the Open approach. Similarly, farmers and fishermen of the Pabna Irrigation and Rural Development Project forced the project managers to use the sluice gates to let the river water in and emulate flood conditions that prevailed before construction of cordons.

Rivers and Sustainable Development. S. Nazrul Islam, Oxford University Press (2020). © Oxford University Press.
DOI: 10.1093/oso/9780190079024.001.0001

More broadly, a robust civic environment and river movement has emerged in Bangladesh over the last two decades. It has provided a deep and extensive analysis of the river crisis and put forward the demand for abandoning the Cordon approach and for adopting the Open approach. Organizations have been formed at the national level to campaign for protection of Bangladesh's rivers. Under the guidance of these organizations, and sometimes spontaneously, movements to save rivers have emerged at the local level all across the country. Responding to these demands, the government of Bangladesh has also taken some steps for the protection of rivers, forming initially (in 2004) a River Task Force focused on rivers around Dhaka city and later the National River Conservation Commission (NRCC) in 2016, with a more comprehensive, national remit.

Also, as noted in chapter 11, there have been some changes in the thinking of foreign lending agencies, which exert significant influence on the decisions and project proposals of Bangladeshi water authorities. The World Bank's opposition to large-scale gravity flow irrigation projects led to the switch from the pump-gravity to pump-gravity-pump or simply pump method of irrigation in some projects. Second, drainage congestion problems received more attention in the design of cordons. Third, there was some broadening of the objectives of cordon projects. Instead of focusing exclusively on increasing crop output, objectives were formulated in terms of development, poverty reduction, etc., thus allowing more attention to protection of open fisheries, waterways, employment of the rural poor, etc. Fourth, there was more recognition of the necessity of local people's participation in designing and managing water development projects, instead of foisting on them cordon projects from above and commanding them to comply with certain rules and regulations. Local people, who have been practicing the Open approach for thousands of years, have a much better sense of the merits of this approach than do government officials and foreign consultants. Finally, the current prime minister of Bangladesh, Sheikh Hasina, frequently makes pronouncements that agree with the Open approach. For example, she asked rivers to be kept free, appealed to the people to learn to live with floods, instructed the country's engineers not to build unnecessary structures that obstruct water flow, and took the initiative to construct new canals to connect parts of the Dhaka city with the Balu River.

However, the response of Bangladesh's water establishment to the prime minister's directives and exhortations has not always been enthusiastic. The water establishment somehow manages in many cases to ignore these directives and continue to promote the building of structures whenever and wherever it gets an opportunity to do so. Often it implements river protection decisions in such a way as to produce perverse outcomes. The reluctance of the policy establishment to embrace whole-heartedly the Open approach can be seen in the *Bangladesh Delta Plan 2100*, published by the Bangladesh Planning Commission. This plan, finalized in September 2018, fails to take a clear position in favor of the Open approach. Instead, it vacillates between the two approaches, displays confusion and eclecticism, and clings mostly to the Cordon approach. Hence, more efforts are necessary to make Bangladesh adopt the Open approach as a general policy regarding rivers.

The objective conditions for an upswell in public opinion for the Open approach are maturing as the manifestations of the fundamental inappropriateness of the Cordon approach for the Bengal Delta become ever more acute. In addition to flooding, the problem of waterlogging is now spreading all across the delta; lean-season river flows are dwindling; and salinity is intruding further north. Climate change is aggravating these problems with each passing day. It is necessary to explain better how the Open approach can help to confront these problems. It is also necessary to explain how the implementation of the Open approach can actually proceed.

The discussion of this chapter is organized as follows. Section 12.2 discusses the merits of the Open approach in confronting the adverse effects of climate change in the Bengal Delta. Section 12.3 notes efforts toward the Open approach in the Indian part of the Bengal Delta. The rest of the chapter reviews the progress of the Open approach in the Bangladesh part of the delta. Section 12.4 presents the struggle against the Cordon approach-based FAP that was drawn up following the historic flood of 1988. Section 12.5 notes the changes that took place in Bangladesh's water-related institutional set up following the FAP process and the prospects they present for promoting the Open approach. The chapter next presents a few examples of water projects that, to a certain degree, conform with the Open approach. Section 12.6 presents the opening of the Pabna Irrigation Project to revive flood conditions. Section 12.7 reviews the opening of some coastal polders under the new TRM system. Section 12.8 presents some Open approach-conforming irrigation projects, such as the Barisal Irrigation Project and the Matamuhuri Irrigation Project. Section 12.9 offers a brief review of the *Bangladesh Delta Plan 2100*, approved recently by the government. Section 12.10 draws attention to the broad-based civic movement that has emerged in Bangladesh for adoption of the Open approach. Section 12.11 concludes.

12.2 Open approach as a climate change strategy for Bengal Delta

The Open approach can play a particularly useful role in protecting the Bengal Delta against climate change. To see this, it is useful to note first the different adverse effects that climate change is having and will have on the Bengal Delta.

12.2.1 Adverse effects of climate change on the Bengal Delta

Though the Bengal Delta and its people have made almost zero contribution to causing climate change, they are bearing the brunt of its adverse effects and will do so even more in the future. It is true that "Mitigation is the best adaptation!" However, even if future GHG emissions are brought down to zero, the world cannot escape the effects of the carbon dioxide that has already accumulated in the atmosphere.[1] As a result, considerable climate change is now a fact of life and is going to intensify in the coming decades. Deltas are likely to be worst affected. The following first lists and then provides a brief description of the major ways in which climate change is affecting the Bengal Delta.

- Submergence
- Salinity intrusion
- Aggravation of seasonality of river flow
- Aggravation of extreme weather events
- Aggravation of diseases and epidemics

12.2.1.1 Submergence
The *Climate Change 2007: Synthesis Report* (IPCC 2007, p. 47) notes that sea level has risen, due to climate change, at an average rate of 3.1 mm per year during 1993–2003 and may rise in the future by several meters as a result of the complete melting of the Greenland and Antarctica ice sheets and the North Pole ice cap. This would result in the submergence of a large part of the Bengal Delta, as most of this delta lies within 10 m from the current sea level. The situation of Bangladesh is particularly serious, as the country is densely populated with about 165 million people in 2019 and projected to grow to 220 million by 2050 (UNSD 2019). According to geological surveys, a rise of sea level by 1 m would submerge about 15 percent of Bangladesh's landmass and convert tens of millions of Bangladeshis into "climate refugees," destabilizing the country, South Asia, and the world as a whole. A similar situation is faced by the Indian part of the Bengal Delta, particularly its southern part, including the districts of 24-Parganas (both North and South) and Medinipur (both East and West).

12.2.1.2 Salinity intrusion
Even the part of the Bengal Delta that will escape submergence will be affected by salinity intrusion that will harm agriculture, drinking water, and the local flora and fauna. According to the Soil Resources Development Institute (SRDI), the area of Bangladesh affected by salinity has increased from 83.3 million hectares in 1973 to 105.6 million hectares in 2009 and is continuing to increase. Salinity is now spreading to non-coastal areas as well, increasing the difficulty of living

and forcing people to leave their homes and migrate. According to a joint study by the International Food Policy Research Institute (IFPRI) and Ohio State University, conducted in 2018, between 15,000 to 30,000 families are migrating from Chittagong and Khulna districts each year due to rising soil salinity.[2]

12.2.1.3 Aggravation of seasonality of river flows

Climate change is exacerbating the already extreme seasonality of Bengal rivers, in several ways. First, warming temperatures cause more evaporation of water from the Indian Ocean, resulting in more moisture in monsoon winds, and hence more precipitation in summer. Second, warming temperatures are also leading to more melting of the Himalayan snows and glaciers. These two processes are increasing the peak season (summer) water flow of Bengal rivers. Third, receding Himalayan glaciers are reducing the winter flow of Bengal rivers. Meanwhile, by diverting the winter season flow while letting the summer season flow pass through, the diversionary structures built by India on Bengal rivers are further exacerbating the seasonality of river flows in the Bangladesh part of the Bengal Delta.

12.2.1.4 Aggravation of extreme weather events

Sitting at the apex of the Bay of Bengal, which itself is a triangular extension of the Indian Ocean, the Bengal Delta is a frequent destination of cyclones and their accompanying tidal bores.[3] Climate change is making both more frequent and menacing. Climate change is also causing rainfall pattern to be erratic, with untimely excessive rainfall and lack of rain during times when it is generally expected. These changes are causing floods and droughts at unusual times of the year.

12.2.1.5 Aggravation of diseases and epidemics

People living in the Bengal Delta, located in the tropics, are already vulnerable to the spread of diseases—both known and new, and in particular to water- and vector-borne diseases. Global climate change is likely to make the problem worse. Incidence of diseases such as dengue, malaria, diarrhea, and pneumonia has increased following climate related extreme weather events, resulting in increased health costs for many households.[4]

12.2.2 Open approach to confront climate change effects in Bengal Delta

It so happens that the Open approach, together with the cross-sectional variant of the Ecological approach adopted for the upper reaches of rivers, can help the Bengal Delta withstand many of the adverse effects of climate change. The following provides a brief review of how this can happen.

12.2.2.1 Counteracting submergence and salinity intrusion

As noted earlier, geological studies show that sedimentation in the past has raised the elevation of Bangladesh's coast by about 2 mm per year. This is close to the rate at which sea level is rising along Bengal Delta coast due to global climate change. By ensuring unfettered flow of sediment to the coast, the Open approach can therefore countervail the submergence process to a significant degree. Currently, the Cordon approach is hampering the process of increasing coastal elevation. It is causing the sediment to either settle on riverbeds or pass deep into the sea because of the more confined nature of the flow. Deposition of sediment deep into the sea may help delta formation processes in the long term, but it does not help Bengal Delta much with regard to confronting current sea level rises. Furthermore, the Bay of Bengal has, what is known as, the "swath of no ground," a deep canyon where sediments can get lost forever, so to speak. Needless to say, for the countervailing effect of the Open approach to be more effective, it has to be combined with adoption of the cross-sectional version of the Ecological approach for upstream reaches of the rivers. As noted earlier, upstream diversion of rivers has reduced the volume of sediment reaching the coast from about 2 billion tons to about 1 billion and is decreasing it further.[5] Similarly, the Open approach, in combination with the cross-sectional version of the Ecological approach applied upstream, can help the Bengal Delta to withstand submergence and salinity intrusion by restoring the original, natural volumes of freshwater flows to the coast.

12.2.2.2 Counteracting seasonality of river flows

The Open approach can help to counteract and cope with aggravation of the seasonality effect of climate change in two ways. First, by allowing river overflows to spread over floodplains, it creates the physical capacity to deal with increased summer flows caused by climate change. Second, by allowing to spread over floodplains, the Open approach also helps to retain more summer water, both in surface waterbodies and in underground aquifers. A part of this retained water can then flow back to the river and augment its dry season flow.

12.2.2.3 Coping with extreme weather events

The Open approach can help coastal people to face extreme weather events, such as cyclones and tidal bores. Currently, coastal polders create a false sense of security, often leading to lack of preparedness on the part of people living inside them. The Open approach will persuade people to build on higher ground, with stronger materials, and with better emergency facilities, including drinking water supply. Second, the cordons currently have become a cause of prolonged waterlogging inside the polders where both rainwater and seawater carried by tidal bores get trapped. The Open approach does away with this problem.

Third, the Open approach can help coastal people withstand cyclones by better protecting the Sundarbans, the mangrove forests that serve as a barrier between them and the sea. Currently, decrease in sediment and freshwater volumes due to the Commercial and Cordon approaches is leading to the decay of these forests, as pointed out by the recent UNESCO team that visited the area in connection with the controversy regarding the Rampal thermal power station that Bangladesh and India together are building near these forests. The recent experience of hurricanes, such as Aila, Nargis, etc., illustrated again how the Sundarbans save the coastal population from direct hits by cyclones and tidal bores.

12.2.2.4 Counteracting diseases and epidemics

The Open approach can help to mitigate the aggravation of diseases and epidemics by restoring the general cleansing effect of regular river and tidal flows. It can be particularly effective in this regard by helping coastal areas to avoid prolonged waterlogging that currently follows cyclones, due to the ob-struction to drainage caused by the polders.

12.3 Open approach in the Indian part of Bengal Delta

Responding to the crises created by the Cordon approach, more people in both parts of the Bengal Delta are now advocating for the Open approach. The discussion (in chapter 11) of the Ajay embankments of the Indian part of the Bengal Delta noted scholars expressing the view that "periodic spilling is not only normal for a stream but it is also a very welcome process. Flood(ing) is a natural process of river which cannot be controlled, and it is helpful for the floodplain areas also" (Roy and Dutta 2012, p. 6). Elaborating further, these authors argue against "structural flood mitigation strategies" because of their negative effects on floodplains. Referring to the Kunar River, Roy and Dutta note that "embankment increases the chance of flood rather than mitigate flood." They find that "embankments are liable to failure, and when they fail, the damage can be much greater than if there were no embankment (p. 1)." Other scholars from the Indian part of the Bengal Delta, who have also advocated for more nature-conforming approaches in dealing with rivers, include Agarwal and Narain eds. (1997); Bandyopadhaya and Parveen (2002); Bandyopadhyay and Jana (2010); Bhattacharya (2011); Biswas (2015); Chakraborty (2010); Chandra (2012); Gain and Giupponi (2014); Ghosh (2013); Ghosh, Mandal, and Banerjee (2015); Ghosh and Mistri (2015); Mitra and Mukhopadhaya (2005); Mukherjee and Saha (2016); Nandy (2001); Rudra (2000, 2002, 2010), Rudra, Panda, and Romshoo (2012); Sarkar (2006), and Sharma (2002). Earlier, we saw Somanathan (2013) doubting embankments as an effective flood control strategy, and Mishra (2001,

2005, 2015) advocating strongly for the Open approach in the context of the Kosi floodplains in Bihar.

At the same time, there is some ambivalence among some river scholars and activists of the Indian part of the Bengal Delta about letting rivers free. Some of them find it hard to embrace the Ecological and Open approaches whole-heartedly, even though they intellectually comprehend their rationale. The same applies to some Indian river scholars and activists from other parts of the country. This is, in part, because of India's upstream location, which allows it to divert river flows away from Bangladesh. As noted earlier, India has already taken advantage of its upstream location to divert river water away from Bangladesh by constructing barrages, such as the Farakka on the Ganges River and Gajoldoba on the Teesta River. Being beneficiaries of these diversions, it is difficult for many Indian scholars and activists to be consistent with regard their support for the Ecological and Open approaches.

By contrast, the people, river scholars, and activists of Bangladesh gravitate naturally to the Ecological and Open approaches. For several decades now, they have been conducting a spirited campaign for the Open approach and have achieved some successes. The following provides a brief review of these efforts.

12.4 Struggle against FAP in Bangladesh

In the wake of the 1988 flood, there was a push by the Government of Bangladesh (GoB) and foreign lending agencies toward a major embankment construction plan. The original idea, contained in the 1964 Master Plan, of double embanking major rivers, was revived and promoted in several studies commissioned to suggest what was to be done following the 1988 flood. Although there were differences in viewpoints among these studies, GoB—in collaboration with some major aid agencies—formulated a Flood Action Plan (FAP) that was geared toward further embankment construction. However, many river scholars and activists, as well as NGOs and civil society organizations and their international allies put up a stiff resistance against this plan, so that construction of major embankments could not proceed.

12.4.1 GoB and UNDP study

Some of the studies conducted in the wake of the 1988 flood were under-taken by GoB, with support from the UN Development Programme (UNDP). Immediately after the 1988 flood, GoB carried out a rapid study and published a preliminary report, titled *National Flood Program*. Proceeding from this brief report, GoB undertook two more comprehensive studies—*Flood Policy Study* and *Flood Preparedness Study*—with UNDP assistance and carried out by a multidisciplinary and multisectoral team consisting of both Bangladeshi and foreign professionals. These studies were completed in early 1989. The *Flood Policy Study* (GoB and UNDP 1989) followed the Cordon approach. It set the long-term objective as "contiguous flood protection on both sides of each of the three main rivers", namely the Ganges, Brahmaputra, and Meghna (World Bank 1990, p. 26). It concluded that "embankments will form the basis for an effective flood protection program. They are to be built to provide a controlled environment in which social and economic development can be undertaken with confidence" (ibid.).

Furthermore, the GoB and UNDP study called for dividing protected areas into locally managed "compartments" to provide controlled flooding and drainage and to contain the area of flooding in case of an embankment breach. It was claimed that such compartments together will form a "contiguous flood control system" along main rivers and will provide greater benefits than a "stand-alone embankment scheme." As we shall see, this idea of compartmentalization proved to be an important flash point for later debate.[6] The proposed GoB and UNDP plan also suggested controlled flooding and drainage for the basins of minor rivers too, "to complement the main stem development."

GoB and UNDP (1989) set six objectives, including the objective of "creating flood free land to accommodate the increasing population."[7] It also developed 11 guiding principles, some of which were actually contradictory to each other (see World Bank 1990, p. 3). For example, on the one hand, it contained principles calling for "*controlled flooding*, wherever possible and appropriate" (Principle 1) and of "reduction and distribution of (sediment) load on the main rivers through diversion of flows into major distributaries" (Principle 7). On the other hand, it included principles calling for "safe conveyance of the large cross-boundary flow to the Bay of Bengal by channeling it through the major rivers with the help of embankments on both sides" (Principle 5). (NDC 1993, p. 27; GoB and UNDP 1989). Despite the call for "controlled flooding," the main emphasis of the GoB and UNDP study was on "safe conveyance" and double embankment.[8]

12.4.2 French study

Some serendipitous incidents played an important role in the process leading to the FAP. Danielle Mitterrand, wife of the then French President, Francois Mitterrand, was visiting Bangladesh when the 1988 flood hit and submerged even the diplomatic quarters of the city, so that she got stuck in the flood and experienced Bangladesh's flood problem on a first-hand basis. Upon returning to

Paris, she apparently persuaded her husband to take some initiative to solve the problem. Accordingly, the French government sent a team of engineers to study the problem and suggest necessary measures. The French team, working jointly with BWDB, prepared *Prefeasibility Study of Flood Control in Bangladesh* (French Engineering Consortium 1989). The solution in this report was based on the idea of double embankments that would confine river flows to river channels only.[9] In particular, it suggested the following:

- Embankments along the three major rivers, the Brahmaputra-Jamuna, the Ganges-Padma, and the Meghna.
- Embankments along the main tributaries, the Teesta, Dharla, Dudhkumar, Atrai, Kangsa, Titas, and Gumti rivers, in order to counter the floods in these rivers and the backwater effect during high stages in the major rivers.
- Embankments along certain distributaries, the Old Brahmaputra, the Dhaleswari, and the Arial Khan, in order to alleviate the discharge flowing in the major rivers (World Bank 1990, p. 27).

Thus, the French proposal advocated a stronger version of the Cordon approach. It estimated the total length of necessary main embankments to be 3,350 km, of which 2,300 km would be new and the remaining 1,050 km would be reinforced existing embankments. The height would be designed for a 100 year return flood and would average 4.5 m with a maximum of 7.4 m. For 18 major towns, these would be designed for a 500-year return flood. There would also be river training to prevent bank erosion and to offer protection to urban areas facing such erosion, such as Sirajganj and Chandpur. The proposal included regulatory structures to control desired flooding of the protected areas and ensure drainage through either gravity or pumps.[10]

12.4.3 JICA study

The Japan International Cooperation Agency (JICA) also conducted a study, *Report of the Survey of Flood Control Planning in Bangladesh*, which came out in May 1989. This study also advocated the Cordon approach and creation of flood-free conditions to enhance agriculture, accommodate increasing population, and allow development of commercial and industrial enterprises.[11] The main long-term goal was to contain the three major rivers through double embanking.[12]

The report considered other options of flood prevention, such as creation of reservoirs, dredging, creation of large diversionary channels, etc., but ruled them out because of land scarcity, ineffectiveness, etc.[13] Hence, it settled for embankments. However, in doing so, the JICA study, unlike the UNDP and French studies, argued for a less hurried approach. It cautioned against construction of long and continuous river embankments without careful examination of their technical feasibility and economic viability. It rather recommended a "stage-wise" and "cellular" implementation process, beginning with construction of embankments in limited areas, in combination with existing or planned road embankments. It argued that a series of such "cellular blocks" will be similar in effect to continuous river embankment, with interior polders serving as secondary embankments.[14] Thus, the JICA proposal was similar to the GoB/UNDP compartmentalization proposal, except for its emphasis on a staged implementation, more intensive use of existing embankments and roads, and complementary non-structural measures, such as flood forecasting, warning, and preparedness.

Along with this long-term plan, comprising both structural and non-structural measures, the JICA study also put forward an immediate action program, with several components. The first included projects that could offer quick benefits. Among these were flood protection and drainage of Dhaka city, improvement of the flood forecasting and warning system, and strengthening of the flood evacuation and relief operations. The second included preparatory works necessary for implementation of the long-term flood control plan. These included topographic mapping, longitudinal and cross-sectional survey of rivers, detailed study, analysis, and research, covering a period of four years.[15]

12.4.4 Eastern Waters Study

While the studies reviewed above advocated for the Cordon approach, including double embanking and creation of flood-free conditions, the Eastern Waters Study (EWS) (Rogers, Lydon, and Seckler, 1989), commissioned by United States Agency for International Development (USAID), struck a different cord. It basically argued against the Cordon approach and for the Open approach. Authored by a team led by Harvard professor of environmental engineering, Peter Rogers, EWS argued that, in view of the huge physical dimensions of Bengal rivers, their extreme seasonality of flow, and the alluvial nature of the soil through which they pass, it is impractical to think of containing their flows within their channels: the physical force of Bengal rivers is so great that no amount of engineering works can withstand it. They cited instances (noted in chapter 11) of devouring of large chunks of the Meghna-Dhonagoda project embankment and area in a matter of days by the Meghna River during the 1988 flood. Based on their analysis, EWS concluded, "The search for embankment protection from floods in the great delta is almost certain to lead to waste and

disappointment" (p. 74). EWS also took note of the beneficial functions of regular river inundation and the importance of their continuation. It noted that
it was quite possible to escape the intricate ways in which the floodplain agriculture, economy, and mode of living depend on these functions. Referring to
the cavalier proposals by foreign consultants for double embankment and safe
passage of outside water to the Bay of Bengal, EWS pointed to the necessity for
"awareness of being outsiders" (p. 73).

Not surprisingly, the EWS recommendations were reportedly not welcome to
Bangladesh's water establishment and other foreign agencies, who were eager to
embark on multi-billion-dollar embankment construction projects. Apparently
under pressure from these quarters, USAID distanced itself from the EWS.[16]

12.4.5   World Bank's coordinating role and formulation of FAP

Facing these different studies with varying recommendations, GoB requested
the World Bank "to coordinate international efforts" aimed at mitigating
Bangladesh's flood problem, and the World Bank agreed in June 1989 to do
so. Following up, the World Bank convened a meeting in Washington DC on
July 11–13, on the eve of the G7 meeting held on July 14–16 in Paris, where
concerns were expressed regarding Bangladesh's flood problem and the need for
coordinated action by the international community was stressed.[17] Attended
by leading figures of all the studies described above, the Washington meeting
decided to focus on a short-run plan for the next five years as the first step in
the long-term flood control program. This plan would include a few "ready
to go projects" and "studies leading to further projects in the near-term and
subsequent plan periods." It would also include "supporting technical, socio-
economic, and environmental studies." (NDC 1993). The World Bank embarked
on preparing such a plan, assisted by a team that included key experts involved
with the above studies.

The preliminary draft of this plan was prepared in August 1989, and it was
revised through discussion with GoB officials in September and October, in
meetings held in Dhaka and Washington DC, respectively. The process led to
the formulation, in November 1989, of the FAP, which was supposed to have
four phases, each ranging five years and together covering the twenty-year period 1990–2010, with the first phase (1990–1995) comprising 26 components
(Table 12.1). This plan was placed in December 1989 before the aid consortium
meeting in London, where it was approved.

As can be seen from Table 12.1, most of the components of FAP first phase were
focused on studies, which were expected to help to identify, plan, and design actual
(embankment) construction projects, to be taken up during future phases. A good

Table 12.1 Components of the Flood Action Plan (Phase I)

FAP Number	Title	Donor(s)
1	Brahmaputra Right Hand Embankment Strengthening	IDA
2	Northwest Regional Study	UK/Japan
3	North Central Regional Study	EC/France
3.1	Jamalpur Priority Project	EC/France
4	Southwest Area Water Management Study	UNDP/ADB
5	Southeast Regional Study	UNDP
6	Northeast Regional Study	Canada
7	Cyclone Protection Study	EC
8a	Greater Dhaka Protection Project	Japan
8b	Dhaka Integrated Town Protection Project	ADB/Finland
9a	Secondary Towns Protection Project	ADB
9b	Meghna Left Bank Protection Project	IDA
10	Flood Forecasting and Early Warning Project	UNDP/ADB/Japan
11	Disaster Preparedness Programme	UNDP
12	FCD/I Agriculture Review	UK/Japan
13	Operation and Maintenance Study	UK/Japan
14	Flood Response Study	United States
15	Land Acquisition and Resettlement Project	Sweden
16	Environmental Study	United States
17	Fisheries Study and Pilot Project	UK
18	Topographic Mapping	France/Finland/Switzerland
19	Geographical Information System	USA
20	Compartmentalization Pilot Project	Netherlands/Germany
21/22	Bank Protection, River Training, and Active Flood Plain Management Pilot Project	Germany/France
23	Floodproofing Pilot Project	United States
24	River Survey Programme	EC
25	Flood Modeling Management Project	Denmark/Netherlands/France/UK
26	Institutional Development Programme	UNDP/France

number of components were focused on supporting measures, such as mapping, forecasting, surveys, modeling, institutional development, management, environment, etc. (NDC 1993, p. 27). Only two involved new construction. Among these were Dhaka Protection (FAP 8) and Secondary Town and Meghna Left Bank protection (FAP 9a and 9b). Several pilot projects were also included (e.g. FAP 20–23). Of these the Compartmentalization Pilot Project (FAP 20) later proved to be important, as we shall see.

GoB and the World Bank agreed to coordinate the FAP. Its first phase budget was modest, around $150 million, though the estimates for the cost of future construction of embankments and other infrastructure ran from $4 to 10 billion. However, it was emphasized that such estimates were premature because no decisions had been made regarding FAP activities beyond 1995.[18] Implementation of distinct FAP components was supervised by foreign and Bangladeshi consulting agencies. BWDB served as the executing agency in those cases where the study activities included construction (as in FAP 20).[19]

12.4.6 Opposition to FAP

Though the first phase of FAP did not include large embankment construction projects, except for Dhaka city, it was clear to all that as a whole it was based on the Cordon approach, and the regional study projects, pilot projects envisaged under FAP 20, and phase I data gathering projects were all geared toward the goal of double embanking Bangladesh's rivers.[20] In other words, FAP represented re-imposition of the Cordon approach, promoted earlier by the IECO Master Plan.

While it was possible to impose the IECO Master Plan on Bangladesh without any public discussion in 1964, this was not possible with FAP. Times were different, and there was much more awareness and expertise both within and outside Bangladesh to let FAP pass without any challenge. A large number of independent water experts, economists, and other public intellectuals joined the discussion and opposed the Cordon approach embodied in the idea of double embanking rivers. Instead, they argued for the Open approach.

One of the studies that offered a comprehensive critique of FAP was *Floodplains or Flood Plains?* (Hughes, Adnan, and Dalal-Clayton 1994). In this study, the authors provided a detailed analysis of the vital role that regular river overflows played in the ecology, economy, and life of Bangladesh's floodplains and the harm that would result from embankments aimed at preventing these overflows. It noted that floodwater was an integral part of the floodplain system and characterized floodwater as a "prime common good," which creates a wide range of "secondary" (common property) resources.[21] It provided the insightful comparison between cordon projects and "enclosures" of England aimed at depriving people of common property resources.[22] The study concluded that FAP "might create as many problems as it aims to solve,"[23] and noted that the ultimate solution to the floodplain problems created by cordons was "continuation of the natural processes" of regular inundation and sedimentation[24] and "allowing natural forces to operate without hindrance."[25]

Needless to say, the analysis and conclusions presented in Hughes, Adnan, and Dalal-Clayton(1994) agree well with the Open approach. Other studies and publications opposing FAP include Boyce (1990), Custers (1992), Haskonig (2003), Islam (1990), and NDC (1993). And, of course, there was the EWS (Rogers et al. (1989) that opposed the Cordon approach even before FAP was formulated.

The opposition to FAP, however, was not limited to experts and intellectuals. Much wider sections of Bangladesh civil society engaged themselves with the issue and opposed FAP. By the 1980s, various NGOs—including their associations, such as the Association of Development Agencies in Bangladesh (ADAB) and the Coalition of Environmental NGOs (CEN)—had emerged in Bangladesh, and many of them took up opposing FAP as an important cause. They engaged in different forms of citizen mobilization, including lobbying in both national and international arenas. Representatives of some of these NGOs and others travelled to Paris and London to have their views heard by the G-7 summit and the London meeting of the aid consortium, respectively.[26] Partly in response to the activities of Bangladeshi NGOs, European environmentalists also picked up the FAP issue. A conference on FAP was held in May 1993 in Strasbourg under the auspices of the Green Party of the European Parliament. Following the conference, the International Flood Action Plan Campaign Coalition was formed by international NGOs and other concerned organizations. The FAP issue was also raised in the 1992 United Nations Conference on Environment and Development, held in Rio de Janeiro, Brazil.[27] Thus, the FAP issue witnessed considerable internationalization, so that donor countries had to take note of their domestic public opinion in considering their support for this plan.

12.4.7 FAP 20—Compartmentalization or embankments within embankment

As noted above, FAP Phase I projects—apart from the Dhaka city embankment project—were mostly focused on data gathering and studies. However, these included several so-called pilot projects (FAP 17, 20, 21, 22, and 23). Of these,

FAP 20 was of particular significance, because it was aimed at testing out the idea of compartmentalization, which was to serve as the model to be followed throughout the floodplains of Bangladesh. FAP 20 was therefore considered the cornerstone of FAP as a whole.[28]

Compartmentalization, as noted earlier, entailed dividing up an area that was already behind an (main) embankment into different "water management units" through construction of auxiliary embankments. The apparent goal was to provide additional and differential level of protection to different units, depending on their physical and land-use characteristics. It was thought that such compartmentalization will also help to confine flood to specific units, in case of a failure of the main riverbank embankment. The claim was that compartmentalization will provide a more secure environment for intensive agriculture and integrated rural and urban development.[29] Clearly, compartmentalization was another Dutch idea to be imposed on Bangladesh. It was not a coincidence that FAP 20 was financed by the Netherlands (together with Germany).

However, compartmentalization amounted to doubling down along a path that was already misdirected. It meant additional obstructions to water flow on floodplains. Of course, the compartmentalization plans included inlet structures at the upstream side and outlets at the downstream side for drainage, etc. However, given the experience of Bangladesh, these regulators were a sure recipe for additional drainage congestion. Needless to say, compartmentalization meant separating floodplains even more from the nurturing functions of river overflow. Also, construction of auxiliary embankments meant more wastage of scarce cultivable land and more disruption of open fisheries and waterways.

Originally, FAP 20 envisaged implementation of the compartmentalization idea in two areas of Bangladesh, Tangail and Sirajganj, on the left- and right-hand banks of the Jamuna River, respectively. Later on, the Sirajganj site was given up, and the project went ahead only in Tangail. Figure 12.1 shows the map of Tangail district and the compartmentalization project area.[30]

12.4.8 Opposition to FAP 20

While the FAP study and data gathering projects were not visible to the public, pilot projects, such as FAP 20, were visible, and they had noticeable effects on the life of the local population. As a result, these projects became the focal points of opposition to FAP on the ground.

It may be recalled here that there were disagreements among lending agencies about the overall objectives of FAP, with some viewing it as a flood-control project and others as a development project.[31] FAP 20 was also controversial. In particular, there were disagreements regarding the role of local public participation

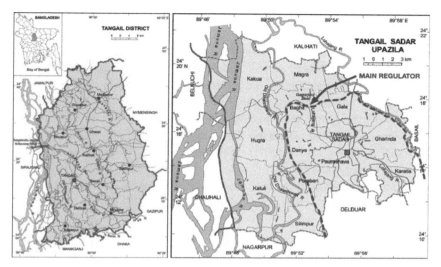

Figure 12.1 Map of FAP 20 (Tangail)—A pilot project of compartmentalization
Source: Crelis F. Rammelt, Zahed Md. Masud, and Arvid Masud (2018).

in the project. Apparently, the Dutch viewed the project as a way of introducing not only their physical model of polders but also their institutional model of polder management. Accordingly, they were of the view that the very design of the compartmentalization project was to be determined on the basis of consultation with local people. Similarly, the physical/structural steps, outlined in endnote 31, were to be complemented by steps of an institutional nature geared at more participatory decision making and management.[32] It may be noted that within the compartments there are important conflicts of interests (between small and large farmers, inhabitants of dry and wet lands, city dwellers, boat people, and fishermen), and it is likely that those with greater influence will prevail. From this perspective, FAP 20 was to determine both whether participation and organization of the people could be realized and how the interests of fisheries, the environment, agriculture, and industry should be integrated into the formulation of technical proposals. By contrast, the Panel of Experts (PoE) of the Flood Plan Coordination Committee (FPCO) thought that experts were enough to design the compartmentalization idea embodied by FAP 20 and viewed people participating only in operation, maintenance, and evaluation.[33]

In any case, a lot of opposition, from many sources, built up against FAP 20. As noted earlier, local people who were to lose their land for auxiliary embankment construction for uncertain benefits were not encouraged about the project.

Compartmentalization was also a direct threat to the fishing communities who depended on open fisheries for their livelihood. The same was true for boatmen and others who depended on waterways. Farmers who were already moving toward winter *boro* as the main crop didn't see much value in the flood protection that compartmentalization was promising to offer and were instead worried about the disruption that the auxiliary embankments would cause. Unlike in previous decades, when local people could not ventilate their grievances, this time around they could rely on the NGOs and civil society organizations to help them in this regard. They voiced their opposition in the local consultations that were held as part of the process of implementation. They also actively resisted the actual physical work of the project. In several cases, opposition turned violent. In face of the virulent opposition, implementation of FAP 20 was in jeopardy.[34]

12.4.9 Disarray of FAP

With its cornerstone, FAP 20, thwarted by the public, the future of FAP as a whole became uncertain. As noted earlier, there were divergent views among its protagonists regarding FAP from the very beginning.[35] Broadly speaking, these views may be classified into the following three strands:

(i) *Flood elimination in earnest*: This is the view advocating "safe passage of transboundary" water to the Bay of Bengal, implying no overflow and no flood, and seriously meaning it.

(ii) *Flood elimination in practice*: This is the view advocating flood control but actually meaning or accepting flood elimination. Advocates either anticipated or at least were not perturbed by transmutation of flood control into flood prevention. As we saw earlier, many donors were using the terms "flood control" and "flood prevention" interchangeably.

(iii) *Flood control in earnest*: This is the view advocating flood control as opposed to flood elimination. Protagonists of this view were arguing for embankments as a way of minimizing the damage that excessive and untimely river overflow might cause to crop, life, and properties. (As noted earlier, embankments aimed at regulation of timing and control of river overflow can be a part of the Open approach too).

These three different viewpoints were not discussed and debated openly and thoroughly among various protagonists of FAP, who therefore all could rally behind the FAP idea of double embankment, because the exact nature of these embankments was still not determined. Accordingly, they could all support the Phase I projects, which were supposed to provide the knowledge base for answering the questions regarding feasibility and detailed specifications of the future embankment projects. They could also support pilot projects of embankment construction, such as under FAP 20, because these had an exploratory nature. Furthermore, the portfolio of projects of FAP Phase I included pilot projects not only for embankment construction but also for flood proofing (FAP 23). Finally, all the donor money was not pooled to support all the projects. Instead, donors had the option of picking particular projects they wanted to support. As a result, they could choose projects that were more to their liking.

However, all this careful orchestration and coordination by the World Bank faced trouble when the general FAP idea of double embankment and the implementation of the particular idea of compartmentalization (under FAP 20) faced stiff opposition from Bangladesh civil society and the affected people. As a result, future phases of FAP never materialized, and no new embankments were constructed along major rivers in the past decades, except for the embankments along segments of the left bank of the lower Meghna as part of the M-D and CIP, described earlier. These projects however, as we saw, were conceived long before FAP and were part of the 1964 Master Plan.[36]

12.5 Institutional changes and prospects of Open approach in Bangladesh

Despite the disarray it ultimately faced, the FAP process provided an impetus toward institutional changes in Bangladesh's water management. Since its formation in 1959, BWDB (previously EPWAPDA) has been in charge of both policy formulation and implementation in the area of water development. However, many observers felt a conflict of interest in this combination of tasks, because it had the potential to bias policies toward structural interventions, which resulted in a larger budget for the organization. Also, with time, there was more recognition of the fact that water development is a multidisciplinary affair, which therefore requires participation of professionals from different disciplines and representation of people from different walks of life. Achieving such a broader framework for policy formulation was difficult within the setup of BWDB, which was dominated by civil engineers.

Partly in response to these issues, some effort at separating policy formulation from implementation started in the 1980s. In 1983, GoB formed the Master Plan Organization (MPO), tasked with formulation of a national water plan and water development framework for 1985–2005 that could replace the Master Plan of 1964, which was clearly out of date half a century after its formulation. In the same year, the National Water Council was also established to "supervise the planning of water resources." There was also a National Flood Council, chaired by the prime minister and its Implementation Commission, along with the River Research Institute, and two trustee-based organizations, namely Centre for Environmental and Geographical Information Services (CEGIS) and the Water Modeling Institute (WMI).

In this background, the FAP process led to the creation of a new institutional set up, with the Flood Plan Coordination Organization (FPCO) as the apex body chaired by the minister of water development and flood control. Under it was the Panel of Experts (PoE), composed of foreign and Bangladeshi experts and financed mostly by UNDP. In addition, there were the Technical Committee (TC) and Review Committee (RC). An important feature of this set up was that it was put directly under the water ministry instead of the BWDB, though the latter continued to be the agency "responsible for implementation of large water works." Thus FAP advanced further the process of separating the responsibility of water policy formulation from that of water project implementation.

FAP was however a project and hence the bodies set up for FAP had a temporary character. After FAP (Phase I) was formulated, FPCO, which became redundant, was therefore merged with MPO in 1992 to form the Water Resources Planning Organization (WARPO), entrusted with the task of preparing "a national water resources master plan, to assist in formulation of water policy, and to manage data regarding water use." Thus, the mandate of WARPO was to continue the planning task that EPWAPDA/ IECO performed earlier and was entrusted to MPO in the early 1980s. In short, formation of WARPO created a certain latitude in water policy discussion that was not present before.

Unfortunately, WARPO has not quite lived up to the expectations. Unlike BWDB, which was dominated by (mostly civil) engineers, WARPO was supposed to be a multidisciplinary organization. In reality, however, it was staffed mostly by engineers coming from BWDB, with a few exceptions. As a result, it failed to conduct a critical evaluation and deep analysis of the country's past water development strategy and projects. Instead, it followed, by and large, the Cordon approach in preparing the National Water Management Plan (WARPO 2001a and 2001b). Meanwhile, there were plans to downsize BWDB substantially, from more than 20,000 to about 8,000 staff. Also, part of the responsibility for project implementation was shifted to the Local Government Engineering Division (LGED), an implementation agency of the Ministry of Local Government and Rural Development (LGRD). Under the new arrangement, LGED was entrusted the task of managing FCDIs with command areas less than 1,000 hectares. Despite being under the local government ministry, LGED is an engineer-dominated agency, and, in the view of observers, has done serious damage to the country's natural drainage system by its indiscriminate and ill-judged construction of rural infrastructure in recent decades.

Thus, the water-related institutional set up of Bangladesh has undergone some changes over time. Some of these changes created more scope for a shift from the Cordon to Open approach. However, this scope still remains largely unutilized due to lack of fulfillment of complementary conditions. Meanwhile, some progress toward the Open approach can be seen along other directions. For example, some of the Cordon projects had to be modified, under pressure from local residents, along the ideas of the Open approach. The following discussion reviews two such examples.

12.6 Opening up of the Pabna Irrigation and Rural Development Project

The Pabna Irrigation and Rural Development Project (PIRDP) is an example of a Cordon project that had to be modified to allow the restoration of annual river inundations. As noted in chapter 11, PIRDP was conceived as a partial cordon project, with many openings in the embankments. Unlike some other cordon projects, where openings are mainly for drainage, the openings of the PIRDP were meant to let river water both flow in and drain out.

Under the business-as-usual (BAU) scenario, these sluice gates would have become dysfunctional, converting the partial cordon into a full cordon. However, regular river inundation was so important for the living and livelihood of the local people that an opposition grew to elimination of flooding in the name of flood control. This was because people in the project area, being part of the Chalaan Beel, were used to deep annual submergence, and hence prevention of river overflow altogether would have been too much of a disruption for both the local economy and ecology. The local people therefore demanded the use of the openings to replicate the annual flooding that used to occur before construction of the embankments. The demand was stronger because the pump-cum-gravity irrigation arrangement of the project was not effective either, as noted in chapter 11.[37]

Responding to this demand, the project managers introduced a "new mode of water management," under which the process of regular inundation by river overflow was restored. According to ADB (1996, p. 9), "while not envisaged under the Project, this mode of water management has benefited a large portion of farmers in the Project area." The new mode of water management allowed flooding inside the embankment to be less deep and of shorter duration as compared to that witnessed by areas outside (see Table 12.2).

The experience of PIRDP in allowing controlled inundation led the ADB to question the water establishment's preference for flood elimination and replacement of natural irrigation and drainage by pumps to transfer water from

Table 12.2 Comparative flood indices inside and outside PIRDP's Talimnagar and Baulikhola sluice gates in 1995 and 1996.

Sluice Gate	Baulikhola		Talimnagar	
Position	Inside	Outside	Inside	Outside
Year: 1995				
Flood Index (foot-days inundated)	454	667	420	632
Maximum Floodplain Depth (ft)	5.9	9.6	5.8	8.6
Days Floodplain Inundated	111	121	109	129
Year: 1996				
Flood Index (foot-days inundated)	435	535	398	532
Maximum Floodplain Depth (ft)	6.5	8.6	6.4	8.3
Days Floodplain Inundated	97	107	90	115

Source: MRAG (1997, Table 2.3).

and to rivers across the embankments.[38] In fact, the ADB report cautioned that the "new mode of management" allowing controlled inundation could be jeopardized if the gravity-cum-pump method of drainage took greater hold in the project area, because the latter would drain way much water from areas where crops could be grown under rainfed conditions.[39]

The PIRDP experience therefore shows how, under pressure from local people, the water authorities were forced to open up a Cordon project to restore annual river inundation. It helps to illustrate the distinction between flood control and flood elimination. It also shows how discontinuous embankments can be a part of the Open approach under which connections between floodplains and river channels are preserved while avoiding the extreme outcomes.

12.7 Opening up of coastal polders—Tidal River Management (TRM)

Some of the coastal polders provide another example of the opening up of cordons, under pressure from local people, to end the problem of waterlogging. This modification was later adopted formally as TRM.

As noted earlier, the water establishment's traditional response to the polder crisis has been mainly to adopt rehabilitation projects aimed at repairing embankments, restoring drainage canals, removing silt to make flap gates operable, etc. However, these measures provided at best some temporary relief and did not improve the situation in the long run, as was seen from the experience of the KJDRP. Meanwhile, people living in some of polders in lower Jessore and upper Khulna, fed up with perennial waterlogging, came up with their own solution to the polder crisis. From their own experience they knew that tidal water coming in during high tides was laden with sediment, which was deposited on the plains. At low tide, relatively clear water flowed out, scouring canals and riverbeds (to regain sediment equilibrium) and thus keeping drainage channels open (Figure 12.2).

Armed with this indigenous knowledge, local people in several polders took matters into their own hands, cut the embankments, let high tide flow in, channeling it to local beels (lakes or depressions), where the sediment was deposited. During low tides, the sediment-free water, on its way out, washed away some of the sediment from the riverbed, unclogging drainage outlets and mitigating the waterlogging problem. In view of the effectiveness of this system, the local people wanted it to be adopted as a general strategy to alleviate the waterlogging problem of the polders.[40]

Accustomed to taking ideas only from foreign experts, the water authorities of Bangladesh were not eager to learn from local people and admit to them the failure of their projects. Also, they did not have much material interest in a solution that did not require a lot of construction and budget. As a result, many years passed before the local people's idea was accepted by the authorities.[41] Eventually, under pressure from the local public, the idea was formulated as the TRM project

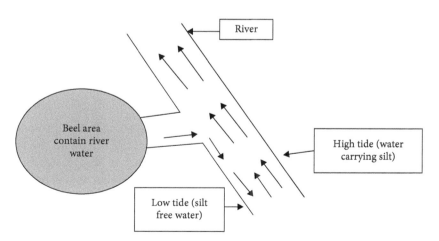

Figure 12.2 The idea of Tidal River Management
Source: Shampa and Ibne Mayaz Md. Pramanik (2012)

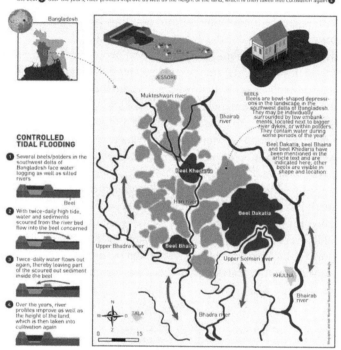

Controlled flooding in beels/polders with Tidal River Management in the Bangladesh delta

Several beels/polders in the southwest delta of Bangladesh face water logging as well as silted rivers ❶ This hampers effective drainage of water as well as agricultural production. Tidal River Management (TRM) involves the temporary (usually several years) removal of an embankment section adjacent to a beel or polder. During the years of tidal in- and outflow, the area is not suitable for agriculture but may be used for aquaculture. With twice-daily high tide, water and sediments scoured from the river bed flow into the beel concerned ❷ Twice-daily water flows out again, thereby leaving part of the scoured out sediment inside the beel ❸ Over the years, river profiles improve as well as the height of the land, which is then taken into cultivation again ❹

Figure 12.3   Application of the Tidal River Management idea
Source: Lock Weijts and Martijn van Stavernan.

and adopted for implementation in several of the polders in the greater Khulna-Jessore area (Figure 12.3).

However, TRM could not provide a general solution to the polder crisis. Under this approach, the saline tidal overflow was to be confined to particular low-lying depressions, i.e., beels. There was no plan to allow a general overflow of the tidal water. Thus, the applicability of the TRM approach was contingent on the availability of such beels, and once a particular beel was filled up by sediment, it was necessary to find another. Thus, availability of beels acts as a constraint on the applicability of the TRM system. Another constraint relates to the availability of land through which the tidal saline water was to reach the beels. To the extent that beels were generally under public ownership, using them for public purposes posed less of a problem. However, in places where new channels were to be constructed for the tidal flow to reach the beels, the ownership of land to be used for such channel construction became a problem.

One can see the evidence of path-dependence in this predicament. Prior to the polder construction, all the enclosed land was exposed to river and tidal flows and there was no hesitance in accepting tidal flows. However, after many decades of a saline-free environment, it has now become difficult for the people to accept general inundation and temporary eight-month long (*ashtomashi*) dams as the solution. (See ADB 2007, pp. 8, 10, and 14 for more discussion on these issues.) Thus, though the TRM has shown a way out of the waterlogging problem faced by the polders, its wider application faces some problems owing to path dependence. A general program of de-polderization in Bangladesh will have to deal with this problem.

12.8 Open approach conforming irrigation projects

Another dimension along which the Open approach is making progress in Bangladesh is through adoption of irrigation projects that do not require construction of cordons. The following reviews two such examples.

Earlier, we noticed that the World Bank disagreed with the IECO Master Plan's idea of large embankment projects aimed at flood protection as an immediate necessity for raising agricultural output. Instead it argued, based on the LWRSS that it conducted during 1970–1972, for greater attention to the area that is normally not flooded and raising productivity of agriculture in this area through small-scale irrigation based on existing surface waterbodies and using LLPs. We also saw that this idea was adopted, if grudgingly, in modifying the irrigation method of some of the cordon projects, such as CIP. In case of the Barisal Irrigation project, the change was more radical.

12.8.1 Barisal Irrigation Project

The Barisal Irrigation Project was originally conceived by the Master Plan as another cordon project, and it had two phases, as can be seen from Figures 12.4 and 12.5. However, it was changed into an irrigation project without embankments and was mostly confined to what was originally Phase I. Aimed at expansion of surface water irrigation using LLPs, the project was consciously viewed as a departure from conventional FCDI project philosophy. It was not a coincidence that the changed project was formulated by the planning commission (and

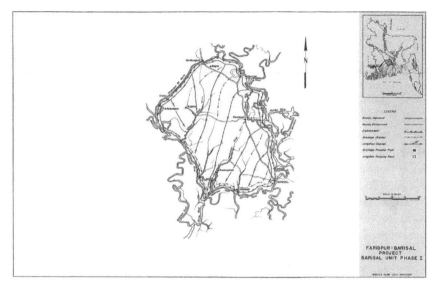

Figure 12.4 Map of Barisal Irrigation Project – Phase I
Source: IECO (1964b), p. 279.

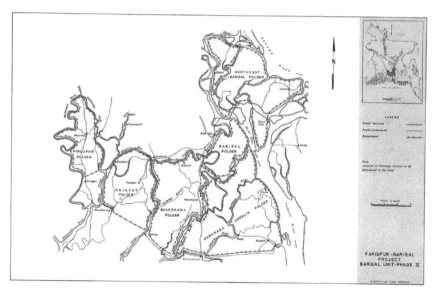

Figure 12.5 Map of Barisal Irrigation Project – Phase II
Source: IECO (1964b), p. 283.

not by BWDB) in 1973, soon after the LWRSS completed. It was appraised in October–November of the same year.[42]

The project has a command area of 83,772 hectares, located in four upazillas that surround the district city of Barisal.[43] It aimed at providing irrigation to 57,062 of 59,490 hectares of cultivable land lying in the command area. It was expected that this would expand cultivation of HYV boro from 5,261 to 45,326 hectares and would allow a shift from local to HYV aman in 35,208 hectares. Together, the project was expected to bring about a huge increase in agricultural output.

The expansion of irrigation was planned to be achieved through the use of primary and secondary pumps. About 100 primary pump stations, each with a capacity of 0.7 cumec, were to lift water from perennial channels (rivers, canals, etc.) to creek storage ponds. From these ponds (or, where possible, from perennial channels), about 2,500 secondary pumps, each with a capacity of 0.06 cumec and serving from 16.2 to 24.3 hectares of land, were to lift water to farmers' fields. About 750 control structures, with gates to allow drainage and passage for boats, were to control water levels in about 100 primary irrigation units. The project also involved formation of pump groups to ensure the cooperation among farmers, needed for effective operation and utilization of the irrigation facility (IBRD-IDA 1975, p. 9).[44]

As already noted, the project was prepared directly by the planning commission, and it required participation of both BWDB and BADC for implementation. BADC was in charge of irrigation using LLPs in the country from the very beginning (i.e., the 1950s), and so it had considerable experience in fielding these pumps and forming the cooperatives necessary for their proper utilization. In fact, BADC was already engaged in LLP-based irrigation activities in the area before the Barisal Irrigation Project was conceived. Therefore, enlisting its participation in this project was quite logical.[45] Thus, the Barisal Irrigation Project provides an example of irrigation expansion without necessitating construction of embankments.[46]

12.8.2 Rubber dam projects—Matamuhuri Irrigation Project

Rubber dam projects are another example of irrigation that does not require cordons or frontal intervention of permanent nature. Bangladesh has implemented several such projects in recent years. One of these is the Matamuhuri Irrigation Project, conceived as a pilot project and based on a rubber dam constructed on Bhola khal (canal), in Chokoria and Pekua upazillas (sub-districts) of Cox's Bazar zilla (district) (Figure 12.6). Being close to the sea, it has the additional objective of preventing saline intrusion. BWDB conducted the feasibility study

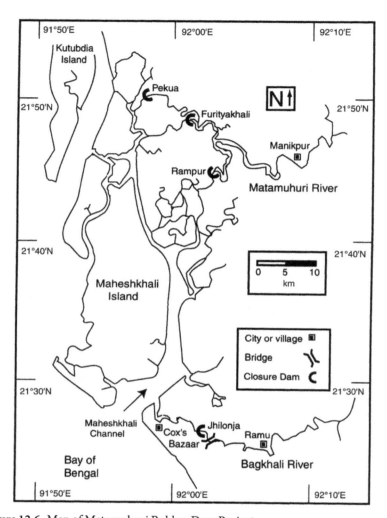

Figure 12.6 Map of Matamuhuri Rubber Dam Project
Source: Brian Smith, Benazir Ahmed, M. Edrise Ali, and Gill Braulik (2001).

in 1995–1996 and submitted a project covering a gross command area of 27,764 hectares. In view of its high estimated cost, the government decided to go for a pilot project of reduced size, involving construction of a single rubber dam with related structures. The modified project had a total command area of 14,000 hectares, of which 5,000 hectares are cultivable, and 3,815 hectares are irrigable. Prior to this project, local people, collecting money from the beneficiaries, used to construct earthen cross dams in the dry season and remove them before the rainy season (known as *ashtomashi* dams). Thus, the rubber dam project was replacing a water management method that was already practiced by the local population. In fact, it serves as an example of return to a pre-industrial method but with incorporation of new industrial-era technology.

In addition to an 84.4 m long rubber dam at Pekua on Bhola Khal, the project involved remodeling and rehabilitation of pre-existing regulators and sluices and construction of some new ones; improvement of canals and irrigation channels; and construction of shrimp inlets, a pump station, and some related structures. The rubber dam itself required construction of a concrete base slab on which the rubber bag was installed. The height of the rubber bag, upon inflation, was 4 m. Since the sill level was -1 m, the top level of the rubber bag could reach a height of 3 m, allowing a retention level of 2.7 m. For efficient irrigation, a water level of 2 m was required. A footbridge was also constructed at the site.[47]

The project was completed in 2005 and, according to BWDB, was successful. Irrigation expanded to about 4,000 hectares, the cropping intensity increased from 178 to 205 percent, and the crop production increased by 12,013 metric tons. BWDB further claims that the sweet water part of the river flow could be kept separate from the saline part, allowing optimal use of both.[48]

These examples show that irrigation in the Bengal Delta does not have to rely invariably on cordons. The Open approach can ensure necessary irrigation through optimum utilization of river inundation, even using embankments with adequate and efficient regulators, as is the case with the modified PIRDP. The Open approach can create more scope for LLP-based irrigation by retaining monsoon water in surface waterbodies in floodplains. It can increase the potential of tube-well irrigation by enhancing recharge of groundwater aquifers. It can also make use of rubber dams both to enhance irrigation and to make optimum use of saline and freshwater in coastal areas.

12.9 *Bangladesh Delta Plan 2100*

Despite the projects described here, the progress of the Open approach in Bangladesh remains partial. This ambiguity found reflection again in the *Bangladesh Delta Plan 2100* (GoB and GoN 2017), prepared recently under the auspices of the GoB Planning Commission. This large document, henceforth referred to as BDP 2100, was prepared over several years (during 2014–2019), with financial and technical assistance from the Netherlands.[49] Six out of the eight consulting firms that formed the consortium to prepare this plan were Dutch.

Hence, it is not unreasonable to assume that the Dutch thinking held sway in for-mulating the plan.

The BDP 2100 had the scope and potential to break out of the Cordon ap-proach and embrace wholeheartedly the Open approach. This was expected particularly in view of the fact that the Dutch themselves have undergone some rethinking since the 1990s about the right approach to rivers. As we saw in chapter 8, they initiated the "Room for Rivers" project based on the realization that floodplains need to be made available to rivers. It is somewhat disappointing that this rethinking— adopted by the Dutch in their own country—did not find adequate reflection in the BDP 2100, for which they were the lead consultants.

It is true that the BDP 2100 recognizes two different, possible strategies re-garding rivers in Bangladesh. One was, what it calls, the strategy of Optimal Water Control (OWC), which, according to it, has the following features:

i) focuses on *interventions* as soon as knowledge, institutional capacity, and finance permits;

ii) provides a basic *safety against floods* for all, allowing regular floods to enter by controlled gates;

iii) aims for being independent of India by means of its *own barrages* and flow diversion projects.

From its more detailed presentation, it further transpires that OWC clings to the proposition of double embankment of the major rivers of Bangladesh and bases itself on rosy pictures of a "fully regulated polder system." It also suggests prob-lematic pictures of "periodically bringing sediment to closed polders from out-side."[50] It also focuses on barrage construction, apparently going further in this regard than even the IECO Master Plan and FAP. Thus, OWC promotes not only the Cordon approach but also the frontal version of the Commercial approach.

The second strategy of BDP 2100 is what it calls Adaptation by Design (ABD), which has the following characteristics:

i) tries to *avoid interventions* where possible by looking for alternatives, but does not exclude interventions;

ii) focuses on *local solutions* and strengthening of local communities and governance;

iii) puts great efforts in *cooperation and transboundary negotiations*;

iv) uses natural processes where possible (*Building with Nature*); and

v) is based on *differential flood zoning*, rather than full embankment of the major rivers.

It is clear that the ABD is closer to the Open approach to rivers. This can be seen in its suggestion to "avoid interventions," to avoid "full embankment," to "build with nature," and to heed views of local communities.[51]

The recognition of the two alternative strategies is a welcome feature of BDP 2100, and it was expected that it would offer a thorough discussion of the pros and cons and arrive at a clear position regarding them. Unfortunately, it offers instead a confusing discussion, leaves the issues basically unsettled, and adopts an eclectic position that ultimately gravitates to the Cordon and Commercial approaches (see Islam 2019 for details). This becomes clear from its choice of projects. For example, the portfolio of 15 projects that it suggests in Volume I for immediate execution put the Ganges Barrages at number one and the Brahmaputra Barrage at number three. The more elaborate list of about 90 projects suggested in Volume II is again dominated by proposals for barrages and embankments.

It is therefore disappointing that after so many decades of adverse consequences of the Cordon approach, after a considerable shift toward the Open approach globally, and after the shift toward the Open approach in their country by the Dutch themselves, the BDP 2100 fails to make a clear departure from the Cordon approach. This is disconcerting because, according to its promoters, BDP 2100 is to set the direction of land and water development in Bangladesh for the next 100 years, so the strategy adopted by BDP 2100 is likely to have a long-term impact.

However, as was the case with FAP, the civil society of Bangladesh has engaged itself in the discussion of BDP 2100 and has already pointed out its limitations and offered concrete suggestions about how this plan should be reworked to make it more appropriate for the deltaic environment of Bangladesh.[52] It remains to be seen whether the government heeds the suggestions from the civil society, which is now playing an important role. The next section offers a review of the new activism of Bangladesh civil society toward the Open approach.

12.10 Renewed movement for Open approach

Efforts by the civil society and local people, as noted earlier, played an important role in the progress toward the Open approach achieved so far in Bangladesh. Its evidence could be seen in the resistance against FAP, in the campaign of the people of the Chalaan Beel for opening the embankments of PIRDP, and in the initiative of the people of southwest Bangladesh to open up the polders to let river water in.

Toward the end of the 1990s, civil society's support for the Open approach attained a more sustained and comprehensive character; it became part of the broader movement for protection of the environment, which had deteriorated in the 1980s and early 1990s due to unfettered industrialization efforts. Rivers, the backbone of the Bangladesh environment, were a prime victim of this process.

Not surprisingly, protection of rivers became the foremost task of the Bangladesh environment movement.

The current civic movement began with the formation of pro-environment groups, such as the *Poribesh Rokkhya Shopoth* (POROSH) in Dhaka city. This civic effort received a boost with the formation of the Bangladesh Environment Network (BEN), with the goal of mobilizing the expertise and resources of non-resident Bangladeshis (NRBs) to work together with those inside Bangladesh to protect the country's environment. To bring pro-environment forces of the country together—so they could attain a critical mass necessary to influence policies—BEN in 1999, proposed to hold a broad-based conference inviting all pro-environment organizations, experts, and activists of the country. Later known as the first International Conference on Bangladesh Environment (ICBEN), this conference was held in January 2000. The success of ICBEN led to the formation of *Bangladesh Poribesh Andolon* (BAPA)[53] in July 2000 as a united platform of pro-environment organizations and individuals. Since then, BAPA and BEN have been working together for protection of Bangladesh's environment.

Given the importance of rivers in Bangladesh's environment, BAPA and BEN focused special attention on them and initiated the movement to protect rivers, beginning with the *Buriganga Bachao Andolon* (Movement for Protection of the Buriganga River). Similar movements sprang up in other districts, focused on local rivers. To bring these river movements together and set a common and effective direction for them, BAPA and BEN convened the National Conference on Rivers in 2006, which adopted the *Resolution on Rivers*, based on the Open approach. It also formed the *Jatiyo Nodi Rokkha Andolon* (National Movement to Protect Rivers), to bring local river protection movements under one umbrella.

The river movement, under the auspices of BAPA and BEN, has had considerable success. It made the citizenry aware of the plight of the rivers. The media took note and started publicizing the pollution and encroachment of rivers, made easier by the Cordon approach, which views rivers as an "outsider" and not an integral part of floodplains. Facing public pressure, the government formed the River Task Force, focused initially on the four rivers that surround the capital city of Dhaka.[54] The judiciary also joined the effort, and the High Court, responding to public interest litigation suits filed by pro-environment organizations, directed the government to demarcate river boundaries and to set up pillars along these boundaries in order to stop any further encroachment.[55]

In time, BAPA, BEN, and other environmental organizations persuaded the River Task Force to extend its attention to rivers outside Dhaka. One of these is the Baral River of North Bengal (northwest Bangladesh), mentioned earlier in chapter 5. Taking off from the Padma (Ganges) River near Charghat, Rajshahi, this river flows east about 200 km to meet the Jamuna (Brahmaputra) River at Baghabari near Sirajganj. Many rivers of North Bengal, such as the Atrai, Korotoya, Ichamati, Gumani, Hoorsagar, etc., flow from north to south to meet the Baral River and sustain the Chalaan Beel, also mentioned earlier. Unfortunately, construction of ill-advised sluice gates cut the Baral River off from the Padma River, an outcome that was aggravated by the reduction of flow in the Padma River due to India's Farakka Barrage. Meanwhile, BHRE, embankments of PIRDP, and several other sluice gates constructed on the Baral River reduced the flow of the Jamuna and other north-south running rivers mentioned above. As a result, the Baral River, that used to be the region's major waterway along which large ships once plied, has been strangled.[56]

To resuscitate the Baral River, BAPA and BEN helped local people to launch the *Baral Bachao Andolon* (Save the Baral River Movement) and wage a persistent campaign, including numerous rallies, meetings, representations, processions, and petitions, culminating in a 200 km long human chain along the bank of the entire river. Thanks to this vigorous campaign, the River Task Force ultimately issued a directive to free the Baral River. As a result, some earthen dams across the river were removed and some parts of the river were re-excavated. However, the concrete sluice gates are yet to be removed. Despite its incomplete nature, the victory in reviving the Baral River is evidence of the need for a popular movement to save the rivers and adopt the Open approach.

Responding to the river movement, the government of Bangladesh has formed in 2017 the National River Conservation Commission (NRCC), with a nation-wide remit. While the River Task Force is an administrative body, the NRCC is a constitutional body with greater stature, though lacking executive power. Despite the limitation, the NRCC in recent years has displayed some vigor. Meanwhile, the High Court of Bangladesh, responding to another public interest suit brought by pro-river organizations, issued a landmark verdict in July 2019 declaring rivers "living entities" having the status of a "legal/juristic person," with inherent rights to exist and prosper. It also appointed the NRCC as the custodian of rivers and directed it to carry out its role in cooperation with other government agencies.[57] Earlier, Prime Minister of Bangladesh, Sheikh Hasina called for living with floods and urged the country's engineers to refrain from constructing unnecessary structures in rivers.[58] Needless to say, these exhortations too, despite not being too effective yet, reflect the Open approach to rivers.

In short, there now exists a robust environment movement in Bangladesh that has taken up protection of rivers as its major goal. Its efforts have already led to some progress toward the Open and Ecological approaches to rivers, and the stage has been set for further progress in this direction.

12.11 Conclusions

The Open Approach is more suitable for the Bengal Delta than for any other part of the world. The people of this delta have been following this approach from time immemorial. The Cordon approach was imposed on it during the 1950s and 1960s, largely through external influence. The experience of more than half a century however has convinced many of the people of the Bengal Delta of the need to discard the Cordon Approach and adopt the Open Approach. This conviction is spreading in both parts of the Bengal Delta. In the Indian part, river scholars have analyzed the experience of the Cordon approach applied to the Ajay and other rivers and have come to the conclusion that "periodic spilling is not only normal for a stream but it is also a very welcome process." Needless to say, this is a basic message of the Open approach. Support for the Open approach also found expression in the resistance to the Cordon approach-based FAP that some donors and collaborating Bangladeshi agencies tried to impose on Bangladesh in the wake of the historic flood of 1988. It found expression in the insistence of the people of Chalaan Beel that the PIRDP be modified. It was manifested by the initiative of the local people of the Khulna-Jessore area to adopt the TRM method. The water authorities themselves adopted several irrigation projects that do not require construction of cordons.

A robust civic movement, advocating the Open approach, has now developed in Bangladesh. It is leading a nationwide movement to protect rivers from cordons, frontal interventions, pollution, and encroachment. The media has come out in support of this movement and is offering constant exposure on the plight of the rivers of the country. Responding to the movement, the government has formed the River Task Force and the NRCC. The judiciary has also taken up the cause and directed the government to demarcate the boundaries of rivers to protect them from encroachment. More recently, it has declared rivers "living entities" having the status of "legal/juristic" person. The prime minister of Bangladesh, Sheikh Hasina, has been continually urging her agencies not to construct unnecessary river-intervening structures and the people to live with floods. As a result of the above developments, some on-the-ground progress has also been achieved. The fabled Baral River of North Bengal has been partially opened up, as has been the case with the Turag River near Dhaka city. Upon the insistence of the prime minister, a new canal is being constructed to connect parts of East Dhaka with the Balu River. The government is also displaying more activism in freeing riverbanks from encroaching structures.

However, the reluctance among government circles to embrace the Open approach wholeheartedly continues. This reluctance can be seen in the BDP 2100, prepared recently by the Planning Commission of Bangladesh, with financial and technical assistance from the Netherlands. Despite showing awareness about the possibility of contrasting strategies toward rivers, BDP 2100 adopts a confusing eclectic position, which in terms of project choice falls back largely on the Cordon approach. It thus ends up proposing more embankments and barrages, instead of a comprehensive plan of de-polderization.

The movement for the Open approach in Bangladesh therefore has a long way to go. Especially important is clarifying to the public how the Open approach can help in solving various problems, such as waterlogging, bank erosion, catastrophic flood, encroachment, pollution, etc. A second important task is to explain how to transition from the Cordon to the Open approach can take place, given the problems of path dependence. It is clear that de-polderization of Bangladesh has to be a gradual process, with attention to the specifics of particular polders. However, climate change has made this transition both imperative and urgent. As the saying goes, nothing concentrates the mind more than a death sentence, and climate change is indeed close to a death sentence for a low-lying area such as the Bengal Delta. Once the gravity of the situation is made clear, it should be easier to mobilize the people to carry out this transition, despite the short-term dislocations it may entail. However, with well crafted policies, this transition can be a win-win process.

Notes

1. Development and application on a large scale of technologies for removing carbon dioxide from the atmosphere can make a change in this regard. However, the prospects of that happening soon are still not that bright. For discussions on the climate change challenges and tasks for Bangladesh and the Bengal Delta in general, see Bhattacharya, Ahmed, Islam, and Matin eds. (2009). See GoB (2008) for Bangladesh government's plan of action regarding climate change. See also Nianthi and Husain (2004) for impact of climate change on rivers in the particular context of IRLP.
2. See Chen and Mueller (2018) and Haider (2019).
3. Cyclones are often accompanied by tidal bores that sweep the coastal areas, devastating life and property. The cyclone that hit in November 1970 is estimated to have killed almost half a million people. As an indication of what is to come, cyclones have become more frequent in recent years.
4. See Kabir, Rahman, Smith, Lusha, and Milto. (2016), https://www.ncbi.nlm.nih.gov/pmc/articles/PMC4821870/.
5. The restoration of the original, natural volume of sediment will therefore require cooperation between India and Bangladesh (and also Nepal, Bhutan, and China), on the basis of the Ecological approach. In particular, it will require India to remove frontal intervention structures to resume the natural flows of the rivers. Also of note in this regard is the fact that while the Commercial approach is likely to reduce sediment volume, by trapping more of it behind dams and barrages, more deforestation and topsoil exposure occurring in the catchment areas are likely to increase sediment volume. The net result of these two counteracting trends will depend on their relative strengths. For the countervailing effect of sedimentation on submergence resulting from sea level rise, see Bhattacharya (2009).
6. Compartments were "to provide water control (controlled flooding and controlled drainage) to protected areas and also to contain areal extent of flooding in the event

of any embankment breach. Compartments will also provide focal points for flood preparedness and warning activities. Over the plan period, compartments will build up into contiguous flood control systems along the main rivers with wider benefits than a stand-alone embankment scheme and at a pace consistent with institutional capacities. Other components of the Plan include active floodplain management and the development of minor rivers for controlled flooding and drainage to complement the main stem development" (World Bank 1990, p. 26). For the background of 1987 and 1988 floods, see Brammer (1990). See also Dewan, Mishigaki, and Komatsu (2003).

7. These six objectives were: (i) safeguard life and livelihoods; (ii) improve agro-ecological conditions to increase crop production; (iii) enhance development of public facilities, commerce, and industry; (iv) minimize potential flood damage; (v) create flood-free land to accommodate the increasing population; (vi) meet the needs of fisheries, navigation, communications, and public health (World Bank 1990, p. 26).

8. According to World Bank (1989), "The concept of controlled flooding was not new when officially introduced in the UNDP Flood Policy Study (GoB and UNDP 1989). Experience with it was not yet available, however; hence the UNDP study also suggested a pilot project. This was adopted in the Flood Action Plan in December 1989."

9. "The solution presented in this report for protection against floods was essentially based on the extensive confinement of the main rivers and distributaries by longitudinal embankments. This represents a drastic departure from the previous policy which relied mainly on individual polders" (World Bank 1990, p. 27).

10. See World Bank (1990, p. 27) for details of the French proposal.

11. See Japanese Flood Control Experts (1989). The major objectives of the study were: (a) minimizing the potential flood damages; (b) creating flood-free lands to accommodate the increasing population; (c) enhancing agricultural land use to facilitate the adoption of high-yield crop varieties, and (d) enhancing conditions for effective development of commercial and industrial enterprises (World Bank 1990, p. 29).

12. The essential target of the long-term flood control plan is to mitigate flooding in the central part of the country by means of appropriate measures against flooding by the three major rivers (World Bank 1990, p. 29).

13. See World Bank (1990, p. 30) for details.

14. See World Bank (1990, p. 30); in addition, the study argued for flood forecasting and warning, and flood fighting.

15. See World Bank (1990, p. 30).

16. During personal communications, Prof. Peter Rogers informed me that the hostility of the Bangladesh water establishment and their foreign backers to EWS was so great that, during the FAP controversy, he was once not allowed to fly from Nepal to make a scheduled visit to Bangladesh.

17. The G-7 meeting also took note of the different studies mentioned earlier.

18. See NDC (1993, p. 30).

19. See NDC (1993, p.30).

20. NDC (1993), however, thinks that the UNDP/GoB proposal was a compromise between the two, urging for combining the benefits of flooding with the advantages

of protection against flooding. It further thinks that in the end the GoB and UNDP proposal of controlled flooding and drainage prevailed within the establishment and became the basis of FAP. The GoB and UNDP proposals also recognized the importance of giving attention to people's participation; social, environmental, and economic aspects; and operation and maintenance (NDC 1993, p. 26). See Wescoat, Chowdhury, Parker, Khondker, James, and Pitman (1992) for five views of FAP.

21. See Hughes et al. (1994, pp. 9–10).

22. See Hughes et al. (1994, p. 2).

23. See Hughes et al. (1994, p. 66).

24. See Hughes et al. (1994, p. 22).

25. See Hughes et al (1994, p. 69).

26. As noted earlier, the G-7 meeting was held in Paris in July 1989, and the London meeting of the aid consortium was held in December 1989 to consider financing of FAP.

27. See ADAB (1992). For a broader discussion of floods in Bangladesh, see Hofer and Messerli (2006).

28. According to NDC (1993), "FAP-20 is meant to provide data and guidelines for the FAP as a whole. Hence FPCO and donors regard the project as a *cornerstone* of the FAP." More concretely, "FAP-20 was supposed to provide: (i) feasible systems of water management for the development of protected areas; and (ii) guidelines for the development of protected areas in FAP generally" (author's emphasis).

29. See WARPO (2018, FAP-20, p. 6)."

30. Compartmentalization in Tangail area offered the the following alternatives:

   I(a): Full compartmentalization: All peripheral structures gated with the exception of the ungated or partly gated, navigable inlet structures Lohajang and Sadullapur; limited peripheral road (20 km); southern embankment as feeder road with bridges/ungated structures; flow regulators between Lohajang and sub-compartments 9, 10, and 11 only.

   I(b): The same as I(a), but also gated, unnavigable inlet structures Lohajang and Sadullapur; including mooring facilities.

   II(a): The same as I(a), but also including flow regulators between Lohajang and remaining sub-compartments.

   II(b): The same as Ia, but also including flow regulators between Lohajang and remaining sub-compartments.

   III(a): The same as II(b), but with closed/regulated structures along southern embankment, with exception of the open Lohajang outlet.

   III(b) The same as IIIa, but with complete peripheral road (60 km) (WARPO, 2018, FAP-20).

31. "The most important lesson to be learned from the progression of FAP-20 is that FAP objectives need clarification. Essentially, the FAP is viewed in two opposed ways: the FAP as a flood-control project and the FAP as a development project. As FAP-20 (and the FAP generally) moves forward, it will be difficult to combine the different points of view into a single work plan" (WARPO 2018, p. 43).

32. FAP 20 concludes that most drainage problems in flood-control projects are due to faulty planning, design, and construction. As a rule, drainage problems are badly

underestimated. Compartmentalization will probably improve drainage during the monsoon, but the impact will remain limited because of the adverse hydrological conditions downstream (NDC 1993).

33. See NDC (1993, p. 4) for details.

34. The French plan was thwarted, but the tendency toward building cordons and structures persisted, partly because of the big budget that such projects entail.

35. This was with even without considering the views of EWS, which was not considered as a FAP protagonist.

36. Even the regional studies were not discussed properly. For details regarding the disarray of FAP see NDC (1993, pp. 47–57). For discussion of the FAP experience and its aftermath, see also Brammer (2010). See Chowdhury and Rasul (2012) for a discussion of equity and social justice in water governance in Bangladesh.

37. See ADB (1996, pp. 9–13).

38. See ADB (1996, p. 2).

39. As ADB (1996, p. 13) put it, "It seems however that expansion of large-scale surface irrigation would affect this delicate balance as increased drainage in low lands would drain valuable water from watersheds cropped under rainfed conditions."

40. Worried by the situation, some experts indeed called for at least partial opening up the polders. Brammer (2014), recognizing the necessity for restoring the sedimentation process to counteract sea level rise, suggests modification of polders in a way "that allows tidal water to enter them *at appropriate times of the year* in order to deposit sediment at sufficient rates to raise the land levels in parallel with a rising sea-level and local land subsidence rates (a process known in England as 'warping')" (emphasis added). However, he does not spell out what these "appropriate times" are, nor how opening up during these times only will be sufficient for enough sediment to flow in to counteract sea level rise and subsidence.

41. See ADB (2007, p. vi). See also Karim and Mondal (2017).

42. The discussion of the Barisal Irrigation Project presented in this section is based largely on IBRD-IDA (1975).

43. These upazillas are Kotwali, Babuganj, Jhalkathi, and Nalchiti.

44. The World Bank estimated the EIRR to be about 28 percent. See IBRD-IDA (1975, pp. 9–25) for more details.

45. See IBRD-IDA (1975, pp. 7–9) for further details.

46. See IBRD-IDA (1975, p. 25).

47. The project involved: a) construction of a 84.4 meter long rubber dam at Pekua on Bhola Khal; (b) remodeling of three pre-existing regulators into irrigation intakes/inlets; (c) improvement of 25 km of canals and irrigation channels; (d) rehabilitation of 10 pre-existing sluices; (e) replacement of one pre-existing sluice; (f) construction of two new drainage sluices; (g) re-excavation of 5 km of canal; (h) re-sectioning of 10 km of existing embankment; (i) construction of 2.8 km of retired embankment; (j) construction of 5 shrimp inlets. A footbridge and a pump station were constructed at the rubber dam site.

48. BWDB mentions the following as one of the benefits of this project: "After completion of the project, sweet water of upstream and saline water of downstream of rubber

dam would not mix, hence optimum benefit taken from sweet water and saline water individually." Also, "Protect wetland saline water intrusion and promote accretion of land" is presented as an important objective of this project.

49. BDP 2100 draws upon 23 background papers. See Choudhury, van Scheltinga, van der Bergh, Chowdhury, de Heer, Hossain, and Karim (2012) for information on preparation of BDP 2100. See also GoB and GoN (2014).

50. See BDP 2100 (pp. 244–245) for details of the OWC.

51. See BDP 2100 (pp. 245–247) for details regarding the ABD strategy.

52. See, for example, Islam (2019) for a review of BDP 2100. See also Khalequzzaman (2016, 2019). See Hossain and Sakai (2008) for discussion on severity of flood and its remedial approach. See GoB (1995) for Bangladesh government's plan of action for environment management.

53. In English, Bangladesh Environment Movement.

54. These are the Buriganga River, Turag River, Balu River, and Shitalakkhya River.

55. The Task Force also took the initiative to introduce a circular river route around Dhaka city after removing encroaching structures and reducing pollution.

56. Taking advantage of the situation, unscrupulous local individuals and agencies built several earthen cross-sectional dams, converting the river into a series of ponds. In some parts, the encroachment has proceeded to such an extent that the river no longer exists.

57. See news report in https://www.ndtv.com/world-news/bangladesh-high-court-declares-countrys-rivers-legal-persons-2063061. See also https://www.daily-sun.com/post/403747/2019/07/01/High-Court-declares-all-rivers-as-living-entities.

58. See, for example, *The Daily Star*, August 28, 2017 and *The Dhaka Tribune*, January 25, 2018.

13

Rivers, policies, and interests

13.1 Introduction

Different approaches and policies adopted regarding rivers can have very different outcomes for the rivers themselves and the economy and environment of a country. The examination of the experience of river-related policies across the world presented in the previous chapters demonstrates that the Ecological and Open approaches to rivers are more conducive to sustainable development than the Commercial and Cordon approaches. Nevertheless, it remains the case that the latter approaches continue to dominate policymaking in most countries, particularly in developing countries. How can this paradox be explained, and how can the policy bias toward Commercial and Cordon approaches be ended?

It is important to note in this regard that policy choices are not random occurrences. Instead, they depend on the relative strength of socio-political forces that rally behind different policy options. Generally, there are two types of factors that determine why particular social groups or individuals advocate particular policies. The first of these concerns knowledge (i.e., the epistemological or gnosological aspect of the issue). Thus, people may advocate a particular policy because, as per their knowledge, that policy is the most rational. The second type of factors concerns material interests. Thus, people may advocate a particular policy because that policy serves their material interests. Sometimes and for some people, these two factors may work in the same direction. However, it is also possible for some people to advocate a policy because of material interests even though they know that it is not rational.

To understand the role of material interests in policy choices, it is useful to distinguish between end-beneficiaries and immediate-beneficiaries. The former includes those who benefit from the end-results of a project. It is generally the case that end-beneficiaries are numerous but the size of their per-capita benefit is small. By contrast, the immediate-beneficiaries are generally small in number but the per-capita size of their benefit is large. The theory of public action, developed by Olson (1971) and others, suggests that the beneficiaries who are small in number and whose per-capita benefit is large find it easier to organize and mount joint action to advocate and press for policies that serve their interests.

Rivers and Sustainable Development. S. Nazrul Islam, Oxford University Press (2020). © Oxford University Press.
DOI: 10.1093/oso/9780190079024.001.0001

By contrast, beneficiaries who are numerous and whose per-capita benefit is small find it difficult to organize and mount joint action for policies that are favorable to their interests.

From this perspective, the Ecological and Open approaches are at a disadvantage as compared to the Commercial and Cordon approaches. The beneficiaries of Ecological and Open approaches are generally numerous with small per-capita benefit. In fact, most of the beneficiaries of these approaches are members of future generations, who cannot play any role in the present! This asymmetry with regard to the configuration of material interests is an important reason why it proves difficult to dislodge the Commercial and Cordon approaches from their dominating position and replace them with the Ecological and Open approaches.

In determining the river policies of developing countries, another factor that plays an important role is external influence, i.e., influence coming from developed countries. For a long time, this influence was mostly biased toward the Commercial and Cordon approaches, and it worked through both the knowledge and material interests channels. It is only recently that external influence has favored the Ecological and Open approaches in some instances. However, on balance, the external influence is still heavily biased toward Commercial and Cordon approaches. Countering this external influence is therefore an additional task for many developing countries.

To counter the knowledge and material interests obstacles (of both domestic and external origin), it is important to move river-related policy discussion from the small and closed circles of bureaucrats, technocrats, politicians, and representatives of external lending agencies to the broader public arena, where all stakeholders can engage in the discussion and decisions may be reached through reasoning and in a transparent manner. This shift of venue and modus operandi of the decision-making process has broader effects. As Sen (1999) emphasizes, public discussion not only has an instrumental value (helping to reach better policies) but also has an intrinsic value as a practice in democracy.

However, for public discussion to be effective, it needs to be grounded on information and logic, and it is a duty of scholars to help the public by providing these inputs. It is hoped that the information, analysis, and conceptualization offered in this book will help to encourage a more informed and reasoned public discussion regarding river-related policies and approaches all across the world.

The discussion of this chapter is organized as follows. Section 13.2 analyzes the determinants of river-related policies, including both the knowledge and material interests dimensions. Section 13.3 discusses the problem of external influence on the river-related policies of developing countries. Section 13.4 notes the role of public discussion in helping countries adopt correct approaches and policies toward rivers. Section 13.5 concludes.

13.2 Determinants of river-related policies

Policies in general, river-related policies among them, are not random occurrences. Instead they are the outcomes of struggles among different socio-political forces that rally behind different policy options. It is the relative strength of these forces that determines which of the contending policies gets adopted.

The factors that generally determine why particular social groups or individuals advocate particular policies may be classified into two categories. The first is related to knowledge (i.e., the epistemological or gnosological factors). Thus, people may advocate a particular policy because, to the best of their knowledge, that is the right policy. The second category of factors influencing people's choice of policies is related to material interests. Thus, people may support a particular policy because it serves their material interests. Sometimes, these two factors may work in the same direction. Thus, people may support a policy because it is in their material interests and also because they think it to be the right policy. In other cases, they may run in opposite directions and people decide to go by the one which appears more important for them. Thus, people may support a policy that is in their material interests, even if they know that it is not the right policy from the viewpoint of greater interests of the society.

This policy choice framework also applies to policies regarding rivers. Some people may support the Commercial and Cordon approaches because, as per their knowledge, these are the right approaches. Indeed, at a superficial level, it may indeed appear that damming and embanking rivers are the right ways to deal with them. By comparison, it requires more knowledge to realize that the Ecological and Open approaches are better. It may be possible that some people support the Commercial and Cordon approaches because they think that these are right approaches and also because they hope to benefit materially from them. However, in some cases, a person or group of people may support these approaches because of their material interests even though they know that they are not the right ones from a broader and longer perspective.

To see how material interests play their role in policy choice, it is useful to make a distinction between end-beneficiaries and immediate beneficiaries. The former comprises those who hope to benefit from a project after it has been completed. For example, in case of a dam project, all who are likely to benefit from the electricity generated by the dam or from the irrigation water to be provided by its reservoir are end-beneficiaries, because their benefit starts only after the construction of the dam has ended and it has begun to operate successfully. Usually, end-beneficiaries are numerous and the per person size of the benefit is relatively small.

By contrast, immediate-beneficiaries are those who are likely to benefit from the adoption and the process of implementation of the project itself, and do not have to wait for its completion and/or success. They will benefit even if the project fails to yield the expected results. In case of the above example of a dam project, immediate-beneficiaries comprise consultants, contractors, suppliers, bureaucrats, politicians, and other such persons.[1] The immediate-beneficiaries generally prove to be smaller in number, but the per person size of their benefit is generally large.

It is true that some immediate-beneficiaries may also be end-beneficiaries, though the converse is generally not the case. For example, the reputation of the designer and construction firms of a dam may get a boost if the dam proves to be successful in the end. Similarly, the re-election prospects of a politician may improve if the dam that she worked to get financed yields good benefits to the people of her constituency.

However, it is the immediate interests that generally dominate the behavior of the immediate-beneficiaries. This is particularly the case for projects that have long gestation periods. Bureaucrats are generally not in the same position and same location after a few years. Similarly, politicians' time horizon may be shorter than the completion time of the project. Thus, in championing a project, neither bureaucrats nor politicians may be particularly concerned about the end result. Similarly, the designers and builders can always blame other people and other factors for the failure or poor performance of a project, meanwhile collecting the money they were contracted for.

In his seminal book, *The Logic of Collective Action*, Mancur Olson (1971) shows that beneficiaries who are small in number but for whom the per-capita size of the benefit is large find it easier to organize and mount joint action to get policy options favored by them adopted. This happens for several reasons. First, it is always easier to organize when the relevant group is small in size than when it is large in size. Second, the small size of the group also makes it easier to prevent piggy-backing and shirking by individual members of the group. Third, since the per-capita size of the benefit is large, the potential beneficiaries prove to be eager to band together and be active. By comparison, beneficiaries who are large in number and whose per-capita benefit is small in size find it difficult to get organized and mount joint action to get policy options favored by them adopted. The large number of members of the group makes organizing costly and monitoring shirking behavior difficult. Also, there is less incentive on the part of these beneficiaries to get organized and devote time, energy and resources for joint action, because the size of the per-capita benefit is small.

It is clear that the Commercial and Cordon approaches generally beget large projects involving immediate beneficiaries with large per-capita benefits. As a result, powerful lobbies of immediate-beneficiaries develop in favor of these approaches. By contrast, the Ecological and Open approaches suffer a double jeopardy. First, these approaches have less propensity to generate large river intervention projects. They do not rule out construction of structures entirely, as noticed earlier. However, the wide ranging and persistent activities associated with the Ecological and Open approaches and the need for professionals, organizations, and companies in designing and implementing these activities – in close cooperation with and active role of the local population – is often not readily evident. As a result, these approaches cannot generate powerful lobbies of immediate-beneficiaries. Second, the end-beneficiaries of the Ecological and Open approaches are generally more diffused than those of the Commercial and Cordon approaches. In fact, a large part of the end-beneficiaries of the Ecological and Open approaches are the members of future generations, who are currently not present to voice their opinion and uphold their interests. Meanwhile, the current generation often proves myopic and cannot protect the interests of future generations, to the extent necessary.[2]

There is therefore an important asymmetry between the Commercial and Cordon approaches, on the one hand, and the Ecological and Open approaches on the other, with regard to material interests. The force of material interests works more strongly in favor of the Commercial and Cordon approaches. It is this asymmetry regarding material interests that is an important reason behind the continued dominance of the Commercial and Cordon approaches in policy-making circles in most countries.

The Ecological and Open approaches face knowledge-related difficulties too, as noted earlier. For example, it is easier to claim that dams will generate electricity and provide water for irrigation. Similarly, it is easy to claim that cordons will create flood-free conditions, allowing more settlement and increased agricultural production. By contrast, it is difficult to show immediately that the morphological changes brought about by dams will be harmful for the river or that artificial separation of floodplains from river channels will harm both. Part of the difficulty in this regard lies in the fact that while the benefits of the Commercial approach materialize more or less immediately and are palpable, the benefits of the Ecological approach materialize in the long run and are more diffuse in nature. Seeing the benefits of the Ecological approach therefore requires more mental effort and knowledge.

Thus, both the knowledge barrier and the barrier of material interests work against the Ecological and Open approaches. The advocates of these approaches therefore have a harder task for them.

13.3 External influence on river policies of developing countries

Another factor that plays an important role in determining the river policies of developing countries is external influence, namely influence from developed countries. This influence works through both the knowledge and material interests channels.

The knowledge channel of this influence is clear from earlier chapters. As noted there, modern science and engineering originated in early industrializing countries and then diffused to developing countries. Many experts of developing countries received higher education and degrees from colleges and universities of developed countries, where they were trained in the paradigms that were dominant in those countries. Upon returning to their home countries, they propagated these paradigms to their students, colleagues, policymakers, and all with whom they had pertinent interaction. Even those who didn't receive higher education abroad were indoctrinated by these paradigms through textbooks, reference books, and curricula that came from developed countries. Since paradigms regarding rivers in developed countries were dominated by the Commercial and Cordon approaches, experts and policymakers of developing countries were indoctrinated to think these approaches the only modern way to deal with rivers.[3]

External influence also works through the material interest channel. River-related projects in developing countries were generally undertaken by the government (i.e., the public sector) and were commonly financed by multilateral lending agencies dominated by developed countries. As noted earlier, until recently, the World Bank played a big role in water development projects in developing countries. Regional multilateral banks, such as the ADB, also played a large role. The bilateral lending offered by developed countries was also often directed toward water development projects. Needless to say, this financing from developed countries came together with the paradigms that dominated in these countries. Politicians, bureaucrats, experts, consultants, designers, constructors, suppliers, and all other immediate-beneficiaries of water projects in developing countries were eager to accept these paradigms because of the financing that came with them.

It is also clear from the above that the knowledge and material interests channels of external influence are intertwined. A particular way in which this intertwining works is through the technical assistance (TA) that generally forms an important component of the budget of water projects financed by multilateral and bilateral lending from developed countries. Under this component, experts from developed countries are hired to design and supervise the implementation of projects. In the case of Bangladesh, as we saw, a foreign consultancy firm was hired to formulate the country's Master Plan for water development. In addition, an entire team of foreign consultants was placed, on an in-house basis, at the implementing agency, the EPWAPDA and later the BWDB. All these measures were financed through foreign lending. Thus, the confluence of both knowledge and material interests channels in the external influence favoring the Commercial and Cordon approaches was not limited to only an upper level of decision making, focused on setting the direction and choice of projects. Instead, this confluence was true for more on-the-ground levels of designing concrete projects and implementing them through hired foreign consultants.

Earlier, we noted that many immediate-beneficiaries can absolve themselves from the risk of failure or lack of performance of the projects they promote, inspired by the Commercial and Cordon approaches. This is particularly the case with multilateral lending agencies, the repayment of whose loans is made by the country's central bank from its overall foreign exchange earnings, and not from revenues generated by the projects financed by them. In other words, multilateral lending agencies have created for themselves a largely risk-free environment for their lending operations by delinking repayment of the loans from earnings of the projects financed by them. The individual officials of these agencies who approve these projects also do not face much, if any, consequence of failure or lack of performance. As a result, they have an incentive to maximize their lending by promoting Commercial and Cordon approaches that lead to large-budget projects for structural intervention in rivers.

This does not mean that contrary views, supporting the Ecological and Open approaches, do not get transmitted from developed to developing countries at all. As noted earlier, these approaches have now made some inroads in the academic and policymaking circles of many developed countries. As a result, sometimes these views indeed get conveyed to developing countries. However, as of now, they are generally not backed up by lending such as that routinely offered for water projects conforming with Commercial and Cordon approaches. As noted in chapter 12, during the FAP debate in the wake of the 1988 flood in Bangladesh, the EWS, financed by USAID, came out with recommendations that went against proposals—such as double embanking of all major rivers of the country—that followed from the Commercial and Cordon approaches. However, reportedly under pressure from the water policy establishments, USAID later apparently disassociated itself from the EWS recommendations and followed the "official" line favoring the Commercial and Cordon approaches. This experience illustrates the outcome whereby the lobby working for the Commercial and Cordon approaches in many developing countries—created originally, to a large extent, by the material and intellectual influences of developed countries—has now become so entrenched and powerful that it is able to resist external influences that are contrary to its material interests and understanding.

Thus, domestic and external interests combine to create powerful lobbies in support of the Commercial and Cordon approaches. Adoption of the Ecological and Open approaches will therefore have to overcome both knowledge-related and material interest-related hurdles. The question is how this can be done.

13.4 Role of public discussion in adoption
of correct river policies

One of the important ways in which river-related policies can be corrected is by moving the discussion of these policies from the narrow and closed circles of foreign advisers, bureaucrats, politicians, and technocrats[4] to the public arena, where all stakeholders can engage in an open and transparent discussion, and policy choices can be made through reasoning and consent.

In his discussion, Sen (1999, 2009) notes that democracy should not be equated to mere elections and competition among political parties. Instead, the essence of democracy lies, to a large extent, in reaching decisions of public importance through open public discussion and collective reasoning. In fact, he defines democracy as "governance by discussion". In this regard, Sen distinguishes two different values of public discussion. One is its instrumental value, which refers to the fact that public discussion allows the community to reach better policies. The other is its intrinsic value, as a process of the realization of democracy, which demands participation of the citizenry on an equal footing in making decisions that concern them. The importance of public discussion for better policy making has been emphasized by other researchers, such as Sunay (2012), in a general context, McDonnell (2018), in the particular context of climate policies, and Goodman (2004), in the context of the New Deal era policies.[5]

The necessity of open, public discussion is particularly acute regarding river-related policies. First of all, rivers affect the lives of large multitudes of people, and in some countries, all the people, even directly. Second, rivers and river policies are truly multidimensional. Even among experts, river policies cannot therefore be the subject matter of just one particular discipline. Until recently, however, river policymaking has been dominated mainly by civil engineers, so far as experts are concerned. This has been a consequence of the Commercial and Cordon approaches, which have generally reduced the issue of river policies to determining which particular structural interventions to make, more prosaically, which dams, barrages, and embankments are to be constructed. This identification of river policies with structural interventions made river policies almost the exclusive province of civil and structural engineers.

However, as is clear from the discussion in previous chapters, any intervention in rivers has long-run consequences for their hydrology and morphology. Accordingly, experts in geology, hydrology, geography, soil science, and other related physical sciences deserve to have important roles in river-related policymaking. Also, river policies affect directly agriculture, fisheries, transportation, and other production sectors. Experts of these areas also need to have a role. Financial and economic analysis has a central role in judging the efficiency of investments, making economists indispensable in the policymaking process. Alongside the physical and economic aspects, the social and human consequences of river policies deserve due attention. Accordingly, experts from various social sciences and humanities also have a role in river policymaking.

Apart from experts from different disciplines, it is important to let the people of the river valleys themselves to have an important say regarding policies affecting their own life and rivers. After all, it is they who will be most directly affected by these policies. They may not have the formal expertise, but they have the life experience and the accumulated wisdom about the rivers that are the target of the policies. As noted earlier, one of the consequences of the Commercial and Cordon approaches has been a top-down approach to river policymaking. This has been the corollary of the identification of river policies with structural interventions and hence of re-garding them as a matter of engineering. To the extent that the common people of river valleys were not engineers, it was thought that they did not have the necessary qualifications to evaluate and judge the merits of proposed projects and hence to be engaged in the policy discussion. As a result, river-related projects have generally been imposed on the local people from above. The proposed process of arriving at river policies through public discussion can go a long way to end this undesirable sit-uation and make river policymaking a bottom-up process, entailing a combination of public wisdom and synthesis of expertise from a wide range of disciplines.

For this public discussion process to be effective and successful, it is im-portant to create and make available a common conceptual framework and a necessary pool of information for all the participants of the discussion to use, so that they can talk to each other instead of past each other. In other words, a common set of concepts and vocabulary needs to be created so that the views and arguments of one is understood by others. A particular responsibility in this regard therefore falls on scholars with public interests in mind to lay out the alternative policies and their merits and demerits in a way that is under-standable by the common public. It is encouraging that in recent decades, through the efforts of many individual scholars, as well as of groups, bodies, and organizations, the river-related conversation has entered the public arena.[6] The information, analysis, and conceptualization offered in this book hope-fully will add further to the creation of a common pool of concepts, termi-nology, information, and arguments necessary for a fruitful discussion of policies regarding rivers.

13.5 Conclusions

The necessity of sustainable development is now recognized by all nations. The world community adopted in 2015 the *2030 Agenda* for sustainable develop-ment formulated in the form of the 17 SDGs and 169 targets. There is a con-certed effort to achieve these goals. Rivers have a nexus role to play in sustainable development. Sustainability of a country's agriculture, fisheries, water transportation, settlement, water supply, public health, ecology, hydrological balance, and much more depends on correct choice of policies toward rivers.

Unfortunately, river policies so far have not always been conducive to sustainable development, and it is necessary to correct these policies. Climate change has made correction of river policies more urgent. Path dependence and irreversibility of many consequences of bad river policies have added their weight to this urgency. Barriers arising both from lack of necessary knowledge and from material interests hamper the right policy choice. However, with the passage of time, what was once long-run transmutes into short-run. The adverse consequences of the Commercial approach to rivers could be ignored when this approach was originally adopted because these consequences were, In many respects, long-run. However, after several centuries and many decades, these long-run consequences have now surfaced with greater force. Changes are indeed taking place. Many old dams have been demolished and many new, planned dams have either been canceled or postponed. The drawbacks and ineffectiveness of embankments are getting more recognition. It is now necessary to transform these beginnings into a broad agenda and campaign for a switch from Commercial and Cordon approaches that have outlived their times and embrace the Ecological and Open approaches whose time has come.

Notes

1. River-intervention projects are generally big ticket, publicly funded projects. So, politicians and bureaucrats can hope to benefit a lot—both legally and illegally—by presiding over such large-budget projects. Similarly, the designers, constructors, and suppliers get paid no matter how successful the project proves to be at the end.
2. One commonplace evidence of this failure is the application of a positive (instead of zero) discount rate in evaluating future benefits in project evaluation.
3. Some authors suggest that developing countries' inclination toward dams, barrages, and cordons is a legacy of colonial rule (see for example, Issacman and Issacman 2013). There may be some truth to this view. The process of (deformed) industrialization in today's developing countries was indeed initiated during the colonial rule. However, the source of the Commercial approach is more general, and the approach continues to hold sway even many decades after the end of colonial rule. This is partly because of the overall dominance of the developed industrial economies in the contemporary world.
4. In Japan, the $6 billion Nagara Estuary Dam, completed in 1994 and widely seen as serving no useful purpose, has become a notorious symbol of the so-called "Iron Triangle" formed by the bureaucrats, politicians, and contractors (McCully 2001, p. 251).

5. See Lister (2008) and Sunay (2012) for more discussion on the role of public debate and reasoning in social choice and democracy. See Goodman (2004) for discussion of the role of public discussion in American democracy during the progressive and New Deal eras and McDonnell (2018) for discussion of the role of public debate in the choice of climate related policies in the Netherlands.

6. As mentioned at the beginning, the report, *Dams and Development*, by the WCD (WCD 2000) played an important role in this regard. Patrick McCully's book, *Silenced Rivers* (McCullly 2001), advanced the discussion further. The regular publications and occasional supplementary reports of International Rivers and SANDARP (South Asian Network on Dams, Rivers, and People) are informing and sensitizing large sections of people about issues concerning rivers and river policies.

Glossary of Terms

The following provides brief definitions or explanations of some of the technical terms used in the discussion of this book. Many of these follow the *Highways in the River Environment (HIRE)* manual of the Federal Highway Administration (FHWA 1990).

Abrasion Removal of riverbank material through the rubbing of sediment, debris, or ice against the bank.

Afflux Backwater or height by which water levels are raised because of an obstruction or constriction such as a bridge or a barrage.

Aggradation (of riverbed) Raising of the riverbed due to sediment deposition.

Alluvial Soil and rock material deposited from river flow.

Alluvial river A river whose channel is entirely in alluvium, so that no bedrock is exposed during low flow and is not likely to be exposed by erosion during major flow.

Alluvial fan A landform shaped like a fan and created by the sediment deposited by a river emerging from a high-slope valley onto a low-slope plain.

Alluvium Unconsolidated clay, silt, sand, or gravel deposited by a river in its bed, floodplain, fan, or delta.

Anabranched river A river whose flow is divided during normal or low flows by large islands or bars.

Avulsion A sudden change in the course of a river often by breaching its bank during high flows, occurring most commonly in floodplains, deltas, and alluvial fans.

Capillary action The process through which water from the ground table reaches the surface. As this water disappears through evapotranspiration, the salt contained in it remains as a residue and accumulates in the soil. Since soil salinity makes it more difficult for plants to absorb soil moisture, these salts must be leached out of the plant root zone by applying additional water. This water in excess of plant needs is called the leaching fraction.

Dyke A term used in England and elsewhere to refer to river embankments, often of shorter length.

Ephemeral river A river that does not flow continuously for most of the year.

Estuary Tidal reach at the mouth of a river.

Evapotranspiration The sum of transpiration by plants and evaporation from soil.

Flap gates A gate that prevents water from flowing back into the source of discharge.

Flood Overflow of river water onto its bank and beyond.

Floodplain The land beyond the riverbank created over time through the deposition of sediment by the river overflow.

Flume An open or closed channel to convey water. A large flume is often called an aqueduct.

Funneling effect The upstream broadening of the river when obstructed by a barrage or dam. This happens particularly in alluvial rivers, which can erode their banks easily.

Gradient Slope of the riverbed or the land surface.

Hydraulic potential The potential for generating electricity due to differential in heights of water levels.

Hydrology The scientific study of the movement, distribution, and quality of water on Earth, including the water cycle, water resources, and environmental watershed sustainability.

Island (in a river) A permanently vegetated area that divides the river and remains above water during normal river flow. By contrast, *bars* or *sandbars* are not vegetated.

Lateral erosion Erosion in which lateral component dominates in the removal of material, as opposed to *scour*, in which the vertical component dominates.

Leaching Removal of salt that has accumulated in the soil due to irrigation and other reasons through the flow of freshwater.

Levee A term popularly used in the United States to refer to embankments that are constructed to contain the river within its channel so it does not overflow onto the floodplain. Levees are longer than dykes.

Meandering river A river whose flow has regular concave and convex curves (loops), a pattern known as sinuosity (from sine curves).

Morphology Dimensions, shape, and pattern of river flow.

Mud A soft mixture of silt and clay.

Natural levee A low ridge—often called the bank—along a river, formed over time by deposition of sediment during river overflows. These are different from man-made levees.

Nonalluvial rivers Rivers whose channel lies entirely in bedrock.

Orifice flow Flow of water into an opening that is submerged. The flow is controlled by pressure forces.

Oxbow The horseshoe or bow shaped loop that the river has abandoned.

pH The reciprocal of the logarithm of the hydrogen ion concentration. The concentration is the weight of hydrogen ions, in grams per liter of solution. For example, neutral water has a pH value of 7 and a hydrogen ion concentration of 10^{-7}.

Revetment A river bank lining designed to prevent or halt bank erosion.

Riprap A well-graded mass of durable stone, or other material, that is specifically designed to provide protection from flow-induced erosion.

River training Engineering works built along the bank of a river—either submerged during normal flow or rising above—to direct or lead the river flow into a prescribed channel.

Runoff That part of precipitation that runs off the surface of a drainage area and flows into rivers.

Sand Granular material that is smaller than 2.0 mm and coarser than 0.062 mm in diameter.

Scour The displacement and removal of riverbed material by river flow. The scope of scour is considered to be local, and is distinguished from general riverbed degradation or headcutting.

Seepage Slow movement of water through small cracks and pores in the river bank material.

Shoal A submerged sand bar.

Silt Sediment that is finer than 0.062 mm and coarser than 0.004 mm in diameter.

Sinuosity The ratio between thalweg length and the valley length of a sinuous river.

Sloughing The shallow, transverse movement of a soil mass down a riverbank as a result of unstable conditions at or near the surface (also called clumping). Conditions leading to sloughing include: bed degradation, attack at the bank toe, rapid drawdown, and slope erosion to an angle greater than the angle of repose of the material.

Spur A structure, permeable or impermeable, projecting into a channel from the bank for the purpose of altering flow direction, inducing deposition, or reducing flow velocity along the bank.

Stage Height of river water surface above a reference elevation.

Tailwater Depth of flow in the river directly downstream of a river intervention structure.

Thalweg Line following the deepest part of a river.

Total sediment load Sum of suspended load and bed load or the sum of bed material load and wash load of a river.

Upper bank The portion of a river bank that lies above the average water level of the river.

Wash load Suspended material of very small size (generally clay and colloids) originating primarily from erosion on the land slopes of the drainage area and present to a negligible degree in the bed.

Watershed The catchment area for rainfall which is delineated as the drainage area producing runoff. Usually it is assumed that the base flow in a river also comes from the same area.

Weir A low-height dam over which river water can flow on a regular basis.

References

Adel, Miah M. (2001) "Effects on Water Resources from Upstream Water Diversion in the Ganges Basin." *Journal of Environmental Quality* 30: 336–368.

Adel, Miah M. (2017) "Upstream Natural Resource Piracy-Caused Universality—Lost Economic Health Index GDP." *American Research Journal of Biosciences* 3: 1–32.

Adhikari, Jagannath (2017), "Devastating Himlayan floods are made worse by an international blame game," *The Conversation*. Available at http://theconversation.com/devastating-himalayan-floods-are-made-worse-by-an-international-blame-game-83103.

Agarwal, A., and S. Narain, eds. (1997) *Floods, Floodplains, and Environmental Myths. State of India's Environment—A Citizen's Report*. New Delhi: Centre for Science and Environment.

Ahmed, Feroze M, Qazi Kholiquzzaman Ahmad, and Md. Khalequzzaman (editors). (2004) *Regional Cooperation on Transboundary Rivers: Impact of the India River-Linking Project*. Dhaka: BAPA, BEN, BEA, IEB, BUET, DU, and BWP, BGS, BUP, BNGA, ASB.

Alley, Kelly D. (2004) "The Making of a River Linking Plan in India: Suppressed Science and Spheres of Expert Debate." In Ahmed, Ahmad, and Khalequzzaman (eds.), *Regional Cooperation on Transboundary Rivers*, pp. 1–15 .

Anand, Anisha (2013), Embankments no solution to floods: experts. *Times of India*, October 23, Available at https://timesofindia.indiatimes.com/city/patna/Embankments-no-solution-to-floods-Experts/articleshow/24562812.cms

Arrow, K., B. Bolin, R. Costanza, P. Dasgupta, C. Folke, C. S. Holling, B. O. Jansson, S. Levin, K. G. Meller, C. Perrings, and D. Pimentel (1995) "Economic Growth, Carrying Capacity, and the Environment." *Science* 268: 520–1.

ADB (Asian Development Bank) (1990) *Project Completion Report of the Meghna-Dhonagoda Irrigation Project in Bangladesh*. PCR:BAN 21177.

ADB (Asian Development Bank) (1996a) *Project Performance Audit Report on the Northwest Rural Development Project in Bangladesh*. PPA:BAN 15077.

ADB (Asian Development Bank) (1996b) *Project Performance Audit Report on the Flood Rehabilitation Project*. PPA:BAN 21179.

ADB (Asian Development Bank) (1996c) *Project Performance Audit Report on the Pabna Irrigation Rural Development Project in Bangladesh*. Loan No. 378-BAN[SF]. PPA:BAN 11040.

ADB (Asian Development Bank) (1998) *Project Performance Audit Report on the Ganges-Kobadak Irrigation Rehabilitation Project in Bangladesh*. PPA:BAN 15052.

ADB (Asian Development Bank) (2002) *Project Performance Audit Report on the Second Aquaculture Development Project in Bangladesh*. PPA:BAN 18045.

ADB (Asian Development Bank) (2007), Bangladesh: Khulna-Jessore Drainage Rehabilitation Project, Project Performance Evaluation Report, Operations Evaluation Department, Manila.

ADAB (Association of Development Agencies in Bangladesh) (1992) History of ADAB. Available at http://www.adab.org.bd/about-us/history

Ayeb, Habib (2013) "Egypt No Longer Owns the Nile." *Le Monde Diplomatique*. August. https://mondediplo.com/IMG/jpg/Nile-map.jpg.

Azad, A. K., A. Iqbal, and J. Sultana (2004) "Potential Impacts of Indian River-linking Plan on the Ecosystem of the Sunderbans." In Ahmed, Ahmad, and Khalequzzaman (eds.) *Regional Cooperation on Transboundary Rivers*, 168–181.

Azad, Abul Kalam Md., and Ashraful Alam (2004) "Environmental Impact of India's River Linking Project on Bangladesh." In Ahmed, Ahmad, and Khalequzzaman (eds.) *Regional Cooperation on Transboundary Rivers* 210–220.

Baiyu, Gao (2020). "New concerns for transboundary rivers as China discusses diversion," thethirdpole.net, January 14

Bandyopadhaya, J., and S. Parveen (2002) *The Interlinking of Indian Rivers—Some Questions on the Scientific, Economic, and Environmental Dimensions of the Proposal*. Kolkata: CEDP-IIMC Manak Publishers.

Bandyopadhyay, Sujay, and N.C. Jana (2010) "Severity of Embankment Breaching (EMB) in the Lower Ajay Basin: An Environment Threat." *Indian Journal of Geomorphology* 15: 1–18.

Barrow, Christopher J. (1998) "River Basin Development Planning and Management: A Critical Review." *World Development* 26 (1): 171–186.

Barua, Sudipta, and Jacko Van Ast (2011) "Towards Interactive Flood Management in Dhaka Bangladesh." *Water Policy* 13 (5): 693–716.

Beilfuss, Richard D. (1999), "Can this river be saved? Rethinking Cahora Bassa could make a difference for dam-battered Zambezi," Imternational Rivers, Feb 1. Available at https://www.internationalrivers.org/resources/can-this-river-be-saved-rethinking-cahora-bassa-could-make-a-difference-for-dam%E2%80%93battered

Beilfuss, Richard D. (2012) Environmental flows monitoring and evaluation system for the adaptive management of the Zambezi River basin, Working Dcoument of World Wildlife Foundation Project # 9F083801, June

Bhandari, Ratan (2011), "Nepal Dam's Future Uncertain," International Rivers, retrieved on Decemeber 23, 2011.

Bhattacharya, Dipen (2009) "Climate Change Effects in the Deltaic Environment of Bangladesh." In Bhattacharya et al. (eds.) *Climate Change and the Tasks for Bangladesh*, 36–68.

Bhattacharya, Dipen, M. Feroze Ahmed, Nazrul Islam, and Mohd. A. Matin (eds.) (2009) *Climate Change and the Tasks for Bangladesh*. Dhaka: Bangladesh Poribesh Andolon (BAPA) and Bangladesh Environment Network (BEN) .

Bhattacharya, K. (2011) *The Lower Damodar River, India—Understanding the Human Role in Changing Fluival Environment*. New York: Springer.

Bhaumik, Subir (2009), "Tipaimukh Dam: Hidden facts cause concerns," *Bengal Newz*, July 17, available at subirbhowmikscolumn.blogspot.com/2009/07

Biswas, Sumantra, S. (2015), "Multipurpose projects as a flood controller – Is this the reality? A study of the Damodar River of West Bengal, India," *Environmental Science*, Vol. 5, No. 1, pp. 70–79

Bosshard, Peter (2007) "Dams, Rivers, and People in 2016: An Overview." *Dams, Rivers and People*. International Rivers, CA

Bosshard, Peter (2014) "The Hidden Hand Behind 'Natural' Disasters." International Rivers. https://www.internationalrivers.org/blogs/227/the-hidden-hand- \behind-natural-disasters.

Boyce, James K. (1990) "Birth of a Megaproject: Political Economy of Flood Control in Bangladesh." *Environmental Management* 14 (4).No. 4, pp. 419–428

Brammer, Hugh (1990) "Floods in Bangladesh I: Geographical Background to the 1987 and 1988 Floods." *The Geographical Journal* 156 (1): 12–22.

Brammer, Hugh (2010) "After the Bangladesh Flood Action Plan: Looking to the Future." *Environmental Hazards* 9: 118–130.

Brammer, Hugh (2014), Bangladesh's Dynamic Coastal Regions and Sea Level Rise, *Climate Risk Management*, 1:51-62

Branigan, Tania (2008) "China's Mother River: the Huang He River." *Guardian*, November 25.

Braulik, Gill T., Uzma Noureen, Masood Arshad, and Randall R. Reeves (2015) "Review Status, Threats, and Conservation Management Options for the Endangered Indus River Blind Dolphin." *Journal Biological Conservation* 192: 30–41.

Briggs, M. K. and S. Cornelius (1997), "Opportunities for ecological improvement along the lower Colorado River and Delta," *Wetlands*, 18:513-529

Brink, Elizabeth, and Serena McClain (2004) *Beyond Dams: Options and Alternatives*, American Rivers and International Rivers Network. Washington DC: Berkeley, CA.

Brooker, M. P. (1985) "The Ecological Effects of Channelization (The Impact of River Channelization)." *The Geographical Journal*, 151 (1): 63–69.

Brower, David (with Steve Chapple) (2000) *Let the Mountains Talk, Let the Rivers Run*. New York: New Society Publishers.

Brown, A. J., A. Chapman, J. D. Gosden, and F. J. B. Smith (2008) "The Influence of Infrastructure Embankments on the Consequences of Dam Failure." In *Ensuring Reservoir Safety into the Future*, edited by Thomas Telford pp. 1–15. London. https://doi.org/10.1680/ersitf.35225.0007

BWDB (Bangladesh Water Development Board) (n.d.) Completed Projects, Ministry of Water Resources, Government of Bangladesh. Available at https://www.bwdb.gov.bd/index.php/site/completed_project

Centre for Science and Environment (n.d.) *Flood, Floodplains, and Environmental Myths*. State of India's Environment: A Citizens' Report. Centre for Science and Environment, New Delhi. Available at https://csestore.cse.org.in/books/state-of-indias-environment/soe3-595.html

Chakraborty, Abhik (2010) "Suffering with the River: Floods, Social Transition, and Local Communities in the Ajoy River Basin in West Bengal. India," *Geography*, pp. 125–137.

Chandra, S. (2012) *India Flood Management—Damodar River Basin*. http://www.apfm.info/publications/casestudies/cs_india_full.

Changnon, Stanley A. (1996) The Great Flood of 1992: Causes, Impacts, and Responses, Avalon Publishing

Chari, Mridula (2016), "Over 50 years ago, Bengal's Chief Engineer predicted that the Farakka Dam would flood Bihar," Scroll.in, September 1. Available at https://scroll.in/article/815066/over-50-years-ago-bengals-chief-engineer-predicted-that-the-farakka-dam-would-flood-bihar

Chen, Joyce and Valerie Mueller (2018), "Coastal climate change, soil salinity, and human migration in Bangladesh," *Nature Climate Change*, Vol. 8, pp. 981–985

Choudhury, Giasuddin Ahmed, Catharien Terwisscha van Scheltinga, Dick van den Bergh, Farook Chowdhury, Jaap de Heer, Monowar Hossain, and Zahurul Karim (2012) *Preparations for the Bangladesh Delta Plan*. Wageningen, Alterra, Alterra Report ISSN 1566–7197.

Choudhury, Nurunnabi M. (2004) "Indian River-linking Scheme and Bangladesh Response." In Ahmed, Ahmad, and Khalequzzaman (eds.) *Regional Cooperation on Transboundary Rivers* 458–479.

Choudhury, Nusha Yamina, Alak Paul, and Bimal Kanti Paul (2004), "Impact of coastal embankment on flash flood in Bangladesh: a case study," *Applied Geography*, 24:241-258

Chowdhury, Arif Md., Yahya Khairun, Md. Salequzzaman, and Md. Mizanur Rahman (2011), "Effect of combined shrimp and rice farming on water and soil in Bangladesh," *Aquacult Int*, 19:1193-1206

Chowdhury, AKM Jahir Uddin, and Golam Rasul (2012) "Equity and Social Justice in Water Resource Governance: The Case of Bangladesh." *South Asia Water Studies*, 2 (2): 45–58. www.sawasjournal.org.

Chowdhury, Jahir Uddin and Arpana Rani Dutta (2004) "Effect of Transfer of Brahmaputra Water by Indian RLP on Saline Water Intrusion." In Ahmed, Ahmad, and Khalequzzaman (eds.), *Regional Cooperation on Transboundary Rivers*, 143–158.

Clymer, Adam (1977), Carter's Opposition to Water Projects Linked to '73 Veto of Georgia Dam," *The New York Times*, June 13.

Crelis F. Rammelt, Zahed Md. Masud, and Arvid Masud (2018), "The Waterways of Tangail: Failures to Learn from Flood-Control Efforts in the Brahmaputra Basin of Bangladesh," *Water Alternatives* 11 (1): 106–124

Cusick, Daniel (2019), "No end in sight for record Midwest flood crisis," E&E News, June 16. Available at https://www.scientificamerican.com/article/no-end-in-sight-for-record-midwest-flood-crisis/

Custers, Peter (1992) "Banking on a Flood-Free Future? Flood Mismanagement in Bangladesh." *The Ecologist* 22 (5): 241–247.

Custers, Peter (1993) "Bangladesh's Flood Action Plan: A Critique." *Economic and Political Weekly* July 17–24, 1501–1503.

Dai, Aiguo and Kevin E. Trenberth (2002), "Estimates of freshwater discharge from continents: Latitudinal and seasonal variations," *Journal of Hydrometereology*, Vol. 3, pp. 660–687

Das, Partha, Dadul Chutiya, and Nirupam Hazarika (2009) Adjusting to Floods on the Brahmaputra Plains Assam, India, International Centre for Integrated Mountain Development (ICIMOD), Kathmandu, Nepal.

Dasgupta, Partha, Amartya Sen, and Stephen Marglin (1972), *Guidelines for Project Evaluation*, UNIDO, New York: United Nations

David, Laszlo (1986) "Environmentally Sound Management Of Water Resources." Keynote paper presented at the symposium on Conjuctive Water Use, Proceedings of the Symposium, IAHS Publication No. 156.

Davidson, Bruce (1969), *Australia—Wet or Dry? The Physical and Economic Limits to the Expansion of Irrigation*. Melbourne: Melbourne University Press.

De Brujin, K. M. (2005), *Resilience and Flood Risk Management—A System Approach Applied to Lowland Rivers*. Amsterdam: IOS Press.

Deb, A.K. (1998), "Fake Blue Revolution: Environmental and Socio-Economic Impacts of Shrimp Culture in the Coastal Areas of Bangladesh." *Ocean and Coastal Management*, Vol. 4, pp. 63–88.

Deng, Gang (2000), "Yellow River Upstream Important to West-East Power Transmission," People's Daily, Decmeber 14, 2000

Department of Fisheries (DoF) (2013), *Fishery Statistical Yearbook of Bangladesh, 2013-2014*, Dhaka: Government of Bangladesh

Deutsches Komitee fur Katasrophenvorsorge e. V. (DKKV) (2004) "Flood Risk Reduction in Germany—Lessons Learned from the 2002 Disaster in the Elbe Region." February,

DKKV Publication 23e, German Committee for Risk Reduction, Bonn. http://www. dkkv.org/de/publications/resource.asp?ID=94.

Devkota, Lochan, Alessandra Crosato, and Sanjay Giri (2012) "Effect of the Barrage and Embankments on Flooding and Channel Avulsion Case Study Koshi River, Nepal." *Journal of Rural Infrastructure Development* 3 (3): 124–132.

Dewan, A. M., Makoto Mishigaki, and Mitsuru Komatsu (2003) "Floods in Bangladesh: A Comparative Hydrological Investigation on Two Catastrophic Events." *Journal of the Faculty of Environmental Science and Technology* 8 (1): 53–62 .

Dhungel, Dwarika N., and Santa B. Pun (2004) "Impact of the Indian River-Linking Project on Nepal: A Perspective of Nepalese Professionals." In Ahmed, Ahmad, and Khalequzzaman (eds.) *Regional Cooperation on Transboundary Rivers*. 27–45.

Diganta (2009a), "Dam on Brahmaputra: consequence and reality check," The New Horizon, June 25. Available at https://horizonspeaks.wordpress.com/2009/06/28/ dam-on-brahmaputra-consequence-and-reality-check/

Diganta (2009b), "Tipaimukh Dam FAQ: Effects and Politics," The New Horizon. Available at http://www.horizonSpeaks.wordpress.com/2009/06/08/tipaimukh-faq

Dikshit, Sandeep (2016), "One River, two countries, too many dams," *The Hindu*, June 13.

Diwan, Vijay (2004) "Is Maharashtra Inter-basin Water Transfer Proposal Viable?" In Ahmed, Ahmad, and Khalequzzaman (eds.) *Regional Cooperation on Transboundary Rivers* 124–130.

Dixit, Ajaya (2009) "Kosi Embankment in Nepal: Need for a Paradigm Shift in Responding to Floods." *Economic and Political Weekly,* February 7, pp. 70–78.

Duflo, Esther and Rohini Pande (2007), "Dams," *Quaterly Journal of Economics*, Vol. 122, Issue 2, pp. 601–646

Dutta, C. R. (2004) "Is Interlinking of rivers in the Country at all Possible?" In Ahmed, Ahmad, and Khalequzzaman (eds.) *Regional Cooperation on Transboundary Rivers*, 120–123.

Ebrahim, Zofeen (2014), Pakistan floods trigger fresh dam debate, thethirdpole. net, September 16. Available at https://www.thethirdpole.net/en/2014/09/16/ pakistan-floods-trigger-fresh-dam-debate/

EC (European Commission) (2012), A Blueprint to Safeguard Europe's Water Resources, Brussels, 14.11.2012, COM(2012) 673

EEC (European Economic Community) (1992), On the conservation of natural habitats and of wild fauna and flora, Council Directive 92/43/EEC of May 21, 1992, Official Journal of the European Communities, No. L 206/7 (22.7.92)

EEA (European Environment Agency) (2016) Flood Risks and Environmental Vulnerability: Exploring the Synergies between Floodplain Restoration, Water Policies, and Thematic Policies. EEA Report No. 1/2016. Copenhagen, Denmark.

EU (European Union) (2000), "Directive 2000/60/EC of the European Parliament and of the Council establishing a framework for the Community action in the field of water policy," for short Water Framework Directive (WFD), October 23. European Parliament and Council of the European Parliament, Brussels.

EU (European Union) (2007), Directive 2007/60/EC on the assessment and management of flood risks, for short, Flood Directive. November 26. European Parliament and Council of the European Parliament, Brussels.

EU (European Union) (2009) On the Conservation of Wild Birds, Directive 2009/147/EC of the European Parliament and of the Council of 30 November 2009. Official Journal of the European Union, L 20/7 (26.1.2010)

EU (European Union) (2011), *The EU Biodiversity Strategy to 2020*, European Commission, Luxemburg: Publication Office of the European Union

EU (European Union) (2014), Natural Water Retention Measures, Technical Report 2014-082, Luxemburg, Office for Official Publications of the European Communities

Everding, Gerry (2014) "Humans Have Been Changing Chinese Environment for 3,000 Years. Ancient Levee System Set Stage For Massive, Dynasty-Toppling Floods." *The Source*, Washington University in St. Louis, June 18. https://source.wustl.edu/2014/06/humans-have-been-changing-chinese-environment-for-3000-years/.

Faber, S. and C.Hunt (1994), "River management post-1993: The choice is ours," *Water Resourc. Update*, No. 95, pp. 21–25

Fawthrop, Tom (2019), "Myanmar's Myitsone Dam Dilemma," The Diplomat, March 11, Available at https://thediplomat.com/2019/03/myanmars-myitsone-dam-dilemma/

FHWA (Federal Highway Administration) (1990) *Highways in the River Environment*. FHWA-HI-90–016. Washington DC: US Department of Transportation.

Feyen, Luc, and Paul Watkiss (2011) *River Floods. The Impacts and Economic Costs of River Floods in the European Union, and the Costs and Benefits of Adaptation.* Summary of Sector Results from the Climate Cost Project, Funded by the European Community's Seventh Framework Programme. Technical Policy Briefing Note 03. Stockholm Environment Institute (SEI).

Firoze, A. (2003). The southwest coastal region; problems and potential, *The Dailly Star*, XIV, Issue 215

Flood Action Plan (FAP) 12 (1992) *Bangladesh Flood Action FCD/I Agricultural Study Project Impact Evaluation of Meghna Dhonagoda Irrigation Project,* Dhaka: Flood Plan Coordination Organization, Ministry of Irrigation, Water Development, and Flood Control (prepared by Hunting Techncial Services)

Forbes, Vivian (2015) "Yellow River Changing Course." China Water Risk, November 12. http://chinawaterrisk.org/opinions/yellow-river-changing-course/.

French Engineering Consortium (1989) *Prefeasibility Study of Flood Control in Bangladesh*. Paris: Economic and International Department, Ministry of Public Works.

Gain, Animesh K., and Carlo Giupponi (2014) "Impact of the Farakka Dam on Thresholds of the Hydrologic Flow Regime in Lower Ganges River Basin (Bangladesh)." *Water* 6 (8): 2501–2518.

Ganapathy, Nirmala (2016). "India may review water treaty with Pakistan," *The Strait Times*, September 27.

Gatenby, Victoria (2019), "UNESCO labels Sundarbans of Bangladesh 'World Heritage in Danger,'" News/Bangladesh, July 4. Available at https://www.aljazeera.com/news/2019/07/unesco-labels-sundarbans-bangladesh-world-heritage-danger-190704105023538.html

Ganguly, Sitanshu Shekhar (2004) "Environmental Issues in Linking of Rivers." In Ahmed, Ahmad, and Khalequzzaman (eds.) *Regional Cooperation on Transboundary Rivers*, 159–167.

Getzner, M., M. Jungmeier, T. Köstl, and S. Weiglhofer (2011). *Free-flowing sections of the River Mur—Valuation of ecosystem services.* Final report to Nature Stewardship Council, Styria, E.C.O. Institute of Ecology, Klagenfurt.

Ghosh, Debasis, Mrinal Mandal, and Monali Banerjee (2015) "Environmental Impact of Embankment Breaching: A Case Study along Lower Reaches of Ajay River, West Bengal, India." *International Journal of Arts, Humanities, and Management Studies* 1 (9): 44–54.

Ghosh, S. (2013) *Flood Hydrology and Risk Assessment—Flood Study in a Dam-Controlled River of India*. Saarbrucken, Germany: Lambert Academic Publishing.

Ghosh, Sandipan, and Biswaranjan Mistri (2015) "Geographic Concerns on Flood Climate and Flood Hydrology in Monsoon-Dominated Damodar River Basin, Eastern India." *Geography Journal* Volume 2015 |Article ID 486740 | 16 pages | https://doi.org/10.1155/2015/486740.

Goldsmith, E., and N. Hildyard (1984) *The Social and Environmental Impact of Large Dams*. Cornwall UK: Wadebridge Ecological Center.

Gole, Chintaman V. and Shrikrishna Chitale (1966), Inland Delta Building Activities of Kosi River," *Journal of Hydraulic Division*, Vol. 92, Issue 2, pp. 111–126

Goswami, D. C. (1985) "Brahmaputra River, Assam, India: Physiography, Basin Denudation, and Channel Aggradation." *Water Resources Research* 21 (7): 959–978.

GoA (Government of Australia) (2015) Assessment of Environmental Requirments for the Proposed Basin Plan, Murray-Darling Basin Authority. Available at https://www.mdba.gov.au/publications/mdba-reports/assessing-environmental-water-requirements-basins-rivers

GoB (Government of Bangladesh) (1995) *National Environment Management Action Plan (NEMAP)*. Dhaka: Ministry of Environment and Forests.

GoB (Government of Bangladesh) (1996), *Treaty between the Government of the People's Republic of Bangladesh and the Government of Republic of India on Sharing of the Ganga/Ganges Waters at Farakka*, Dhaka: Ministry of Water Resources

GoB (Government of Bangladesh) (2008) *Bangladesh Climate Change Strategy and Action Plan*. Dhaka: Ministry of Environment and Forests.

GoB and GoN (Government of Bangladesh and Government of the Netherlands) (2014) *Inception Report: Bangladesh Delta Plan 2100*. Formulation Project. Dhaka: General Economics Division, Planning Commission.

GoB and GoN (Government of Bangladesh and Government of the Netherlands) (2017) *Bangladesh Delta Plan 2100*. Dhaka: General Economics Division, Planning Commission, GoB.

GoB and UNDP (Government of Bangladesh and United Nations Development Programme) (1989) *Bangladesh Flood Policy Study. Final Report*. Dhaka: UNDP.

GoE (Government of Ecuador) (2008), *Constitution of the Republic of Ecuador, 2008*. Available at http://pdba.georgetown.edu/Constitutions/Ecuador/english08.html

GoN (Government of the Netherlands) (2014) *Draft National Water Plan 2016–2021*. The Hague: Ministry of Infrastructure and the Environment and Ministry of Economic Affairs.

Goodman, David (2004), "Democracy and Public Discussion in the Progressive and New Deal Eras: From Civic Competence toteh Expression of Opinion," Studies in American Political Development, Vol. 18, Issue 2, October, pp. 81–111.

GoWB (Government of West Bengal) (2015) *Annual Flood Report for the Year 2016*. Irrigation and Waterways Directorate, June.

GoWB (Government of West Bengal) (2016) *Annual Flood Report for the Year 2016*. Irrigation and Waterways Directorate, April.

GoWB (Government of West Bengal) (n.d.) *Flood Management, Irrigation and Waterways Department*. http://wbiwd.gov.in/index.php/applications/embankments.

Green, C. H., D. J. Parker and S. M. Tunstall (2000) "Assessment of Flood Control and Management Options." *World Commission on Dams Thematic Review* 4 (4). 149.

Griffith, C. (1994) *The Environmental Kuznets' Curve: Examining Economic Growth and Environmental Degradation.* Environment Dissemination Notes 9. Washington, DC: The World Bank.

Grist Staff (2008), "Fifteen years after the Great Flood of 1993, floodplain development is booming," March 20. Available at https://grist.org/article/gertz2/

Grossman, Gene, and Alan Krueger (1995) "Economic Growth and the Environment." *Quarterly Journal of Economics* May: 353–377.

Grunwald, Michael (2006), Par for the Corps: A flood od bad projects, *The Washington Post*, May 16

Gujja, Biksham, and Srabani Das (2004) "National Civil Society Committee on Interlinking of Rivers in India: Experiences and Lessons in Establishing a National Dialogue." In Ahmed, Ahmad, and Khalequzzaman (eds.), *Regional Cooperation on Transboundary Rivers*, 503–514.

Haider, Reaz (2019), "Climate Change-Induced Salinity Affecting Soil across Coastal Bangladesh," *reliefweb*, January 15, United News of Bangladesh (UNB) and Inter Press Service (IPS). Available at https://reliefweb.int/report/bangladesh/climate-change-induced-salinity-affecting-soil-across-coastal-bangladesh

Handl, Gunther (2012) Declaration of the United Nations Conference on the Human Environment (Stockholm Declaration), 1972 and the Rio Declaration on Environment and Development, 1992. United Nations Audiovisual Library of International Law 1–11. www.un.org/law/av1.

Haq, Syed Azizul (2004) "International River Linking by India: Anticipated Environmental Impact in Bangladesh and Management approach." In Ahmed, Ahmad, and Khalequzzaman (eds.) *Regional Cooperation on Transboundary Rivers*, 182–194.

Haskonig, Royal (2003) "Controlling or Living with Floods in Bangladesh: Toward and Interdisciplinary and Integrated Approach to Agricultural Damage, Agriculture and Rural Development." Working Paper 10. Washington, DC: World Bank.

Hegel, Friedrich G. W. (1991) *Encyclopedia Logic: Part 1 of the Encyclopedia of Philosophical Sciences*, translated by T. F. Geraets, W. A. Suchting, and H. S. Harris, Indianapolis: Hackett

Hendriks, M. J. A., and J. J. Buntsma (2009) "Water and Spatial Planning in the Netherlands: Living with Water in the Context of Climate Change." In *Climate Change in the Water Sector*, edited by F. Ludwig, P. Kabat, H. van Schaik, and M. van der Valk. pp. 143–157. London, UK: Routledge.

Hersher, Rebecca (2018) "Levees Make Mississippi River Floods Worse, but We Keep Building Them." All Things Considered, NPR, May 21. https://www.npr.org/2018/05/21/610945127/levees-make-mississippi-river-floods-worse-but-we-keep-building-them.

Hettige, H., R. E. B. Lucas, and D. Wheeler (1992) "The Toxic Intensity of Industrial Revolution: Global Patterns, Trends, and Trade Policy." *American Economic Review* 82(2): 478–481.

Hofer, Thomas, and Bruno Messerli (2006) *Floods in Bangladesh: History, Dynamics, and Rethinking the Role of the Himalayas.* Tokyo; New York; Paris: United Nations University Press.

Hoggarth-Vicki, Daniel D., J. Cowan, Ashley S. Halls, Mark Aeron-Thomas, J. Allistair McGregor, Caroline A. Garaway, A. Ian Payne, and Robin L. Welcomme (1996), Management guidelines for Asian floodplain river fisheries. Part 2: Summary of DFID research, Department for International Development of the United Kingdom (DFID), MARG Ltd, and Food and Agriculture Organization (FAO), Rome.

Hoggarth- Vicki, Daniel, J. Cowan, Ashley S. Halls, Mark Aeron- Thomas, J. Allistair McGregor, Caroline A. Garaway, A. Iam Payne, Robin L. Welcomme (1996), Management guidelines for Asian floodplain river fisheries, Part 2: Summary of DFID Research. Rome: Department for International Development of the United Kingdom (DFID), MRAG Ltd./ , and Food and Agriculture Organization (FAO). Available at http:// www.fao.org/ 3/ x1358e/ X1358E00.htm#TOC

Hooijer, A., F. Klijn, G. B. M. Perdroli, and A. G. van Os (2004) "Towards Sustainable Flood Risk Management in the Rhine and Meuse River Basins: Synopsis Of The Findings of IRMA-SPONGE." *River Research and Applications* 20 (3): 343–357. doi:10.1002/rra.781.

Hossain, Akhtar A.N.H. (2004) "Indian Inter Basin Water Transfer Link Project: Bangladesh Perspective." In Ahmed, Ahmad, and Khalequzzaman (eds.) *Regional Cooperation on Transboundary Rivers*, 64–73.

Hossain, M. Z., and Sakai, T. (2008) "Severity of Flood Embankments in Bangladesh and Its Remedial Approach, Agricultural Engineering International." *The CIGR Ejournal.* Manuscript LW 08 004 X, 1-11.

Hossain, Mashud M., And Kazi Madina (2004) "Environmental Impact of Indian River Linking Project in Tetulia Distributary." In Ahmed, Ahmad, and Khalequzzaman (eds.), *Regional Cooperation on Transboundary Rivers*, 195–205.

Hossain, Mokaddem K., and Mahbuba Nasreen (2004) "Transboundary Rivers: Socio-Economic and Environmental Challenges in Bangladesh." In Ahmed, Ahmad, and Khalequzzaman (eds.), *Regional Cooperation on Transboundary Rivers*, 279–287.

Hughes, F. (ed.) (2003), *The Flooded Forest: Guidance for policy makers and river managers in Europe on the restoration of floodplain forests*, Department of Geography, University of Cambridge, Cambridge (UK)

Hughes, F., T. Moss, and K. Richards (2008), "Uncertainty in Riparian and Floodplain Restoration," in S. Darby and D. A. Sear (eds.) *River restoration: managing the uncertainty in restoring physical habitat*, Wiley, Chichester

Hughes, Ross, Shapan Adnan, and Barry Dalal-Clayton (1994) *Floodplains or Flood Plans? A review of approaches to water management in Bangladesh.* London and Dhaka: International Institute for Environment and Development (IIED) and Research and Advisory Services (RAS).

Huq, Saleemul, and Mozaharul Alam (2003) Flood Management and Vulnerability of Dhaka City. In Kreimer A., Arnold M., and Carlin A. (eds.) *Building Safer Cities: The Future of Disaster Risk*, Chapter 9, pp. 121–135, Washington DC: The World Bank

Iacurci, Jenna (2014) "Yellow River Flooding in China Caused by Human Intervention, Not Mother Nature." *Nature World News*, June 20, 2014. http://www.natureworldnews. com/articles/7682/20140620/yellow-river-flooding-in-china-caused-by-human-intervention-not-mother-nature.htm.

IBRD-IDA (1963a) *Brahmaputra Flood Embankment Project East Pakistan.* Report No. TO-334b. International Bank for Reconstruction and Development—International Development Association.

IBRD-IDA (1963b) *Chandpur Flood Protection, Drainage, and Irrigation Project.* Report No. TO-334c. International Bank for Reconstruction and Development—International Development Association.

IBRD-IDA (1963c) *Dept. of Technical Operations, Chandpur Flood Protection, Drainage, and Irrigation Project*, Pakistan. Report No. TO-328c.

IBRD-IDA (1965) *Review of the IECO Master Plan for Water and Power Resources Development in East Pakistan.*

IBRD-IDA (1968a) *Agriculture and Water Development Program: Report of the November 1967 Mission.* Vol. I. The Main Report, International Bank for Reconstruction and Development—International Development Association.

IBRD-IDA (1968b) *General Consultants East Pakistan Water and Power Development Authority.* Report No. TO-691a. International Bank for Reconstruction and Development—International Development Association.

IBRD-IDA (1975) *Bangladesh Appraisal of the Barisal Irrigation Project.* Report No. 449a-BD. International Bank for Reconstruction and Development—International Development Association.

Indrajit (2004) "Negotiating Troubled Waters: River linking, Shared Eco-Systems and Regional Diplomacy." In Ahmed, Ahmad, and Khalequzzaman (eds.), *Regional Cooperation on Transboundary Rivers*, 408–421.

IFCDR (Institute for Flood Control and Drainage Research) (1994), Impact of Megha-Dhonagoda embankments, Bangladesh University of Engineering and Technology, Dhaka

IFMRC (Interagency Floodplain Management Review Committee) (1994), Sharing the Challenge: Floodplain Management into the 21st Century. Washington D.C.: US Government Printing Office.

ICID (International Commission on Irrigation and Drainage) (2000) *The Role of Dams for Irrigation, Drainage, and Flood Control.* New Delhi https://www.icid.org/dam_pdf.pdf

ICOLD (International Commission on Large Dams) (1997) *Position Paper on Dams and the Environment, by the International Commission on Large Dams.* Available at https://www.icold-cigb.org/userfiles/files/DAMS/Position_paper.pdf

IECO (International Engineering Company) (1964a) *Master Plan.* Vol. I. Dhaka: East Pakistan Water and Power Development Board (EPWAPDA).

IECO (International Engineering Company) (1964b) *Master Plan.* Vol. II. Dhaka: East Pakistan Water and Power Development Board (EPWAPDA).IPCC (Intergovernmental Panel on Climate Change) (2007) *Climate Change 2007: Synthesis Report.* Fourth Assessment Report, Geneva: WMO and UNEP.

ILO (International Labour Organization) (1957), Indigenous and Tribal Populations Convention, 1957 (N0. 107), Geneva. https://www.ilo.org/dyn/normlex/en/f?p=NORMLEXPUB:12100:0::NO:12100:P12100_INSTRUMENT_ID:312252:NO

IFMRC (1994) (Interagency Floodplain Management Review Committee) *A Blueprint for Change—Sharing the Challenge: Floodplain Management into the 21st Centrury*, Washington D.C. https://fas.org/irp/agency/dhs/fema/sharing.pdf

IPCC (Intergovernmental Panel on Climate Change) (2014), *Climate Change 2014: Synthesis Report.* Fifth Assessment Report, Geneva: WMO and UNEP

IR (International Rivers) (2012a) *The Great New Walls: A Guide to China's Overseas Dam Industry.* Berkeley, CA: International Rivers.

IR (International Rivers) (2012b) "Laos Announces "Postponement" of Xayaburi Dam But Vows to Continue Construction and Resettlement Activities," July 15. Available at https://www.internationalrivers.org/resources/laos-announces-%E2%80%9Cpostponement%E2%80%9D-of-xayaburi-dam-but-vows-to-continue-construction-and

IRN (International Rivers Network) (2007) "Early Adopters: Advances in Flood Management." In Defore the Deluge: Coping with Floods in a Changing Climate, 2007 *Dams, Rivers and People* annual report, pp. 18–21, Berkeley, CA19.

Islam, Aminul Md. and Kuniyoshi Takeuchi (2008) *Flood Hazard Mapping of Dhaka-Narayanganj-Demra (DND) Project Using Geo-Informatic Tools.*Available at semanticscholar.org/

Islam, AZMM (1992), A Study of the Meghna-Dhonagoda Project, Institute of Flood Control and Drainage Research, Bangladesh University of Engineering and Technology, Dhaka

Islam, Fakrul Md. and Yoshiro Higano (1999) "Internaional Environmental Issue between India and Bangladesh: Environmental and Socio-Economic Effects on the Teesta River Area," ERSA conference papers ersa99pa226. European Regional Science Association

Islam, Fakrul Md., and Yoshiro Higano (2002) "Attainment of Economic Benefit through Optimal Sharing of International River Water," *Indian Journal of Regional Science*, 32(2): 1–10

Islam, Nazrul (1990) "Let the Delta Be a Delta! An Essay in Dissent on the Flood Problem of Bangladesh." *The Journal of Social Studies* 48 (April): 18–41.

Islam, Nazrul (1997a) "Notes on Farakka: The Problem of Water Sharing between India and Bangladesh." *The Journal of Social Studies* 76 (April) 1–9. Also in French as "Le barrage du Farakka ou le partage de l'eau entre l'Inde et le Bangladesh." *Alternatives Sud* VIII (2001) 4: 257–264.

Islam, Nazrul (1997b) "Income-Environment Relationship: How Different is Asia?" *Asian Development Review* 15 (1): 18–51.

Islam, Nazrul (1999) "Flood Control in Bangladesh: Which Way Now?" *Journal of Social Studies* 83 (January) 1–31.

Islam, Nazrul (2001) "Open Approach to Flood Control: The Way to the Future." *Futures: The Journal of Forecasting, Planning, and Policy* 33 (8): 783–802.

Islam, Nazrul (2004) "Towards a New Philosophy regarding Rivers." In Ahmed, Ahmad, and Khalequzzaman (eds.), *Regional Cooperation on Transboundary Rivers*, 525–540.

Islam, Nazrul (2006a) "The Commercial Approach vs. the Ecological Approach to Rivers: An Essay in Response to the Indian River Linking Project (IRLP)." *Futures: The Journal of Forecasting, Planning, and Policy* 38 (5): 586–605.

Islam, Nazrul (2006b) "IRLP, or the Ecological Approach to Rivers?" *Economic and Political Weekly* 41 (17 April-May): 17–26.

Islam, Nazrul (2009) "Climate Change and the Tasks for Bangladesh." In Bhattacharya et al. (eds.), *Climate Change and the Tasks for Bangladesh*, 13–35.

Islam, Nazrul (2014), Towards a Sustainable Social Model: Implications for the Post-2015 Agenda, UN-DESA Working Paper No. 136, (ST/ESA/2014/DWP/136), New York: United Nations

Islam, Nazrul (2016) *Let the Delta Be a Delta: The Way to Protect Bangladesh Rivers*. Dhaka: Eastern Academic.

Islam, Nazrul (2018), *Bangladesh Environment Movement: History, Achievements, Challenges*, Dhaka: Eastern Academic.

Islam, Nazrul (2019), *Bangladesh Delta Plan: A Review*, Dhaka: Eastern Academic.

Islam, Nazrul Md. (2000) *Embankment Erosion Control. Towards Cheap and Simple Practical Solutions for Bangladesh*. Coastal Embankment Rehabilitation Project South Khulshi, Chittagong, Bangladesh,

Islam, Nazrul, J. Vincent, and T. Panayotou (1999a) *Unveiling the Income-Environment Relationship: Exploration into the Determinants of Environmental Quality*. Development Discussion Paper No. 701. Cambridge, MA: Harvard Institute for International Development.

Islam, Rafiqul ed. (2004) *Where Land Meets the Sea: A Profile of the Coastal Zone of Bangladesh*. Dhaka: University Press Limited.

Islam, Rafiqul (2006) "Managing Diverse Land Uses in Coastal Bangladesh: Institutional Approaches," in C. T. Hoanh et al. eds. *Environment and Livelihoods in Tropical Castal Zones: Managing Agriculture-Fisheries-Aquaculture Conflicts*. Comprehensive Assessment iof Water Management in Agriculture Series, Vol. 2 (Oxon: CABI) pp. 237–242

Islam, Rafiqul (2007) "Pre- and Post-Tsunami Coastal Planning and Land-Use Policies and Issues in Bangladesh." Proceedings of the worlkshop on coastal area planning and management in Asian tsunami-affected countries, Bangkok: FAO. Available at http://www.fao.org/tempref/docrep/fao/010/ag124e/ag124e_full.pdf#page=66   FAO, Bangkok.

Islam, Saidul Md., and Md. Nazrul Islam (2016) "'Environmentalism of the Poor': The Tipaimukh Dam, Ecological Disasters and Environmental Resistance beyond Borders." *Bandung: Journal of the Global South* 3: 27. https://doi.org/10.1186/s40728-016-0030-5.

Iyer, Ramaswamy (2002) "Linking of Rivers: Judicial Activism or Error?" *Economic and Political Weekly*, Vol, 37, Issue 46, November 16, pp. 4595–4596.

Iyer, Ramaswamy (2004a) "Address at the Plenary Session of International Conference on Regional Cooperation on Transboundary Rivers (ICRCTR)." Dhaka, Bangladesh, December 17.

Iyer, Ramaswamy (2004b) "Beyond Drainage Basin and IWRM: Towards a Transformation of Thinking on Water." Available at https://www.researchgate.net/publication/4926438_BEYOND_DRAINAGE_BASIN_AND_IWRM_Towards_a_Transformation_of_Thinking_on_Water

Jacob, Nitya (2015), "Bhutan's experiments with happiness," DownToEarth, June 28. Available at http://www.downtoearth.org.in/coverage/bhutans-experiments-with--happiness-42467

Jacobs, Jeffrey W. (2017) "River Basin Planning." *Water Encyclopedia*. http://www.waterencyclopedia.com/Re-St/River-Basin-Planning.html, accessed on March 22, 2017.

Jairath, Jasveen (2004) "'Why' Interlinking: Muddy Discourse and Compulsions of Legitimacy." In Ahmed, Ahmad, and Khalequzzaman (eds.)., *Regional Cooperation on Transboundary Rivers*, 102–119.

Japanese Flood Control Experts (1989) *A Preliminary Study on Flood Control in Bangladesh*. Tokyo: Japanese International Cooperation Agency (JICA).

Jha, S. K. (1996) "The 'Kuznets' Curve: A Reassessment." *World Development* 24: 773–780.

JICA (1989), *Report of the Survey of Flood Control Planning in Bangladesh*, Dhaka and Tokyo

Johannessen, Åse., and Jakob J. Granit (2015) "Integrating Flood Risk, River Basin Management and Adaptive Management: Gaps, Barriers and Opportunities, Illustrated by a Case Study from Kristianstad, Sweden." *International Journal of Water Governance* 3: 5–24.

Johannessen, Å., and Thomas Hahn (2013) "Social Learning towards a More Adaptive Paradigm? Reducing Flood Risk in Kristianstad Municipality, Sweden." *Global Environmental Change* 23: 372–381.

Jones, P. H. (1985) *Geology and Groundwater Resources of Bangladesh*. Report Prepared for the World Bank, South Asia Region.

Kabir, Md. Iqbal, Md. Bayzidur Rahman, Wayne Smith, Mirza Afreen Fatima Lusha, and Abul Hasnat Milton (2016) "Climate Change and Health in Bangladesh: A Base-line

Cross-Sectional Survey." *Global Health Action* 9. https://www.ncbi.nlm.nih.gov/pmc/articles/PMC4821870/.

Karim, M. and J. Stellwagen (1998). Shrimp Agriculture. Final Report, Vol. 6, Fourth Fisheries Project, Department of Fisheries, Dhaka, Bangladesh

Karim, Rezaul Md., and Rakhee Mondal (2017) "Local Stakeholder Analysis of Tidal River Management (TRM) at Beel Kapalia and the Implication of TRM as a Disaster Management Approach." *Hydrology* 5 (1):1–6. DOI:10.11648/j.hyd.20170501.11.

Kashem, Abul Md., Abdullah AL Manun Siddiqui, and Mahbubur Rahman (2017), "Coastal zone of Khulna District in Bangladesh: Fisheires land use and its potential," *International Journal of Fisheries and Aquatic Studies*, 5(2): 599-608

Kaufmann, R., B. Davidsotter, and S. Garnham (1995) The Determinants of Atmospheric SO2 Concentrations: Reconsidering the Environmental Kuznets' Curve. Boston: Center for Energy and Environmental Studies, Boston University.

Khalequzzaman, Md. (1994) "Recent Floods in Bangladesh: Possible Causes and Solutions." *Natural Hazards* 9: 65–80.

Khalequzzaman, Md. (2018) "Underlying Causes of Haor Flood in 2017." Presentation at BAPA-BEN Conference on Flood, Waterlogging, and Landslides, January, Dhaka.

Khalequzzaman, Md. (2019) "BDP 2100: Is it Compatible with the Sustainable Development Goals (SDGs)?" Paper presented at the Special Conference on Bangladesh Delta Plan 2100 and Bangladesh Environment, organized by BAPA and BEN, Dhaka, January.

Khalequzzaman, Md., Puneet Srivastava, and Fazlay Faruque (2004) "The Indian River-Linking Project: A Geologic, Hydrologic, Ecological, and Socio-Economic Perspective." In Ahmed, Ahmad, and Khalequzzaman (eds.)., *Regional Cooperation on Transboundary Rivers*, 78–90.

Khan, Shaheen Rafi (1999) The case against Kalabagh Dam, Working Paper Series No. 48, Sustainable Development Policy Institute (SDPI), Islamabad, Pakistan

Khatun, Tajkera (2004) "The Ganges Water Withdrawal in the Upstream at Farakka and Its Impact in the Downstream Bangladesh." In Ahmed et al., *Regional Cooperation on Transboundary Rivers*, 221–249.

Khondker, L., Douglas James, and Keith Pitman (1992) "Five Views of the Flood Action Plan for Bangladesh." Working Paper #77, Natural Hazard Research, March.

Kidder, Tristram R., and Haiwang Liu (2014) "Bridging Theoretical Gaps in Geoarcheology: Archeology, Geoarcheology, and History in the Yellow River Valley, China." *Archeological and Anthropological Sciences*. doi:10.1007/s12520-014-0184-5.

Klijn, F., M. van Buuren, and S. A. M. van Rooji (2004) "Flood-Risk Management Striategies for an Uncertain Future: Living with the Rhine River Floods in the Netherlands? *AMBIO: A Journal of the Human Environment* 33 (3): 141–147.

Krishna, Gopal (2004) "Status of the Writ Petition on 'Networking of Rivers' in India." In Ahmed, Ahmad, and Khalequzzaman (eds.)., *Regional Cooperation on Transboundary Rivers*, 442–457.

Kumar, Madan (2016), "'Remove Farakka Barrage,' Demands Nitish Kumar as Flood Situation in Bihar Turns Grim," *The Times of India*, August 21 (updated on August 22), 2016.

Kusler, J. and L. Larson (1993) "Beyond the Ark. A New Approach to US Floodplain Management." *Environment* 35 (5): 7–11.

Kuznets, S. (1955) "Economic Growth and Income Inequality." *American Economic Review* 45: 1–28.

Kuznets, S. (1963) "Quantitative Aspects of the Economic Growth of Nations: VIII. Distribution of Income by Size." *Economic Development and Cultural Change* XI (II): 1–80.

Lafitte, Raymond (2007), Baglihar Dam and Hydroelectirc Plant: Expert Determination, Executive Summary, Lausanne, Switzerland, Government of Pakistan and Government of India. Available at http://siteresources.worldbank.org/SOUTHASIAEXT/Resources/223546-1171996340255/BagliharSummary.pdf

Le Blanc, David (2015), Towards integration at last? The sustainable development goals as a network of targets, UN-DESA Working Paper No. 141 (ST/ESA/2015/DWP/141), New York: United Nations

Leung, George (1996) "Reclamation and Sediment Control in the Middle Yellow River Valley." Water International. 21 (1): 12–19. doi:10.1080/02508069608686482.

Lister, Andrew (2008) "Public Reason and Democracy." *Critical Review of International Social and Political Philosophy* 11 (3): 273–289.

Little, Ian M. D. and James A. Mirlees (1969), *Manual of Industrial Project Analysis in Developing Countries*. Paris: OECD

Mahanta, Anjana (2013), "The case of Kirichu in the Indo-Bhutan context: Transboundary hydropower projects and downstream flooding," paper submitted for the Sustainable Mountain Development Summit, held at Kohima, Nagaland, India, on September 25-27. Available at https://www.indiawaterportal.org/sites/indiawaterportal.org/files/kurichu.pdf

Mahanta, Anjana (2014), "Damming Bhutan affects India," based on a paper submitted for the Sustainable Mountain Development Summit, held at Kohima, Nagaland, India, on September 25-27. Available at https://www.indiawaterportal.org/articles/damming-bhutan-affects-india

Marris, Emma (2018) "Mississippi River Flooding Worse Now than Any Time in Past 500 Years." *Nature* April 4. https://www.nature.com/articles/d41586-018-04061-z.

McCully, Patrick (2001) *Silenced Rivers: The Ecology and Politics of Large Dams.* London: Zed Books.

McCully, Patrick (2007) *Before the Deluge: Coping with Floods in a Changing Climate*. IRN Dams, Rivers, and People Report. Berkeley, CA: International Rivers Network.

McDonnell, Emily (2018), Democracy: The Importance of Public Debate with Climate Related Topics," The Beam, August 18. Available at https://medium.com/thebeammagazine/democracy-the-importance-of-public-debate-with-climate-related-topics-fae4af876374

McPhee, John (1989) *The Control of Nature*. New York: The Noonday Press, Ferrar, Strauss, and Giroux.

McPhee, John (1998) *Annals of the Former World*. New York: Ferrar, Strauss, and Giroux.

Meadows, Donella H., Dennis L. Meadows, Jorgen Randers, and Willian W. Behrens III (1972), *The Limits to Growth*, New York: Universe Books

Meadows, Donella H., Jorgen Randers, and Dennis L. Meadows (2002), *The Limits to Growth: The 30-Year Update*, White River Junction (Vermont): Chelsa Green

Min, Su, Peng Yao, Zheng Bing Wang, Chang Kuan Zhang, and Marcel J. F. Stive (2016), "Exploratory morphodynamic hindcast of the a evolution of the abandoned Yellow River Delta, 1578-1855 CE," *Marine Geology* 383: 99-119

Mirza, M., Monirul Qader, Ahsan Uddin Ahmed, and Qazi Kholiquzzaman Ahmad (2008) *Interlinking of Rivers in India: Issues and Concerns*. London: CRC Press, Taylor and Francis Group.

Mishra, Dinesh. K. (2001) "Living with Floods: People's Perspective." *Economic and Political Weekly* 36 (2): 2756–2761.

Mishra, Dinesh K. (2005), "Abandoned victims of the Kosi embankments," Infochange Features, January. Available at http://infochangeindia.org/other-nonpublished-118/222-features/6413-abandoned-victims-of-the-kosi-embankments.html

Mishra, Dinesh K. (2015), "Flawed embankment strategy converts Bihar into a watery grave," India waterportal. Available at https://www.indiawaterportal.org/articles/flawed-embankment-strategy-converts-bihar-watery-grave

Mitra, N., and M. Mukhopadhaya (2005) "Flood Plain of Lower Ajay Basin: An Ecological Threat." In *River Floods: A Socio-Technical Approach*, edited by K. M. B. Rahim, M. Mukhopadhya, and S. Sarkar, 159–165. Kolkata: ACB Publications.

Mohammad, Ghulam (1966) "Development of Irrigated Agriculture in East Pakistan: Some Basic Considerations." *The Pakistan Development Review* 6 (3) 315–375. http://www.jstor.org/stable/41257931.

Mowla, Qazi Azizul, and Mohammed Saiful Islam (2013) "Natural Drainage System and Waterlogging in Dhaka: Measures to Address the Problems." *Journal of Bangladesh Institute of Planners* 6 (December): 23–33

MRAG (Marine Resources Assessment Group) (1997), *Fisheries Dynamics of Modified Floodplains in Southern Asia* (R5953), ODA, UK http://www.fao.org/3/x1358e/X1358E02.htm

Mukherjee, Baishali, and Ujwal Deep Saha (2016) "Teesta Barrage Project—A Brief Review of Unattained Goals and Associated Changes." *International Journal of Science and Research (IJSR)* 5 (5): 2027–2032. ISSN online: 2319–7064, Paper ID: Nov163787.

Munoz, Samuel E., Liviu Giosan, Matthew D. Therrell, Jonathan W. F. Remo, Zhixiong Shen, Richard M. Sullivan, Charlotte Wiman, Michelle O'Donnell, and Jeffrey P. Donnelly (2018) "Climatic Control of Mississippi River Flood Hazard Amplified by River Engineering." *Nature* 556: 95–98.

Munshi, Md. Ruhul Amin (2008) *Flood Disaster Mitigation in DND: Land Use Perspective.* M. Disaster Management diss. BRAC University, Dhaka, Bangladesh.

Nandy Ashish (2001) "Dams and Dissent: India First Modern Environmental Activist and His Critique of the DVC Project." *Futures* 33: 709–731

Nanson, G. C. and J. C. Croke (1992), "A genetic classification of floodplains," *Geomorphology* 4(6), 459-486

NAPM (National Alliance of People's Movements) (2004) "Resolutions Adopted at the National Convention on Interlinking of Rivers—Feasibility and Justifiability." Organized by NAPM, in association with Other Organizations, Delhi, December 2–3.

NDC (Netherlands Development Cooperation) (1993) *Flood Action Plan, Bangladesh: A Study of the Debate on Flood Control in Bangladesh.* The Hague: IOV, Netherlands Development Cooperation, Ministry of Foreign Affairs.

Needham, Joseph (1971), Science and civilization in China, Vol. 4, Physics and Physical technology, Part III, Cambridge: Cambridge University Press

Nianthi, Rekha K. W. J., and Zahid Husain (2004) "Impact of Climate Change on Rivers with Special Reference to River-Linking Project." In Ahmed, Ahmad, and Khalequzzaman (eds.)., *Regional Cooperation on Transboundary Rivers*, 263–278.

Nienhuis, P. H., and R. S. E. W. Leuven (2001) "River Restoration and Flood Protection: Controversy or Synergism?" *Hydrobiologia* 444 (1–3): 85–99.

Nikesh, Sai D (2015). "India to aid development in Bangladesh through Tripura, *The Dollar Business*, January 14, https://www.thedollarbusiness.com/news/india-to-aid-development-of-bangladesh- through-tripura/12673

Nishat, A., M. Reazuddin, and others (1999), *The 1998 Flood: Impact on Environment of Dhaka City*, Dhaka: Department of Environment in conjunction with IUCN Bangladesh

Noolkar-Oak, Gauri (2017). *Geopolitics of water conflicts in the Teesta River basin*, Both Ends, The Netherlands.

Olson, Mancur (1971) *The Logic of Collective Action: Public Goods and the Theory of Groups*. Cambridge, MA: Harvard University Press.

Panayotou, T. (1995) "Environmental Degradation at Different Stages of Economic Development." In *Beyond Rio (The Environmental Crisis and Sustainable Livelihoods in the Third World)*, edited by I. Ahmed and J. A. Doeleman. London: Macmillan Press Ltd., for the International Labor Organization (ILO).

Parker, D. J., ed. (2000) *Floods*. Vol. 1. London, UK: Routledge.

Paul, Bimal K. (1995), "Farmers' Responses to the Flood Action Plan (FAP) of Bangladesh: An empirical Study," *World Development*, Vol. 23, Issue 2, February, pp. 299–309

Pearce, Fred (2014) *Downstream Voices: Wetlands Solutions to Reducing Disaster Risks*. Wetlands International, The Netherlands

Pegram, G., Y. Li, T. Le. Quesne, R. Speed, J. Li, and F. Shen (2013) *River Basin Planning: Principles, Procedures, and Approaches for Strategic Basin Planning*. Paris. UNESCO.

Petrow, T., A. H. Thieken, H. Kreibich, C. H. Bahlburg, and B. Merz (2006) "Improvements on Flood Alleviation in Germany: Lessons Learned from the Elbe Flood in August 2002." *Environmental Management* 38 (5): 717–732.

Peyton, Will (2020). "China's water ambitions may leave neighbors dry," *The News Lens*, March 11.

Pinter, Nicholas (2005) "One Step Forward, Two Steps Back on US Floodplains." *Science* (April 8), pp. 207–208.

Pinter, Nicholas (2006), "New Orleans revival recipes (Letter to the Editor)," Issues in Science and Technology, Spring, 22, 3. Available at https://books.apple.com/gb/book/new-orleans-revival-recipes-letter-to-the-editor/id517475044

Pinter, Nicholas, Abebe A. Jemberie, Jonathan W. F. Remo, Reuben A. Heine, and Biran S. Ickes (2008), "Flood trends and river engineering on the Mississippi River system," Geophysical Research Letters, Hydrology and Land Surface Studies, December 12. Available at https://agupubs.onlinelibrary.wiley.com/doi/full/10.1029/2008GL035987

Pinter, Nicholas, Fredrik Huthoff, Jennifer Dierauer, Jonathan W. F.Remo, and Amanda Damptz (2016). "Modeling residual flood risk behind levees, Upper Mississippi River, USA, *Environmental Science and Policy*, Vol. 58 (April), pp. 1310140

Postel, Sandra (1999). *Pillar of Sand—Can the Irrigation Miracle Last?* New York: W. W. Norton & Company

Pottinger, Lori (2003) "Connecting the Drops: Holistic Watershed Approaches Take Some US Cities by Storm." *World Rivers Review*, April., pp. 8–9 (continued in 14).

Pukhan, Samudra Dev (2005) Floods—The annual mayhem in Assam: A Technocrat's Viewpoint. *Ishani* 1(6), p. 6.

Qian, Ning and Dingzhong Dai (1980), "The Problems of River Sedimentation and the Present Status of Its Research in China," In Chinese Society of Hydraulic Engineers (eds.) Proceedings of the International Symposium on River Sedimentation, Beijing: Guanghua Press.

Rahman, Atiur (1995), *Beel Dakatia: environmental consequences of development disaster*, Dhaka: University Press Lmited.

Rammelt, Crelis F., Zahed Md. Masud, and Arvid Masud (2018) "The Waterways of Tangail: Failures to Learn from Flood-Control Efforts in the Brahmaputra Basin of Bangladesh." *Water Alternatives* 11 (1): 106–124.

Ramsar Convention Secretariat (2016), An Introduction to the Ramsar Convention on Wetlands (previously The Ramsar Convention Manual), Ramsar Convention Secretariat, Gland, Switzerland

Rashid, Salim (2001) "Compact Townships as a Strategy for Development." *CTBUH Review* 1 3 (Fall) 1–19.

Rashid, Salim, and Shakil Quayes (2000) "Compact Township, Rural Migration, and Urbanization." In Feroze Ahmed (editor), *Bangladesh Environment 2000*, Dhaka: Bangladesh Poribesh Andolon (BAPA) 380–393.

Reisner, Marc (1993) *Cadillac Desert: The American West and Its Disappearing Water.* New York: Penguin Books.

Revkin, Andrew (2013), "Can Bhutan achieve hydropowered happiness?" *The New York Times* Dot Earth Blog, December 10. https://dotearth.blogs.nytimes.com/2013/12/10/can-bhutan-achieve-hydro-powered-happiness/

Rishaduzzaman (2013), Assignment on Impact of Tipaimukh Dam. February 7, Available https://www.scribd.com/document/124331342/Assignment-on-Impact-of-Tipaimukh-Dam

Rogers, Peter P., P. Lydon, and D. Seckler, (1989) Eastern waters study: Strategies to manage flood and drought in the Ganges-Brahmaputra basin, ISPAN Technical Support Center

Roth, Lawrence H. (2007) "The New Orleans Levees: The Worst Engineering Catastrophe in US History—What Went Wrong and Why." Presentation by Deputy Executive Director, American Society of Civil Engineers. Available at https://biotech.law.lsu.edu/climate/ocean-rise/against-the-deluge/01-new_orleans_levees.pdf

Roy, Arundhuti (1999), The Greater Common Good: The Human Costs of Big Dams. *Frontline*, Vol. 16, Issue 11, April 6.

Roy, Pinaki (2013), "Teesta braces for further blow." *The Daily Star*, May 25, 2013

Roy, Pinaki (2014) "Teesta Waters Go Up and Down." thethirdpole.net, April 24. https://www.thethirdpole.net/en/2014/04/24/teesta-waters-go-up-and-down/.

Roy, Suvendu, and Shyamal Dutta (2012) "Role of Embankment in Flood: A Study of the Confluence Zone of Kunur and Ajay Rivers, Lower Ajay River Basin, West Bengal." *Indian Streams Research Journal* 2 (9): 1–7.

কল্যাণ রুদ্র (২০০০), "গঙ্গা কী অন্য পথ বেছে নবিে?" দেশ, ৯ই ডসিমেম্বর, পৃ ৫৮-৬৮ [Rudra, Kalyan (2000) "Will the Ganges choose an alternative route?" *Desh*, December 9, pp. 58–68]

Rudra, Kalyan (2002), "Floods in West Bengal, 2000—Causes and Consequences." In *Changing Environmental Scenario*, edited by S. Basu, 326–347. Kolkata, India: ACB Publications.

Rudra, Kalyan (2010) *The Encroaching Ganga and Social Conflicts: The Case of West Bengal, India.* West Bengal, India: Department of Geography, Sri Chaitanya College, Habra. Available at https://www.indiawaterportal.org/articles/encroaching-ganga-and-social-conflicts-case-west-bengal

Rudra, Kalyan, Ranjan K. Panda, and Shakil Romshoo (2012) *Living Rivers, Dying Rivers: Rivers of West Bengal, Orissa, & Indus System.* India International Centre.

Available at https://www.indiawaterportal.org/articles/living-rivers-dying-rivers-rivers-west-bengal-orissa-indus-system

Saha, Biswajit, M.H. Tareq, and Provat Saha (2012) "Quantification of the Environmental Degradation due to Urbanization of Dhaka-Naraynaganj-Demra (DND) Project Area in Bangladesh." *Global Journal of Science Frontier Research Environment and Earth Science* 12, (1): 1–10.

Saha, Manoj Kumar (2004) "Indian River Linking Project: Impact on Environment." In Ahmed, Ahmad, and Khalequzzaman (eds.), *Regional Cooperation on Transboundary Rivers*, 206–209.

Sajen, Shamsuddoza (2016), "India's River Linking Project—A Disaster in Waiting," *The Daily Star*, May 26

Saleh, A. F. M., S. M. U. Ahmed, M. R. Rahman, M. Salehin, and M. S. Mondal (2002), "Performance Evaluation of FCD/FCDI Projects During the 1998 Flood," in M. A. Ali, S. M. Seraj, and S. Ahmad (eds), *Engineering Concerns of Flood: A 1998 Perspective*, pp. 253–266, Dhaka: Bangladesh University of Engineering and Technology (BUET). ISBN 984-823-002-5

Sarkar, A. (2006) *River Bank Erosion: Geomorphology and Environment*. Kolkata: acb publications.

Sayers, P., G. Galloway, E. Penning-Rowsell, L. Yuanyuan, S. Fuxin, C. Yiwei, W. Kang, T. Le Quesne, L. Wang, and Y. Guan (2015) "Strategic flood management: ten golden rules to guide a sound approach," *International Journal of River Basin Management*, 13(2), 137-151

Schneider, Kristina (2001), "Water Resources and International Conflicts: Game Theory." Department of Civil Engineering, University of Texas. http://www.ce.utexas.edu/prof/maidment/grad/schneider/WaterResourceManagement/WaterResources.ppt.

Schumm, Stanley A., ed. (1972) *River Morphology*. Stroudsburg, PA: Dowden, Hutchinson and Ross, Inc.

Sein, Thein (2011), "Letter to the Parliament on cancellation of the Myitsone Dam," Government of Myanmar. Available at Eleven Media Group, August 31, 2011

Sejuwal, Kalendra (2017), "Indian dams causing floods in Nepal: Locals." *my Republica*, August 17, https://myrepublica.nagariknetwork.com/news/25778/

Selden, T. M., and D. Song (1995) Neoclassical Growth, the J-curve for Abatement, and the Inverted U-curve for Pollution. Journal of Environmental Economics and Management, Vol. 29, Issue 2, pp. 162–168

Selden, T. M., and D. Song (1994) "Environmental Quality and Development: Is There a Kuznets' Curve for Air Pollution Emissions?" *Journal of Environmental Economics and Management*, 27: 147–162.

Sen, Amartya K. (1999), *Development as Freedom*, New York: Oxford University Press

Sen, Amartya K. (2009), *The Idea of Justice*, Cambridge, Mass: The Belknap Press of Harvard University Press

Shabman, L. (1994), "Responding to the 1993 flood: The restoration option," *Water Resourc. Update*, No. 95, pp. 26–30

Shafiq, N., and S. Bondyopadhayay (1992) *Economic Growth and Environmental Quality: Time Series and Cross-country Evidence*. World Bank Policy Research Working Paper WPS 904. Washington DC: The World Bank.

Shampa, and Ibne Mayaz Md. Pramanik (2012) "Tidal River Management (TRM) for Selected Coastal Area to Mitigate Drainage Congestion." *International Journal of Scientific and Technology Research* 1 (5): ISSB 2277-8616.

Shams, Md. Shamim, and Parvez Sarwar Hossain (2013) "Application of Numerical Model for Drainage Improvement of Dhaka-Narayanganj-Demra (DND) Area of Dhaka City, Bangladesh." International Journal of Scientific and Engineering Research 4 (10): 1–14.

Sharma, Om P., Mark Everard, and Deep N. Pandey (2017), Wise Water Solutions in Rajasthan, Udaypur, India: WaterHarvest, University of West of England, and Water Wisdom Foundation.

Sharma, S. S. (2002) "Floods in West Bengal—Nature, Analysis, and Solution." In Basu, *Changing Environmental Scenario*, 315–322.

Sinha, C. P. (2004), "Interlinking of Indian Rivers," in Ahmed, Ahmad, and Khalequzzaman (eds.) *Regional Cooperation on Transboundary Rivers*, pp. 56–63.

Sinha, Rajiv, Alok Gupta, Kanchan Mishra, Shivam Tripathi, Santosh Nepal, S. M. Wahid, and Somil Swarnkara (2019) "Basin-scale Hydrology Aand Sediment Dynamics of the Kosi River in the Himalayan Foreland." *Journal of Hydrology*, 570 (March): 156–166.

Smith, Brian D., Benazir Ahmed, Muhammad Edrise Ali, and Gill Braulik (2001) "Status of the Ganges River Dolphin or *Shushuk gangetica* in Kaptai Lake and the Southern Rivers of Bangladesh." *Oryx* 35 (1): 61–72.

Smith, Keith and Roy Ward (1998) *Floods—Physical Processes and Human Impacts*. Chichester, UK: Wiley and Sons.

Snow, Edgar (1961), Red Star Over China, New York: Grove Press

Somanathan, E. (2013) "Are Embankments a Good Flood-Control Strategy? A Case Study of the Kosi River," *Water Policy*, Vol. 15, Issue S1, pp. 75–88

Song, D. (1993) *A Simple Static Model for Pollution Kuznets' Curve*. Syracuse, NY: Department of Economics, Syracuse University.

Sparks, R. E. (2006) "Rethinking and then Rebuilding New Orleans." *Issues in Science and Technology*, Vol. 22, No. 2 (Winter), pp. 1–12.

Stern, Nicholas (2007) *The Economics of Climate Change: The Review*. Cambridge: Cambridge University Press.

Stern, Nicholas (2009) *A Blueprint for a Safer Planet: How to Manage Climate Change and Create a New Era of Prosperity*. London: The Bodley Head.

Sultana, Farhana (2004) "Engendering a Catastrophe: A Gendered Analysis of India's River-Linking Project." In Ahmed, Ahmad, and Khalequzzaman (eds.)., *Regional Cooperation on Transboundary Rivers*, 288–305.

Sultana, J., and A. K. Azad (2004)"Regional Management of Common Rivers; Impact of Farakka Barrage on Bangladesh." In Ahmed, Ahmad, and Khalequzzaman (eds.), *Regional Cooperation on Transboundary Rivers*, 250–262.

Sultana, P., P. M.Thompson, and M. G. Daplyn (1995), Rapid Rural Appriasal impact of surface water management projects on agriculture in Bangladesh, *Project Appraisal*, Vol. 10, No. 4. Surrey, England

Sunay, Reyhan (2012), "The importance of public debate in democratic regimes," *European Scientific Journal*, Vol. 8, No. 9, pp. 34–45

Tanabe, Nobuhiru (2014) "Arase Dam: Japan's First Dam Removal Project Underway," *Japan for Sustainability (JFS)* Newsletter No.147 (November)

Tawhid, Khandoker Golam (2004) "Causes and Effects of Waterlogging in Dhaka City, Bangladesh." M. diss., Department of Land and Water Resource Engineering, KTH, Stockholm.

Teclaff, Ludwick A. (1996) "Evolution of the River Basin Concept in National and International Law." *Natural Resources Journal* 36 (Spring) 359–391.

Tewari, Avantika (2017), "Narmada's story of 'Development' and destruction," *Feminism in India*, August 10. https://feminisminindia.com/2017/08/10/narmada-development-destruction/

Thakkar, Himanshu (2006) "What, Who, How, and When of Experiencing Floods as a Disaster." *SANDRP*, November.

Thakkar, Himanshu (2007) "A Dam-Made Disaster: How Large Dams and Embankments Have Worsened India's Floods." In McCully (2007), pp. 14–17.

Thakkar, Himanshu (2015) "Damodar Valley Dams' Role in West Bengal Floods—DVC Dams Could Have Helped Reduce the Floods, They Increased It." *SANDRP*, August 5. https://sandrp.in/2015/08/05/damodar-valley-dams-role-in-w-bengal-floods-dvc-dams-could-have-helped-reduce-the-floods-they-increased-it/.

Thijsse, J. T. (1964) *Report on Hydrology of East Pakistan.* May-October, Dhaka: EPWAPDA (mimeo)

Thijsse, J. T. (1965) *Additional Report on Hydrology of East Pakistan.* March–April, Dhaka: EPWAPDA (mimeo)

Thomas, Maria (2015), "The slow and dangerous death of Indus River Delta," Quartz India, July 9. Available at http//:www.qz.com/india/448049/the-slow-and-dangerous-death-of-indus-river-delta

Thompson, Paul (1990), "Impact of Flood Control on Agriculture and Rural Development in Bangladesh: Post Evalutation of the Chandpur Project," Flood Hazard Research Centre Publication No. 178. Enfield, UK: Middlesex Polytechnique

Thompson, Paul M. and Parvin Sultana (1996), Distributional and Social Impact of Flood Control in Bangladesh, *The Geographical Journal*, Vol. 162, Part 1, pp. 1–13.

Tippett, J., B. Searle, C.Pahl-Wostl, and Y. Rees (2005) "Social Learning in Public Participation in River Basin Management—Early Findings from HarmoniCOP European Case Studies." *Environmental Science and Policy* 8 (3): 287–299.

Tobin, G. A. (1995) "The Levee Love Affair: A Stormy Relationship." *Water Resources Bulletin* 31:3.pp. 359–367

Tockner, K., and J. A. Stanford (2002), "Riverine flood plains: present state and future trends," *Environmental Conservation* 29(03): 308–330

Tregear, T. (1965) *A Geography of China.* London: University of London Press

Tunstall, S. M., C. L. Johnson, and E. C. Penning-Rowsell (2004) Flood Hazard Management in England and Wales: From Land Drainage to Flood Risk Management. Paper presented at the World Congress on Natural Disaster Mitigation, February 19–21, New Delhi, India.

UN (United Nations) (1970) *Integrated River Basin Development.* Report of a Panel of Experts, II. A.4. (E/3066.Rev. 1) New York.

UN (United Nations) (1976) *River Basin Development: Policies and Planning: Proceeding of UN Interregional Seminar of River Basin and Interbasin Development*, Vols. I and II. Natural Resources/Water Series No. 6. New York; Budapest: United Nations.

UN (United Nations) (1977) "Report of the United Nations Water Conference, Mar del Plata, 14–25 March 1977." United Nations Publication, Sales No. E.77.II.A.12.

UN (United Nations) (1982) World Charter for Nature, General Assembly Resolution A/RES/37/7.

UN (United Nations) (1997) *Convention on the Law of the Non-navigational Uses of International Watercourses.* General Assembly Resolution 51/229. Official Records of the General Assembly, Fifty-First Session, Supplement No. 49 (A/51/49).

UN (United Nations) (2000), *United Nations Millennium Declaration,* New York: UN General Assembly Resolution A/RES/55/2

UN (United Nations) (2009) *World Economics and Social Survey: Promoting Development, Saving the Planet.* New York: UN Department of Economic and Social Affaris (Sales No. E09.II.C.1).

UN (United Nations) (2011), *World Economic and Social Survey: The Great Green Technological Transformation,* New York: UN Department of Economic and Social Affairs (Sales No. E11.II.C.1).

UN (United Nations) (2015a) *Transforming our world: The 2030 Agenda for Sustainable Development,* New York: UN General Assembly Resolution A/RES/70/1

UN (United Nations) (2015b) Sendai Framework for Disaster Risk Reduction 2015-2030, UNISDR, Geneva

UN (United Nations) (2015c), *Paris Agreement on Climate Change,* UNFCCC: Paris

UNDP (United Nations Development Programme) (2007) *Human Development Report 2007/2008: Fighting Climate Change: Human Solidarity in a Divided World.* Basingstoke, UK: Palgrave-Macmillan.

UNEP (United Nations Environment Programme) (1985) *Environmentally Sound Management of Inland Waters (EMINWA), Introduction to the Programme.* Manuscript, UNEP, Nairobi.

UNEP (United Nations Environment Programme) (2004), Policy Options of the Madeira River Basin, in Amazon Basin -- GIWA Regional Assessment, pp. 43- 47, UNEP, Nairobi

UNEP (United Nations Environment Programme) (2015) *Nile River Basin: Adaptation to Climate Change Induced Water Stress in the Nile Basin: Summary for Decision Makers.* UNEP, Nairobi. Available at https://www.unenvironment.org/resources/report/nile-river-basin-adaptation-climate-change-induced-water-stress-nile-basin-summary

UNESCO (United Nations Economic, Scientific, and Cultural Organizaiton) (1971), Convention on Wetlands of International Importance especially as Waterflowl Habitat, UNESCO, Paris https://www.ramsar.org/sites/default/files/documents/library/current_convention_text_e.pdf

UNESCO (United Nations Economic, Scientific, and Cultural Organizaiton) (1994).

UNESCO (United Nations Economic, Scientific, and Cultural Organizaiton) (2016), Report on the Mission to the Sundarbans World Heritage Site, Bangladesh, from 22 to 28 March, 2016.

UNFCCC (United Nations Framework Convention on Climate Change) (1992) *United Nations Framework Convention on Climate Change.* Bonn: United Nations.

UNTAA (United Nations Technical Assistance Administration) (1959) *Water and Power Development in East Pakistan.* Prepared for the Government of Pakistan by the United Nations Water Control Mission, New York.

UN -HABITAT (2008) *State of the World Cities 2008/2009, Case Study: Dhaka's Extreme Vulnerability To Climate Change.* Nairobi, Kenya.

UNSD (United Nations Statistical Division) (2019), Statistical Yearbook (Sixty-second issue). Available at http://data.un.org/_Docs/SYB/PDFs/SYB62_1_201907_Population,%20Surface%20Area%20and%20Density.pdf

USACE (United States Army Corps of Engineers) (2009), Louisiana Coastal Protection and Restoration Project (LACPR) Final Report, June

USFCR (United States Foreign Relations Committee) (2007), Multinational Development Banks: Development Effectiveness of Infrastructure Projects, Hearing held on July 12, 2006, Washington DC: US Government Printing Office. Available at www.gpoaccess.gov/congress/indexhtml

USSC (United States Supreme Court) (1972) Sierra Club vs. Morton, 405 U. S. 727 (1972). Available at https://caselaw.findlaw.com/us-supreme-court/405/727.html

Van Alphen, J., and E. van Beek (2006) "From Flood Defence to Flood Management—Prerequisites for Sustainable Flood Management." in Floods, from Defence to Management, edited by J. van Alphen, E. van Beek, and M. Taal. pp. 11–15. London: Taylor and Francis.

Van Leussen, W., G. Kater, and P. P. M. van Meel (2000) "Multi-Level Approach to Flood Control in the Dutch Part of the River Meuse." In New Approaches to River Management, edited by A. J. M. Smits, P. H. Nienhuis, and R. S. E. W. Leuven., pp. 287–305, Leiden: Backhuys Publishers

van Staveren, Martijn (2015), Controlled flooding in deltas, Netherlands National IHP-HWRP Committee, http://www.naturebasedsolutions4water.com/controlled-flooding-in-deltas

Vaughn, Christopher S. (1997) "An Impact on Environmental Changes and People's Perceptions of the Meghna-Dhonagoda Embankment." M.Sc. diss., University of East Anglia, Norfolk, UK.

Vincent, J. (1997) "Resource Depletion and Economic Sustainability in Malaysia." Environment and Development Economics 2 (1): 19–37.

Vis, M., F. Klijn, K. M. De Bruijn, and M. van Buuren (2003) "Resilience Srategies for Flood Risk Management in the Netherlands." International Journal of River Basin Management 1 (1): 33–40.

Vombatkere, Sudhir (2004) "A Short Note on Floods and Droughts: Linking River Basins." In Ahmed, Ahmad, and Khalequzzaman (eds.)s., Regional Cooperation on Transboundary Rivers,pp. 131–137

Vombatkere, Sudhir, and Asha Vombatkere (2004) Flood Control and Engineering Interventions: ILR Concept is Simplistic. paper presented at the International Conference on Regional Cooperation on Transboundary Rivers (ICRCTR), December 16-19, Dhaka, Bangladesh

Wahid, Shahriar M., Garrett Kilroy, Arun B. Shrestha, and Sagar Ratna Bajracharya, and Kiran Hunzai (2017) "Opportunities and Challenges in the Trans-Boundary Koshi River Basin." In River System Analysis and Management, edited by Nayan Sharma. Chapter 18, pp. 341–352 Singapore: Springer Science.

Ward, J. V. (1989), "The four-dimensional nature of Lotic ecosystems," Journal of the North American Benthological Society 8(1), 2-8

WARPO (Water Resources Planning Organization) (2001a) National Water Management Plan. Vol. 1. Summary. Dhaka: Ministry of Water Resources, Government of the People's Republic of Bangladesh.

WARPO (Water Resources Planning Organization) (2001b) *National Water Management Plan*. Vol. 2. Main Report. Dhaka: Ministry of Water Resources, Government of the People's Republic of Bangladesh.

WARPO (Water Resources Planning Organization) (2018) Flood Action Plan Reports - FAP-20, Available at http://www.warpo.gov.bd/site/page/6a82024d-60c0-4fb0-9c02-f685b4380e4f/-Wescoat, James L. Jr., Jahir Uddin Chowdhury, Dennis J. Parker, Habibul Haque Khondker, L. Douglas James, and Keith Pitman (1992), Fiver Views of the Flood Action Plan for Bangladesh, Natural Hazard Research, Working Paper # 77, University of Colorado, Boulder.

White, I., and J. Howe (2003) "Policy and Practice: Planning and the European Union Water Framework Directive." *Journal of Evironmental Planning and Management* 46 (4): 621–631.

Willi, H. P. (2006) "Meeting the Challenge of Flood Protection." *EAWAG News*, November.

WMO (World Meteorological Organization) (2011) *Integrated Flood Management: Partnerships in Weather, Climate, and Water for Development*. Geneva. http://www.wmo.int/pages/prog/dra/documents/IFM_factsheet_EN.pdf.

World Bank (1963) *Brahmaputra Flood Embankment Project*. Washington. DC: Department of Technical Operations.

World Bank (1975) *Appraisal of the Barisal Irrigation Project*. Report No. 449a-BD. Irrigation Division. Washington. DC: South Asia Projects Department.

World Bank (1981) *Project Performance Audit Report: Bangladesh Chandpur II Irrigation Project (Credit 340-BD)*. Report No. 3436. April 30.

World Bank (1989) *Bangladesh Action Plan for Flood Control*. December 1989, Washington DC.

World Bank (1990) "Flood Control in Bangladesh: A Plan for Action." World Bank Technical Paper Number 119, Asia Region Technical Department, Washington, DC.

World Bank (2005) *Project Performance Assessment Report—Coastal Embankment Rehabilitation Project, Bangladesh (Credit 2783-BD)*. Report No. 31565. Washington DC.

WCD (World Commission on Dams) (2000) *Dams and Development: A New Framework for Decision Making (Report of the World Commission on Dams)*. London: Earthscan Publications Ltd.

WCED (World Commision on Environment and Development) (1987), *Our Common Future*, New York and Oxford: Oxford University Press.

Wiering, M. A. and B. J. M. Arts (2006) "Discursive Shifts in Dutch River Management: 'Deep' Institutional Change or Adaptation Strategy." *Hydrobiologia* 565, pp. 327–338.

Yang, Z. S., John Milliman, J. Galler, J. P. Liu, and X. G. Sun (1998), "Yellow River's water and sediment discharge decreasing," Eos Transactions Americal Geophysical Union, 79(48): 589

Zahid, Anwar, and Syed Reaz Uddin Ahmed (2006) *Groundwater Resources Development in Bangladesh: Contribution to Irrigation for Food Security and Constraints to Sustainability*. Report No. H039306, International Water Management Institute (IWMI), Sri Lanka

Zvomuya, Fidelis (2012), "Zambezi River: Drained bone dry," International Rivers, December 1. Available at https://www.internationalrivers.org/blogs/1104/the-zambezi-river-drained-bone-dry

Author Index

Subject Index